ROTATING ELECTRODE METHODS AND OXYGEN REDUCTION ELECTROCATALYSTS

ROTATING ELECTRODE METHODS AND OXYGEN REDUCTION ELECTROCATALYSTS

Edited by

WEI XING

State Key Laboratory of Electroanalytical Chemistry
Changchun Institute of Applied Chemistry
Changchun
PR China

GEPING YIN

School of Chemical Engineering and Technology
Harbin Institute of Technology
Harbin
China

and

JIUJUN ZHANG

Energy, Mining and Environment
National Research Council of Canada
Vancouver, BC
Canada

AMSTERDAM • BOSTON • HEIDELBERG • LONDON • NEW YORK • OXFORD
PARIS • SAN DIEGO • SAN FRANCISCO • SINGAPORE • SYDNEY • TOKYO

ELSEVIER

Elsevier
Radarweg 29, PO Box 211, 1000 AE Amsterdam, The Netherlands
The Boulevard, Langford Lane, Kidlington, Oxford, OX5 1GB, UK
225 Wyman Street, Waltham, MA 02451, USA

British Library Cataloguing in Publication Data
A catalogue record for this book is available from the British Library

Library of Congress Cataloging-in-Publication Data
A catalog record for this book is available from the Library of Congress

ISBN: 978-0-444-63278-4

For information on all Elsevier publications
visit our web site at store.elsevier.com

Printed and bound in Poland
14 15 16 17 18   10 9 8 7 6 5 4 3 2 1

CONTENTS

PREFACE

Rotating disk electrode (RDE) and rotating ring-disk electrode (RRDE) techniques are one kind of the important and commonly used methods in electrochemical science and technology, particularly, in the fundamental understanding of electrochemical catalytic reaction mechanisms such as electrocatalytic oxygen reduction reaction (ORR). The kinetics and mechanisms of ORR catalyzed by both noble metal- and nonnoble metal-based electrocatalysts are the most important aspects in fuel cell and other ORR-related electrochemical technologies. Using RDE and RRDE to evaluate the activities of catalysts and their catalyzed ORR mechanisms is necessary and also one of the most feasible approaches in the development of ORR electrocatalysts.

In developing ORR electrocatalysts, significant challenges exist in achieving high catalyst activity and stability. To facilitate the effort to overcome these challenges, a book with focus on the catalyzed ORR and its associated testing and diagnosis of ORR catalysts is particularly useful. Although all researchers in the area of ORR-related electrocatalysts use RDE/RRDE as routine techniques to evaluate their catalysts and explore the catalyzed ORR mechanisms, based on our observation, however, a fundamental understanding of these methods seems not being fully achieved. Some confusion can be found in the literature when RDE/RRDE methods were used and the data explained. Therefore, a detailed and comprehensive description about these techniques from fundamentals to applications is definitely helpful and may be necessary.

In Chapter 1 of this book, the necessary parameters for both RDE/RRDE analysis in ORR study, such as O_2 solubility, O_2 diffusion coefficient, and the viscosity of the aqueous electrolyte solutions, are discussed in depth in terms of their definitions, theoretical background, and experimental measurements. The effects of type/concentration of electrolyte, temperature, and pressure on values of these parameters are also discussed. To provide the readers with useful information, the values of these parameters are collected from the literature, and summarized in several tables. In addition, the values of both the O_2 solubility and diffusion coefficient in Nafion$^{®}$ membranes or ionomers are also listed in the tables. Hopefully, this chapter would be able to serve as a data source for the later chapters of this book, and also the readers could find it useful in their experimental data analysis.

In Chapter 2, to facilitate understanding and preparing the basic knowledge for rotating electrode theory, both the electron transfer and reactant transport theories at the interface of electrode/electrolyte are presented. Regarding the reactant transport, three transportation modes such as diffusion, migration, and convection are described. A focusing discussion is given to the reactant diffusion near the electrode surface using both Fick's first and second laws. In addition, based on the approach in the literature, the kinetics of reactant transport near and within porous matrix electrode layer and its effect on the electron transfer process is also presented using a simple equivalent electrode/electrolyte interface.

In Chapter 3, to give some basic knowledge and concepts, some fundamentals about the catalyst activity and stability of ORR electrocatalysts, which are the targeted research systems by rotating electrode methods, are presented. A detailed description about the electrocatalysts and catalyst layers and their applications for ORR in terms of their types, structures, properties, catalytic activity/stability, as well as their research progress in the past several decades are also given. Furthermore, both the synthesis and characterization methods for ORR electrocatalysts, and the fabrication procedures for catalyst layers are also reviewed.

In Chapter 4, the fundamentals of ORR including thermodynamics and electrode kinetics are presented. The ORR kinetics including reaction mechanisms catalyzed by different electrode materials and catalysts including Pt, Pt alloys, carbon materials, and nonnoble metal catalysts are discussed based on literature in terms of both experiment and theoretical approaches. It is our belief that these fundamentals of ORR are necessary in order to perform the meaningful characterization of catalytic ORR activity using both RDE and RRDE methods.

In Chapter 5, to give readers the knowledge how to appropriately use RDE in their ORR study, fundamentals of both the electron transfer process on electrode surface and diffusion-convection kinetics near the rotating electrode are presented. Two kinds of RDE and their associated theories of the diffusion-convection kinetics and its coupling with the electron transfer process are presented, one is the smooth electrode surface, and the other is the catalyst-layer coated electrode. For measurements using RDE method, the apparatus of RDE and its associated devices such as rotator and electrochemical cell are described to give the readers the basic sense about this technique. The measurement procedure including RDE preparation, catalyst layer fabrication, current−potential curve recording, the data analysis, as well as the cautions in RDE measurements are also discussed in this chapter.

In Chapter 6, the importance of RRDE fundamentals and practical usage in ORR study is emphasized in terms of both the electron transfer process on electrode surface, diffusion-convection kinetics near the electrode, and the ORR mechanism, particularly the detection of intermediate such as peroxide. One of most important parameters of RRDE, the collection efficiency, is deeply described including its concept, theoretical expression, as well as experiment calibration. Its usage in evaluating the ORR kinetic parameters, the apparent electron transfer, and percentage of peroxide formation is also presented. In addition, the measurement procedure including RRDE preparation, current–potential curve recording, and the data analysis are also discussed in this chapter.

Chapter 7 reviews the applications of RDE and RRDE techniques in ORR research and its associated catalyst evaluation. Some typical examples for RDE and RRDE analysis in obtaining the ORR kinetic information such as the overall electron transfer number, electron transfer coefficiency, and exchange current density are also given in this chapter. It demonstrates that both RDE and RRDE methods are a powerful tool in ORR study, and using RDE and RRDE methods, ORR has been successfully studied on Pt electrode, carbon electrode, monolayer metal catalyst, Pt-based catalyst, and nonnoble metal-based catalysts.

It is our hope that this book could be used as a reference for college/university students including undergraduates and graduates, scientists and engineers who work in the areas of energy, electrochemistry science/technology, fuel cells, and batteries.

We would like to acknowledge with deep appreciation our families for their understanding and support of our work. If technical errors exist in this book, we would deeply appreciate the readers' constructive comments for correction and further improvement.

Wei Xing, PhD
Geping Yin, PhD
Jiujun Zhang, PhD

April 2013

BIOGRAPHY

Dr Wei Xing is a Professor and R&D Director at the Chang-chun Institute of Applied Chemistry, Chinese Academy of Sciences (CIAC-CAS). In 1987, Dr Xing received his PhD in Electrochemistry from CIAC-CAS. Dr Xing has more than 25 years of experience as an electrochemist in the area of oxygen reduction reaction and fuel cells, during which, as one of the key senior researchers, he established, and continues to lead the Laboratory of Advanced Power Sources at CIAC-CAS, which develops novel proton exchange membrane fuel cells (PEMFC) catalysts and technologies. His research is mainly concentrated on the R&D of fuel cell technologies including PEMFCs, direct methanol fuel cells (DMFCs), and direct formic acid fuel cells (DFAFCs), in which cathode catalyst development for oxygen reduction reaction is the major focus. To date, he has published more than 160 referred journal papers, 3 books, and 39 patents.

Dr Geping Yin is a Professor at Harbin Institute of Technology (HIT), China, the Vice-Dean of School of Chemical Engineering and Technology and the Director of HIT's Institute of Advanced Chemical Power Sources. She received her BS and PhD in Electrochemistry from HIT in 1982 and 2000, respectively. She has over 30 years of R&D experience in theoretical and applied electrochemistry, and over 15 years in fuel cells. After completing her BS, she took a Lecture position at HIT in 1988, and was promoted to associate and full professor in 1993 and 2000, respectively. Prof. Yin carried out two terms of visiting scholar research at Tokyo Institute of Technology (1985–1986) and Yokohama National University (2009–2010). Prof. Yin serves on several professional Committees or Associations in China, including the Power Sources Committee of Chinese Institute of Communications, and Chinese Industrial Association of Power Sources. She is also member of Editorial Boards for several journals, such as Journal of Chemical Engineering of Chinese Universities, Electrochemistry, Chinese Journal of Power Sources, Battery, and Carbon. Up to now, Prof. Yin has published more than 160 SCI papers, which have been cited for 3100 times (single most cited for 250 times, H-index 29). Some of these papers were selected as "Hot Papers in Engineering" by ISI Web of Knowledge, the top 50 most-cited articles by Chinese mainland authors published in Elsevier's Environmental Sciences journals, and "one hundred most influential international academic papers" in China by Institute of Scientific & Technical Information of China. Prof. Yin is an active member of Electrochemistry Committee of the Chinese Chemical Society and Lead-acid Batteries Committee of Electrotechnical Society of China.

Dr Jiujun Zhang is a Principal Research Officer and Fuel Cell Catalysis Core Competency Leader at the National Research Council of Canada (NRC) Institute for Fuel Cell Innovation (NRC-IFCI, now changed to Energy, Mining & Environment Portfolio (NRC-EME)). Dr Zhang received his BS and MSc in Electrochemistry from Peking University in 1982 and 1985, respectively, and his PhD in Electrochemistry from Wuhan University in 1988. After completing his PhD, he took a position as an associate professor at the Huazhong Normal University for 2 years. Starting in 1990, he carried out three terms of post-doctoral research at the California Institute of Technology, York University, and the University of British Columbia. Dr Zhang has over 30 years of R&D experience in theoretical and applied electrochemistry, including over 15 years of fuel cell R&D (among these 6 years at Ballard Power Systems and 10 years at NRC), and 3 years of electrochemical sensor experience. Dr Zhang holds several adjunct professorships, including one at the University of Waterloo, one at the University of British Columbia, and one at Peking University. Up to now, Dr Zhang has coauthored more than 300 publications including over 200 refereed journal papers with approximately 6200 citations, 11 edited/coauthored books, 11 conference proceeding papers, 12 book chapters, as well as 50 conference and invited oral presentations. He also holds over 10 US/EU/WO/JP/CA patents, 9 US patent publications, and produced in excess of 80 industrial technical reports. Dr Zhang serves as the editor/editorial board member for several international journals as well as Chief-in-Editor for book series (Electrochemical Energy Storage and Conversion, CRC press). Dr Zhang is an active member of The Electrochemical Society, the International Society of Electrochemistry, and the American Chemical Society.

LIST OF CONTRIBUTORS

Weiwei Cai
State Key Laboratory of Electroanalytical Chemistry, Changchun Institute of Applied Chemistry, Changchun, PR China; Laboratory of Advanced Power Sources, Changchun Institute of Applied Chemistry, Changchun, PR China

Chunyu Du
School of Chemical Engineering and Technology, Harbin Institute of Technology, Harbin, China

Ligang Feng
State Key Laboratory of Electroanalytical Chemistry, Changchun Institute of Applied Chemistry, Changchun, PR China

Yang Hu
Laboratory of Advanced Power Sources, Changchun Institute of Applied Chemistry, Changchun, PR China

Zheng Jia
School of Chemical Engineering and Technology, Harbin Institute of Technology, Harbin, China

Changpeng Liu
Laboratory of Advanced Power Sources, Changchun Institute of Applied Chemistry, Changchun, PR China

Qing Lv
Laboratory of Advanced Power Sources, Changchun Institute of Applied Chemistry, Changchun, PR China

Tiantian Shen
School of Chemical Engineering and Technology, Harbin Institute of Technology, Harbin, China

Fengzhan Si
State Key Laboratory of Electroanalytical Chemistry, Changchun Institute of Applied Chemistry, Changchun, PR China; Laboratory of Advanced Power Sources, Changchun Institute of Applied Chemistry, Changchun, PR China

Xiujuan Sun
State Key Laboratory of Electroanalytical Chemistry, Changchun Institute of Applied Chemistry, Changchun, PR China

Yongrong Sun
School of Chemical Engineering and Technology, Harbin Institute of Technology, Harbin, China

Qiang Tan
School of Chemical Engineering and Technology, Harbin Institute of Technology, Harbin, China

Meiling Xiao
State Key Laboratory of Electroanalytical Chemistry, Changchun Institute of Applied Chemistry, Changchun, PR China; Laboratory of Advanced Power Sources, Changchun Institute of Applied Chemistry, Changchun, PR China

Wei Xing
State Key Laboratory of Electroanalytical Chemistry, Changchun Institute of Applied Chemistry, Changchun, PR China

Liang Yan
State Key Laboratory of Electroanalytical Chemistry, Changchun Institute of Applied Chemistry, Changchun, PR China; Laboratory of Advanced Power Sources, Changchun Institute of Applied Chemistry, Changchun, PR China

Shikui Yao
State Key Laboratory of Electroanalytical Chemistry, Changchun Institute of Applied Chemistry, Changchun, PR China

Geping Yin
School of Chemical Engineering and Technology, Harbin Institute of Technology, Harbin, China

Min Yin
Laboratory of Advanced Power Sources, Changchun Institute of Applied Chemistry, Changchun, PR China

Jiujun Zhang
Energy, Mining and Environment, National Research Council of Canada, Vancouver, BC, Canada

Yuwei Zhang
State Key Laboratory of Electroanalytical Chemistry, Changchun
Institute of Applied Chemistry, Changchun, PR China; Laboratory of
Advanced Power Sources, Changchun Institute of Applied Chemistry,
Changchun, PR China

Xiao Zhao
State Key Laboratory of Electroanalytical Chemistry, Changchun
Institute of Applied Chemistry, Changchun, PR China; Laboratory of
Advanced Power Sources, Changchun Institute of Applied Chemistry,
Changchun, PR China

Jianbing Zhu
State Key Laboratory of Electroanalytical Chemistry, Changchun
Institute of Applied Chemistry, Changchun, PR China; Laboratory of
Advanced Power Sources, Changchun Institute of Applied Chemistry,
Changchun, PR China

OXYGEN SOLUBILITY, DIFFUSION COEFFICIENT, AND SOLUTION VISCOSITY

Wei Xing[a], Min Yin[b], Qing Lv[b], Yang Hu[b], Changpeng Liu[b] and Jiujun Zhang[c]

[a]*State Key Laboratory of Electroanalytical Chemistry, Changchun Institute of Applied Chemistry, Changchun, PR China,* [b]*Laboratory of Advanced Power Sources, Changchun Institute of Applied Chemistry, Changchun, PR China,* [c]*Energy, Mining and Environment, National Research Council of Canada, Vancouver, BC, Canada*

Rotating Electrode Methods and Oxygen Reduction Electrocatalysts. http://dx.doi.org/10.1016/B978-0-444-63278-4.00001-X

1.1. Introduction

Rotating electrode technology including rotating disk electrode (RDE) and rotating ring-disk electrode (RRDE) techniques is one of the important electrochemical measurement methods. Particularly, in studying electrode reaction kinetics and mechanisms, both RDE and RRDE have shown their advantages in measuring reaction electron transfer number, reactant concentration and diffusion coefficient, reaction kinetic constant, and reaction intermediates. In this book, our focus will be on the oxygen reduction reaction (ORR) for fuel cell catalyst evaluation, in which both RDE and RRDE techniques have been approved to be the most power methods in terms of measuring the catalysts' ORR activity as well as their catalyzed kinetics and mechanisms.

The most popular theory for analyzing data collected using RDE and RRDE techniques for catalyzed ORR is called the Koutecky–Levich theory,[1] which gives the relationships among the ORR electron transfer number, O_2 concentration (or solubility), O_2 diffusion coefficient, viscosity of the electrolyte solution, and the electrode rotating rate. By analyzing these relationships, both the ORR kinetics and mechanism can be estimated, from which the activity of the catalysts can be evaluated for further catalyst design and down-selection. Therefore, the O_2 concentration (or solubility), O_2 diffusion coefficient, and viscosity of the measuring electrolyte solution are the most frequently used parameters, and their values must be known in order to do the analysis by Koutecky–Levich theory. Because the measurements are carried out in different electrolyte solutions, at different O_2 partial pressures, or at different temperatures, the values of these three important parameters at various conditions will be given in this chapter. These values will be used in later chapters in this book, and hopefully, these parameters will also be useful for the readers for their study and research.

1.2. Physical and Chemical Properties of Oxygen

Oxygen (abbreviated as O_2) is the first element in Group 16 (VIA) of the periodic table. The O_2 is a diatomic O—O gas and is also called molecular oxygen or dioxygen, which has a molecule weight of 15.99994 g per mole of O_2. In O_2 molecule, two oxygen atoms are connected together through a chemical covalent bond. This bond has a length of 121 pm, and a strength of 494 kJ mol^{-1}.

1.2.1. Physical Properties

O_2 is a colorless, odorless, and tasteless gas under normal conditions. Its content in the atmospheric air is 20.94% by volume or 20% by weight. The density of pure oxygen is 1.429 g dm^{-3} at 273 K and 1.0 atm, which is slightly heavier than air. When cooled below its boiling point ($-183\,^{\circ}$C), O_2 becomes a pale blue liquid; when cooled below its melting point ($-218\,^{\circ}$C), the liquid solidifies, retaining its color. The heat of vaporization is 3.4099 kJ mol^{-1} and the heat of fusion is 0.22259 kJ mol^{-1}. Liquid oxygen is potentially hazardous about flames and sparks in the presence of combustible materials. It can be separated from air by fractionated liquefaction and distillation.

Oxygen exists in three allotropic forms. Allotropes are forms of an element with different physical and chemical properties. The three allotropes of oxygen are normal oxygen, or diatomic oxygen, or dioxygen; nascent, atomic, or monatomic oxygen; and ozone, or triatomic oxygen. The three allotropes differ from each other in a number of ways. First, they differ on the simplest level of atoms and molecules. The oxygen that we are most familiar with in the atmosphere has two atoms in every molecule. By comparison, nascent oxygen has only one atom per molecule. The formula is simply O, or sometimes (O). The parentheses indicate that nascent oxygen does not exist very long under normal conditions. It has a tendency to form normal dioxygen. The third allotrope of oxygen, ozone, has three atoms in each molecule. The chemical formula is O_3. Like nascent oxygen, ozone does not exist for very long under normal conditions. It tends to break down and form normal dioxygen. Ozone does occur in fairly large amounts under special conditions. For example, there is an unusually large amount of ozone in the Earth's upper atmosphere. That ozone layer is important to life on Earth. It shields out harmful radiation that comes from the Sun. Ozone is also sometimes found closer to the Earth's surface. It is produced when gasoline is burned in cars and trucks. It is part of the condition known as air pollution. Ozone at ground level is not helpful to life, and may cause health problems for plants, humans, and other animals. The physical properties of ozone are somewhat different from those of dioxygen. It has a slightly bluish color as both a gas and a liquid. It changes to a liquid at a temperature of $-111.9\,°C$ ($-169.4\,°F$) and from a liquid to a solid at $-193\,°C$ ($-135\,°F$). The density is $2.144\,g\,dm^{-3}$.

Oxygen is part of a small group of gases literally paramagnetic, and it is the most paramagnetic of all. Liquid oxygen is also slightly paramagnetic. O_2 has two states. Mostly, the gas exists as a triplet state but singlet oxygen can also be formed and is more reactive. The electronically excited, metastable singlet oxygen molecules might be involved as the reactive intermediate in dye-sensitized photo-oxygenation.[2] It has two electrons in an unpaired triplet state. Oxygen is the only naturally occurring chemical with this property. The singlet form of oxygen reacts swiftly with almost all compounds.

O_2 not only occurs in the atmosphere but also in oceans, lakes, rivers, and ice caps in the form of water. Nearly 89% of the weight of water is oxygen. It is also the most abundant element in the Earth's crust. Its abundance is estimated at about 45% in the earth. That makes it almost twice as abundant as the next most common element, silicon. For example, O_2 occurs in all kinds of minerals. Some common examples include the oxides, carbonates, nitrates, sulfates, and phosphates. Oxides are chemical compounds that contain oxygen and one other element. Calcium oxide, or lime or quicklime (CaO), is an example. Carbonates are compounds that contain oxygen, carbon, and at least one other element. Sodium carbonate, or soda, soda ash, or sal soda (Na_2CO_3), is an example. It is often found in detergents and cleaning products. Nitrates, sulfates, and phosphates also contain oxygen and other elements. The other elements in these compounds are nitrogen, sulfur, or phosphorus plus one other element. Examples of these compounds are potassium nitrate, or saltpeter (KNO_3); magnesium sulfate, or Epsom salts ($MgSO_4$); and calcium phosphate ($Ca_3\,(PO_4)_2$).

Oxygen is essential and necessary for human life and many processes that occur in living creatures, specifically cellular respiration. For example, O_2 in the air is necessary for humans and animals for breathing, and the small amount of dissolved O_2 in fresh or sea water is sufficient to sustain marine and aquatic life and for the destruction of organic wastes in water bodies. Another example is it is used in mitochondria to help generate adenosine triphosphate during oxidative phosphorylation. One important use of oxygen is in medicine. People who have trouble breathing are given extra doses of O_2. O_2 also has many commercial uses.

The most important use is in the manufacture of metals. More than half of the O_2 produced in the United States is used for this purpose.

1.2.2. Chemical Properties

Oxygen has a high affinity for all chemicals except noble gases. Every element, except fluorine and the noble gases, combines spontaneously with oxygen at galactic standard temperature and pressure. Thus most terrestrial oxygen occurs in form of compound with other elements such as silicates oxides, and water. In this sense, although oxygen is quite stable in nature and its bond dissociation energy is very high, it is a very reactive oxidizing agent. It is the essential element in the respiratory processes of most of the living cells. Oxygen itself does not burn but it supports combustion, combines with most elements, and is a component of hundreds of thousands of organic compounds. The combustion (burning) of charcoal is an example. When oxygen reacts with metals, it forms oxides that are mostly ionic in nature. Rusting is an example. Rusting is a process by which a metal combines with oxygen. For example, iron rust is a reaction product of iron and atmosphere oxygen. Decay is another example. Decay is the process by which once-living material combines with oxygen. The products of decay are mainly carbon dioxide (CO_2) and water (H_2O). Oxygen is rarely featured as the central atom in a molecular structure and can never have more than four elements bonded to it due to its small size and its inability to create an expanded valence shell. When it reacts with hydrogen, it forms water, which is extensively hydrogen-bonded, has a large dipole moment, and is considered as a universal solvent.

In this book, electrochemical ORR is our targeted system. This electrochemical ORR is one of the important chemical reactions of O_2, which is also one of the necessary two reactions in fuel cells and metal–air batteries.

1.3. Oxygen Solubility in Aqueous Solutions

Molecular O_2 has a tendancy to dissolve into liquid media such as both aqueous and nonaqueous solutions. The O_2 solubility can be expressed as gram or mole of O_2 per liter of liquid, or mole of O_2 per cubic milliliter: grams per cubic decimeter ($g\ dm^{-3}$) or moles per cubic decimeter ($mol\ dm^{-3}$), or moles per cubic centimeter ($mol\ cm^{-3}$). In this book, we are only focused on aqueous solutions. For nonaqueous solution, the readers may refer to the relevant books and literature.

1.3.1. Solubility in Pure Water

Oxygen solubility in pure or fresh water at 25 °C and 1.0 atm of O_2 pressure is about 1.22×10^{-3} mol dm^{-3} (the values are varied from 1.18 to 1.25 mol dm^{-3} as reported in different literature).[3,4] In air with a normal composition, the oxygen partial pressure is 0.21 atm, the O_2 solubility would become 2.56×10^{-4} mol dm^{-3}. The solubility of oxygen in water has been the subject of much literature. Currently the oxygen polarographic probe is usually used to measure solubility of O_2 in aqueous solutions.[5]

Normally, oxygen solubility is strongly dependent on (1) the amount of dissolved electrolyte salt(s) (decreases at higher concentration of electrolyte), (2) temperature (decreases at higher temperatures), and (3) pressure (increases at higher pressure). We will give some detailed discussion about these factors and their effects on the O_2 solubility in the following subsections.

1.3.2. Electrolyte, Electrolyte Concentration, and pH Effects on O_2 Solubility

In electrochemical measurements, the solutions used must be ionic conductive except those measurements using microelectrodes. To make an ionic conductive solution, some electrolyte salts are normally dissolved in water according to the designed concentration. These inorganic solutes of major interest are acids, alkalis, and salts. When water is added to one of these electrolytes to form ionic conductive solution, the O_2 solubility in such as solution will be decreased. However, the general pattern of O_2 solubility behavior appears to be complex, and is strongly dependent on the type of the electrolytes and their concentration.[6]

The salting-out effect of most electrolytes on O_2 solubility could be described by a relation originally proposed by Sechenov et al.[5] This relation is given by

$$\log\left(\frac{C_{O_2}^{o}}{C_{O_2}}\right) = KC_{el} \tag{1.1}$$

where C_{O_2} is the oxygen solubility in electrolyte solution with a concentration of C_{el}, $C_{O_2}^{o}$ is the oxygen solubility in pure water, and K is the electrolyte-related O_2 solubility constant. Equation (1.1) indicates that O_2 solubility is decreased with increasing the concentration of electrolyte. For some electrolytes, this theoretic equation has been validated by experimental data with sufficient accuracy, as shown in Figure 1.1(A). However, for the electrolytes containing multiple components, the O_2

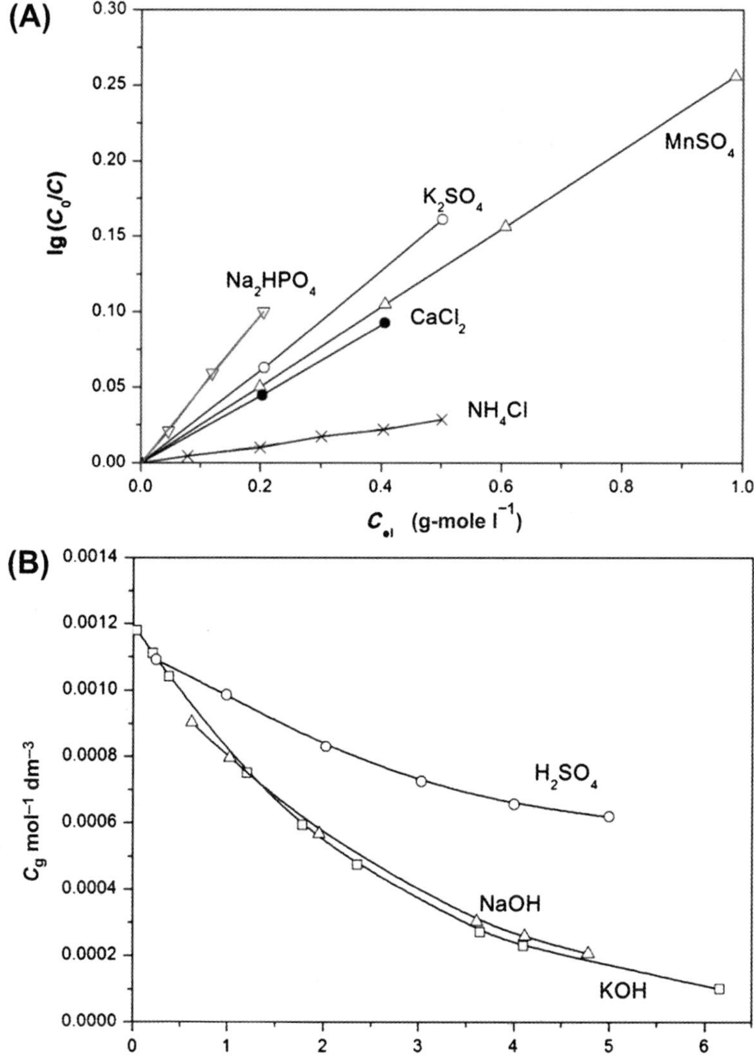

Figure 1.1 (A) Schenov plot according to Eqn (1.1); (B) solubility of O_2 at 25 °C, experimental results for H_2SO_4, NaOH, and KOH.[7−9] (For color version of this figure, the reader is referred to the online version of this book.)

solubility cannot be simply expressed by Eqn (1.1). To address this, John et al.[5] modified Eqn (1.1) into Eqn (1.2):

$$\log\left(\frac{C_{O_2}^o}{C_{O_2}}\right) = \sum_{i=1}^{n} H_i I_i \qquad (1.2)$$

where I_i is the ionic strength of the contributing specific ion i, and H_i is the specific ion-related solubility constant, which can be determined experimentally. If the individual ion's ionic strength can be expressed as $I = \frac{1}{2} C_i Z_i^2$, Eqn (1.2) will become Eqn (1.3):

$$\log\left(\frac{C_{O_2}^0}{C_{O_2}}\right) = \frac{1}{2} \sum_{i=1}^{n} H_i C_i Z_i^2 \qquad (1.3)$$

where C_i is the molar concentration of ion i (moles per cubic decimeter), Z_i is the charge number of ion i, and the sum is taken over all n ions in the solution. If the values of H_i for ionic species could be obtained from the results of relatively simple experiments, Eqn (1.3) can be used in the prediction of oxygen solubility in other electrolyte solutions containing the same ion regardless of the type of compounds from which they were originally formed.[7]

In the study of ORR using rotating electrodes, the commonly employed electrolytes are H_2SO_4, $HClO_4$, NaOH, and KOH. The data of O_2 solubility in these electrolyte solutions are very useful. Figure 1.1(B) gives the plots of O_2 solubility vs electrolyte concentration for these H_2SO_4, NaOH, and KOH electrolyte solutions.[7–9] Unfortunately, it is very difficult to find the solubility of O_2 in $HClO_4$. To be clearer, Table 1.1 lists all O_2 solubility data used for Figure 1.1(B). From Figure 1.1(B), it can be seen that the O_2 solubility falls off with increasing concentration of H_2SO_4, NaOH, or KOH.

Because the commonly used electrolyte solutions in ORR are acidic and alkaline solutions, the pH effect on the O_2 solubility may be considered. Solution pH is determined by the concentration of acid such as H_2SO_4, or alkali such as KOH and NaOH.

Table 1.1. O_2 Solubility in Different Electrolyte Solutions

Electrolyte Concentration (mol dm^{-3})	O_2 Solubility in Aqueous Electrolyte Solution (mol dm^{-3})		
	H_2SO_4	NaOH	KOH
0.05	1.11×10^{-3}	1.10×10^{-3}	1.16×10^{-3}
0.10	1.10×10^{-3}	1.08×10^{-3}	1.15×10^{-3}
0.50	1.05×10^{-3}	0.92×10^{-3}	1.00×10^{-3}
1.00	0.98×10^{-3}	0.81×10^{-3}	0.83×10^{-3}
2.0	0.98×10^{-3}	0.56×10^{-3}	0.55×10^{-3}
3.0	0.72×10^{-3}	0.37×10^{-3}	0.40×10^{-3}
4.0	0.66×10^{-3}	0.24×10^{-3}	0.27×10^{-3}

Note that the O_2 solubility in pure water is taken as $\sim 1.25 \times 10^{-3}$ mol dm^{-3}.

The effect of pH on the O_2 solubility can be understood in terms of the effect of ionic strength, as described by Eqn (1.3). For H_2SO_4, KOH, or NaOH aqueous solution, different pH means different concentration of acid or alkali, then different ionic strengths. According to Eqn (1.3), the logarithm of O_2 solubility is inversely proportional to the ionic strength, the larger the ionic strength, the smaller the O_2 solubility would be. In an electrolyte solution only containing H_2SO_4 solute, increasing pH means the decrease in H_2SO_4 concentration, leading to a decreased solution ionic strength. As a result, the O_2 solubility will be increased. In an electrolyte solution only containing KOH solute, increasing pH means the increase in KOH concentration, leading to an increased solution ionic strength. As a result, the O_2 solubility will be decreased.

1.3.3. Temperature Effect on O_2 Solubility

Oxygen solubility is strongly dependent on temperature. The higher the temperature, the smaller the O_2 solubility. For quantitative expression of O_2 solubility (C_{O_2} with a unit of moles per cubic decimeter) in pure water as a function of temperature in the range of less than 373 K, the following Eqn (1.4) may be used[10,11]:

$$C_{O_2} = \frac{55.56 P_{O_2}}{\exp\left(3.71814 + \frac{5596.17}{T} - \frac{1049668}{T^2}\right) - P_{O_2}} \tag{1.4}$$

where P_{O_2} is the partial pressure above the solution (atm), and T is the temperature (K). Equation (1.4) indicates that the O_2 solubility is decreased with increasing temperature. Table 1.2 lists the experimental values for oxygen solubility at the selected temperatures in different electrolyte solutions. It is shown that the solubility of oxygen is reduced with decreasing temperature between 273.15 and 373.15 K. For a clearer observation, Figure 1.2 shows the plots of C_{O_2} vs T.

Actually, different measurements of C_{O_2} gave slightly different values due to experimental errors. For example, Figure 1.3 shows the O_2 solubility in pure water from different sources.[12]

1.3.4. Pressure Effect on O_2 Solubility

In general, O_2 solubility in aqueous solution is governed by Henry's law, which states that at a constant temperature, the amount of a given gas that dissolves in a given type and volume of liquid is directly proportional to the partial pressure of that gas in equilibrium with that liquid. For O_2 solubility, Henry's law can be put into mathematical terms (at constant temperature) as

Table 1.2. O_2 Solubility at Different Temperatures and 101.325 kPa Pressure in Different Aqueous Solutions[6,12–14]

	O_2 Solubility in Aqueous Electrolyte Solution (mol dm^{-3})			
Temperature (K)	Pure Water	H_2SO_4 (1.0 mol dm^{-3})	NaOH (1.0 mol dm^{-3})	KOH (0.93 mol dm^{-3})
273	2.19×10^{-3}	1.77×10^{-3}	1.42×10^{-3}	1.47×10^{-3}
283	1.70×10^{-3}	1.41×10^{-3}	1.13×10^{-3}	1.17×10^{-3}
293	1.39×10^{-3}	1.17×10^{-3}	0.92×10^{-3}	0.95×10^{-3}
303	1.21×10^{-3}	0.99×10^{-3}	0.77×10^{-3}	0.81×10^{-3}
313	1.04×10^{-3}	0.87×10^{-3}	0.67×10^{-3}	0.71×10^{-3}
323	0.95×10^{-3}	0.79×10^{-3}	0.61×10^{-3}	0.64×10^{-3}
333	0.88×10^{-3}	0.72×10^{-3}	0.57×10^{-3}	0.58×10^{-3}
343	0.83×10^{-3}	0.69×10^{-3}	0.53×10^{-3}	0.55×10^{-3}
353	0.79×10^{-3}	0.67×10^{-3}	0.52×10^{-3}	0.53×10^{-3}
363	0.77×10^{-3}	0.65×10^{-3}	0.51×10^{-3}	0.53×10^{-3}
373	0.76×10^{-3}	0.65×10^{-3}	0.51×10^{-3}	0.53×10^{-3}

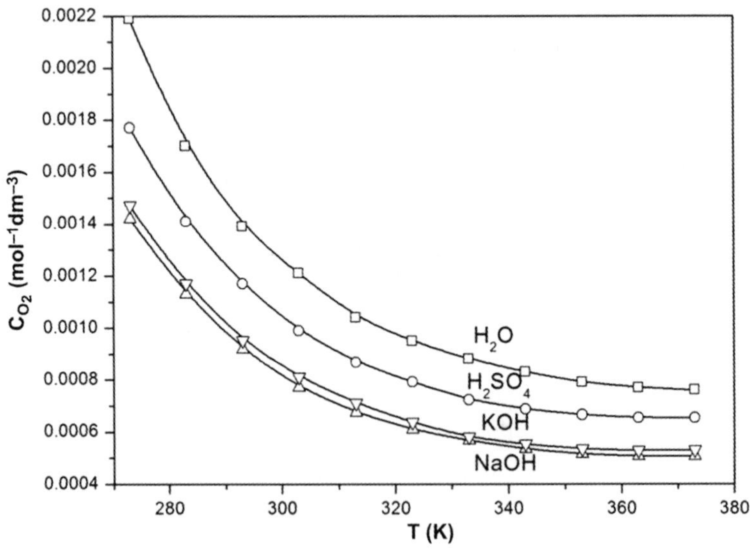

Figure 1.2 O_2 solubility as a function of temperature in different electrolyte solutions. O_2 pressure: 101.325 kPa.[6,12–14]

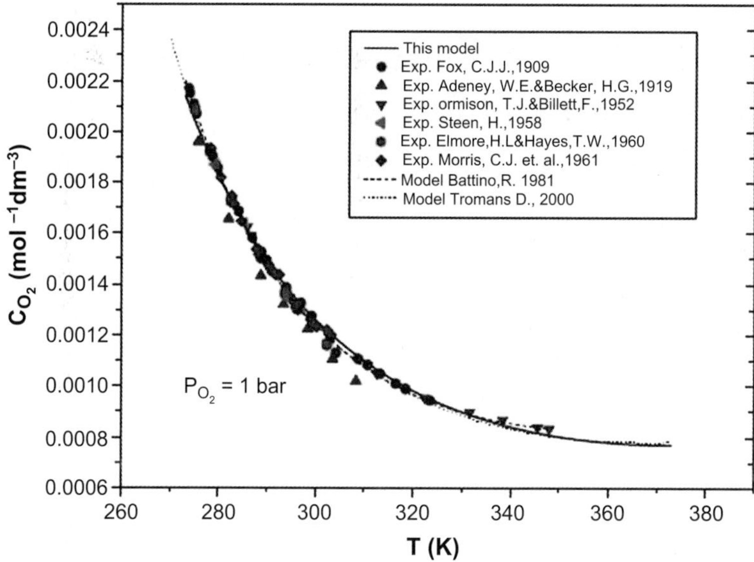

Figure 1.3 Solubility of oxygen in pure water at 1.0 bar.[12] (For color version of this figure, the reader is referred to the online version of this book.)

$$C_{O_2} = K_H P_{O_2} \tag{1.5}$$

where P_{O_2} is the partial pressure of O_2 in the gas above the solution (atmosphere), C_{O_2} is the concentration of the O_2 in the solution (moles per cubic decimeter), and K_H is the constant with the dimensions of pressure divided by concentration (moles per cubic decimeter per atmosphere). In pure water at 298 K, K_H has a value of 769.2 mol dm^{-3} atm^{-1}. It is obvious that the solubility of oxygen increases with increasing the partial pressure of oxygen at a constant rate of K_H. This Henry's law constant is normally dependent on the solute, the solvent, and the temperature. Table 1.3 lists some values of K_H at different temperatures in pure water.

As an example, Figure 1.4 shows the solubility of oxygen in pure water with varying pressure at five different temperatures.[7] It can be seen that the increased rate of oxygen solubility with increasing O_2 partial pressure is different at different temperatures.

1.4. O_2 Diffusion Coefficients in Aqueous Solution

In nature, a substance (a molecule or an ion) dissolved in a liquid such as aqueous solution will flow from a region of its high concentration to a region of its low concentration. To describe

Table 1.3. Henry's Law Constants at Different Temperatures in Pure Water[15]

Temperature (K)	273	283	293	303	313	323	333	353	373
Henry's law constants in pure water (10^3 mol dm^{-3} atm^{-1})	2.18	1.70	1.38	1.17	1.03	0.93	0.87	0.79	0.76

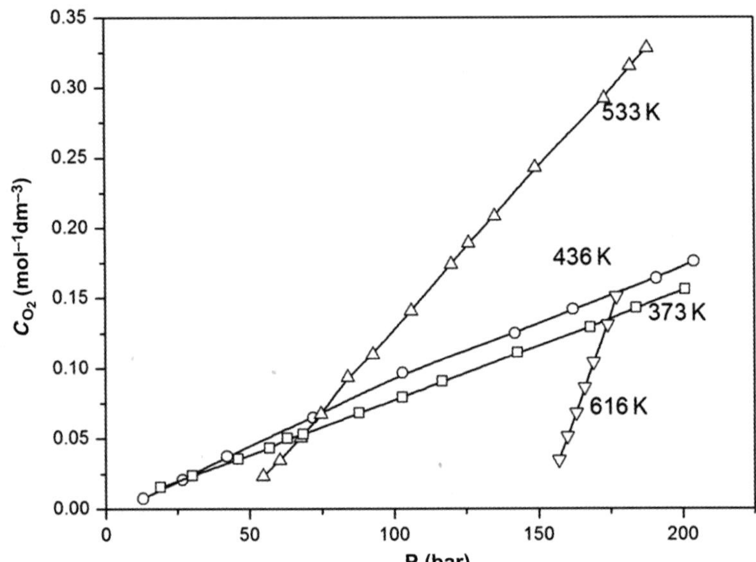

Figure 1.4 Solubility of oxygen in water with varying pressure.[7]

such a natural tendency, Fick's first law is applied, which can be expressed as

$$J = -D\frac{dC}{dx} \tag{1.6}$$

where J is the diffusion flux—the amount of substance that flows through a unit area per unit time (moles per cubic centimeter per second), D is the diffusion coefficient (square centimeters per second), dC is the concentration change of substance (moles per cubic centimeter), and dx is the length of the diffusion path (centimeters). It can be seen that this diffusion coefficient is a parameter determining how fast the corresponding substance can diffuse within the solution. For a deeper description about the diffusion process, please see Chapter two of this book.

In the case of O_2 dissolved in an aqueous solution, Eqn (1.6) will become

$$J_{O_2} = -D_{O_2}\frac{dC_{O_2}}{dx} \tag{1.7}$$

where J_{O_2} is the diffusion flux—the amount of dissolved O_2 that flows through a unit area per unit time (moles per square centimeter per second), D_{O_2} is the diffusion coefficient of O_2 in the aqueous solution (square centimeters per second), dC_{O_2} is the concentration of O_2 (moles per cubic centimeter), and dx is the length of the diffusion path (centimeters). The O_2 diffusion coefficient is one of the necessary parameters in analyzing catalyzed ORR using Koutecky–Levich theory. Its value is strongly dependent on temperature/pressure, the type of the solution, as well as the electrolyte type used. In the following subsections, O_2 diffusion coefficients at different temperatures/pressures in different aqueous electrolyte solutions will be discussed.

1.4.1. O_2 Diffusion Coefficiencies in Pure Water

The estimation of oxygen diffusion coefficient (D_{O_2}) in aqueous solution could be calculated based on the Stokes–Einstein equation using several known parameters such as the molecular weight of water, the absolute temperature, the solution viscosity, and the molar volume of water. For a more detailed discussion, please see the publication from Wilke and Chang.[16] Based on this calculation, different experimental ways have been developed to determine the oxygen diffusion in liquid solutions. For example, Hung and Dinius[17] measured the diffusivities of oxygen dissolved in aqueous solutions by means of a diaphragm cell technique. Holtzapple and Eubank[18] compared three models of diffusion of oxygen through aqueous solutions. Based on mole fraction, chemical potential, or oxygen activity as the driving force, they demonstrated that the three models did not differ significantly in their predictions, except for the extremely high oxygen partial pressure. The mean experimental value of D_{O_2} in pure water is about 2×10^{-5} cm^2 s^{-1} at 20 °C and is sensitive to the temperature. Experimental data for D_{O_2} in pure water by several research groups as examples were shown in Figure 1.5 and Table 1.4.[19] It can be seen that the value of O_2 diffusion coefficient in pure water at 25 °C and 1.0 atm O_2 pressure is varied from 1.9 to 2.3×10^{-5} cm^2 s^{-1}.

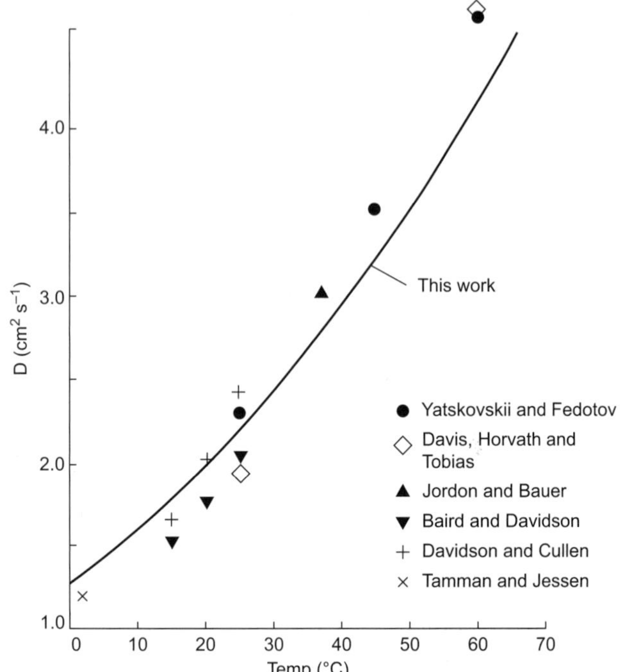

Figure 1.5 Experimental oxygen diffusion coefficients in pure water. Note that the label of Y-axis should be "D, 10^5 cm^2 s^{-1}" rather than "D, cm^2 s^{-1}".[19]

Table 1.4. O$_2$ Diffusion Coefficients in Pure Water at Different Temperatures and 1.0 atm O$_2$ Pressure

Temperature (K)	O$_2$ Diffusion Coefficients in Pure Water (10^5 cm^2 s^{-1})	References
273	1.2	20
288	1.5—1.7	19,21,22
293	2.0	21
298	1.9—2.3	21—23
308	2.6—2.9	19,23
318	3.4	24
333	4.0—4.6	8,24

1.4.2. Electrolyte and Electrolyte Concentration Effects on O_2 Diffusion Coefficient

O_2 diffusion coefficient (D_{O_2} with a unit of square centimeters per second) could be described over the concentration range of electrolytes studied by the Stokes–Einstein equation[25]:

$$D_{O_2} = \frac{kT}{6\pi r\eta} \tag{1.8}$$

where k is the Boltzmann constant (kilogram square centimeters per square second), T is the absolute temperature (Kelvin) is, η is the dynamic viscosity of the solution (kilogram per second per centimeter), and r is the radius of the O_2 molecule (assumed spherical) (centimeters).

From Eqn (1.8), it can be seen that O_2 diffusion coefficient is increased with increasing temperature, and decreased with increasing either the dynamic viscosity of the solution or the radius of the O_2 molecule. Because different electrolyte solutions have different values of dynamic viscosity, the O_2 diffusion coefficient will be strongly dependent on the type/nature of the electrolytes and the electrolyte concentration, which has been proved experimentally. For example, at 25 °C and for KOH concentrations between 1.0 and 8.0 mol dm^{-3}, the O_2 diffusion coefficient changed significantly.[26] Bard et al.[27] calculated D_{O_2} in aqueous solutions containing various concentrations of NaOH (1–12 mol dm^{-3}). The radius of the dissolved O_2 molecule was found to be 2.8 Å, which was independent of NaOH concentration in the concentration range from 2 to 12 mol dm^{-3}. However, the value of D_{O_2} was found to decrease significantly when the solution viscosity was increased in the concentration range from 2 to 12 mol dm^{-3}. The same change of D_{O_2} with increasing concentration was also found in KOH solutions.[26] In addition, D_{O_2} in KOH solutions was found to be always higher than that in the corresponding same concentration of NaOH solutions because of both the lower viscosity of KOH solution and the smaller oxygen radii (1.68 Å) in KOH solution than that in NaOH solution. The calculated and experimental D_{O_2} in NaOH and KOH solutions at 298 K are shown in Figure 1.6(A). Figure 1.6(B) gives the oxygen diffusion coefficients in H_3PO_4 electrolyte with different concentrations.[28] Table 1.5 lists the values of D_{O_2} in different electrolyte solutions for H_3PO_4, $HClO_4$, H_2SO_4, NaOH, and KOH with different concentrations. It can be seen that all D_{O_2} values are decreased with increasing the concentration of electrolytes. As discussed above, this is because the higher the electrolyte

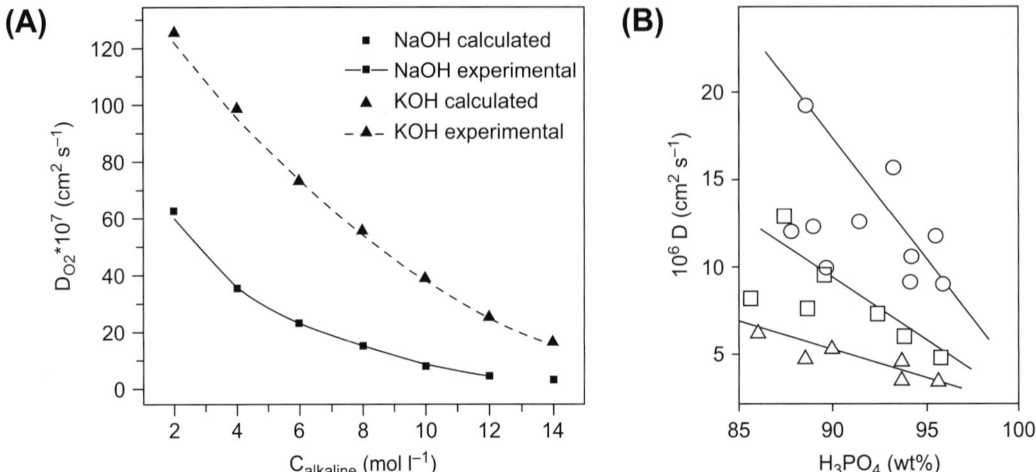

Figure 1.6 (A) Calculated and experimental oxygen diffusion coefficients in NaOH and KOH solutions at 298 K[26]; (B) oxygen diffusion coefficient vs H_3PO_4 concentration. ($\triangle$) 100 °C; ($\square$) 120 °C; ($\bigcirc$) 150 °C.[28]

Table 1.5. O_2 Diffusion Coefficients in Different Electrolyte Solutions

Electrolyte Concentration (mol dm^{-3})	O_2 Diffusion Coefficient in Aqueous Electrolyte Solution (D_{O_2}) ($\times 10^7$ cm^2 s^{-1})				
	H_3PO_4	$HClO_4$	H_2SO_4	NaOH (25 °C)	KOH (25 °C)
0.1	14.6 mol dm^{-3}:	0.1 mol dm^{-3}:	0.5 mol dm^{-3}:	222	190
1	66.6 (100 °C)	167 (25 °C)	140 (25 °C)	165	–
2	82.9 (120 °C)			60.4	126
4				33.8	97.9
6				22.6	73.8
8				15.2	55.4
10				8.08	38.3
12				4.7	24.8
14				2.06	15.5

concentration, the higher the solution viscosity, leading to the reduced D_{O_2} values. It can also be seen that different types of electrolytes can give different D_{O_2} values due to their difference in solution viscosity.

1.4.3. Temperature Effect on O_2 Diffusion Coefficient

As indicated by Eqn (1.8), the value of D_{O_2} is strongly dependent on the temperature, which increases with increasing temperature, as shown in Figure 1.5 for pure water. The variation of D_{O_2} in H_3PO_4 solutions with increasing temperature is also shown in Figure 1.6(B). A typical example is the diffusion coefficient data in 0.1 M NaOH solution from 0 to 65 °C, measured using a rotating Pt electrode.[19] The data were related to temperature as expressed as an Arrhenius form:

$$\log(D_{O_2}) = \log(D_{O_2}^0) - \frac{E_D}{2.303RT} \tag{1.9}$$

where $D_{O_2}^0$ is a constant and E_D is the apparent activation energy, both of which were obtained to be $8.03 \times 10^{-3}\ cm^2\ s^{-1}$ and $3.49\ kcal\ mol^{-1}$, respectively, using Figure 1.7; R is the universal gas constant, and T is the temperature.

1.4.4. Pressure Effect on O_2 Diffusion Coefficient

The pressure effect on O_2 diffusion coefficient (D_{O_2}) is mainly reflected by its effect on the solution viscosity.[29,30] Regarding the effect of pressure on the water viscosity, in general, this effect is insignificant. For instance, in the pressure range of 0–2 atm, the change of relative viscosity of a pure water solution was from

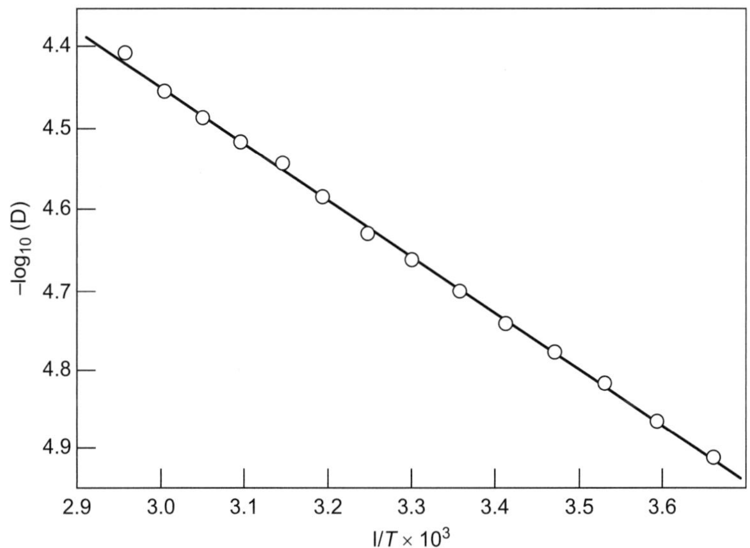

Figure 1.7 Logarithm of the O_2 diffusion coefficient in 0.1 M NaOH solution as a function of the reciprocal of the absolute temperature.[19]

0.93 to 1.05.[31] In a detailed study by Horne et al.,[31] there were two temperature ranges such as below 25 °C, and higher than 25 °C, where pressure had different effects. In the temperature range below 25 °C, the viscosity of water was decreased with increasing pressure at pressure below 1.0 atm, when the pressure was increased to higher than 25 °C, the viscosity was increased with increasing pressure with a turning point at about 1.0 atm.[32] However, in the temperature range higher than 25 °C, the viscosity was increased with increasing pressure from zero pressure to several atmospheres.

As discussed above, the viscosity change of aqueous solution induced by the pressure change would result in a change of O_2 diffusion coefficient. The relationship between D_{O_2} and pressure could be related by a parameter V_a[29]:

$$\log(D_{O_2}) = \log(D_{O_2}^p) - \frac{V_a P}{2.303 RT} \tag{1.10}$$

where $D_{O_2}^p$ is the O_2 diffusion coefficient at zero pressure and T, P is the solution pressure, V_a is the activation volume, and R is the gas constant. When the value of V_a is positive, D_{O_2} will decrease with increasing pressure; and as V_a is negative, D_{O_2} will increase with increasing pressure. For detailed explanation of this active volume and its relationship with pressure, please read the publication by Shimizu et al.[29]

1.5. Viscosity of Aqueous Solution

The viscosity of a liquid is its internal property that offers resistance to flow. Actually a liquid can be also defined as a material that deforms as long as it is subjected to a tensile or shear stress. Under shear, the rate of deformation (or shear rate) is proportional to the shearing stress. According to Newton's idea, the ratio of the stress to the shear rate is a constant, called viscosity, which is independent of the shear rate. For ideal or "Newtonian" liquids, e.g., water, this viscosity is independent of the shear rate. However, the viscosities of many liquids are not independent of the shear rate, which is called "non-Newtonian" liquids. Such "non-Newtonian" liquids can exhibit either shear-thinning or shear-thickening and can be classified according to their viscosity behavior as a function of the shear rate. Generally, viscosity can be defined in two ways:

1. Dynamic viscosity, also absolute viscosity. The usual symbol for dynamic viscosity used by mechanical and chemical engineers is the Greek letter mu (μ). The symbol η is used by chemists, physicists, and the International Union of Pure and

Applied Chemistry. The dynamic viscosity (η) of a liquid is defined as follows (equal to the gradient of the shear stress vs the shear rate curve):

$$\eta = \frac{d\tau}{d\gamma} \tag{1.11}$$

where τ is the shear stress, and γ is the shear rate. The SI physical unit of dynamic viscosity is the pascal-second (Pa s), (equivalent to newton second per square meter, or kilogram per meter per second). The meaning of a viscosity of 1 Pa s is that one liquid is placed between two plates, and one plate is pushed sideways with a shear stress of 1 Pa, within 1 s it moves a distance equal to the thickness of the layer between the plates. Water at 20 °C has a viscosity of 0.001002 Pa s. The centimeter–gram–second (cgs) physical unit for dynamic viscosity is the *poise* (P), named after Jean Louis Marie Poiseuille. It is more commonly expressed, particularly in American Society for Testing and Materials standards, as *centipoise* (cP) (1 P = 0.1 Pa s, and 1 cP = 1 mPa s = 0.001 Pa s).

2. Kinematic viscosity. The kinematic viscosity (v) of a liquid is the viscosity coefficient divided by the density of the liquid (ρ):

$$v = \frac{\eta}{\rho} \tag{1.12}$$

Typically, kinematic viscosity has a unit of square centimeters per second. The cgs physical unit for kinematic viscosity is the stokes (St) (1 St = 1 cm^2 s^{-1} = 10^{-4} m^2 s^{-1}), named after George Gabriel Stokes. It is sometimes expressed in terms of centiStokes (cSt). In US usage, stoke is sometimes used as the singular form.

In electrochemical measurements using rotating electrode technique and data treatment using Koutecky–Levich theory, the most commonly used solution viscosity is kinematic viscosity. Therefore, in the following sections, we will only focus on this kinematic viscosity.

1.5.1. Viscosity of Pure Water

Kinematic viscosity of pure liquid water at different temperatures up to 100 °C is listed in Table 1.6.

1.5.2. Electrolyte, Electrolyte Concentration, and pH Effects on Viscosity

Generally to say, the electrolyte effect on the viscosity of aqueous solution is a complicated matter and several models

Table 1.6. Kinematic Viscosity of Pure Liquid Water at Different Temperatures and 1.0 atm Pressure[33]

Temperature (K)	Kinematic Viscosity (10^2 cm^2 s^{-1})
273	1.787
278	1.519
283	1.307
293	1.004
303	0.801
313	0.658
323	0.553
333	0.475
343	0.413
353	0.365
363	0.326
373	0.290

with hypothesis have been proposed and discussed since 1847. For a detailed discussion, the readers may go to the literature indicated in this subsection. For the usage in electrochemical rotating electrode technique, the viscosity data at different electrolytes may be more useful. Table 1.7 lists the kinematic viscosity data obtained using aqueous solutions containing H_3PO_4, H_2SO_4, $HClO_4$, KOH, and NaOH.

For a clearer observation of the effect of electrolyte concentration on the viscosity, Figure 1.8 shows the plots of kinetic viscosity vs electrolyte concentration for several typical electrolytes. It can be seen that with increasing electrolyte concentration, the kinematic viscosity of the corresponding solution also increase steadily.

The pH effect on viscosity can be seen from Figure 1.8 at different acid or alkali concentrations. For acidic solutions, the higher the acid concentration (the lower the pH), the higher the kinematic viscosity would be. For alkali solutions, the higher the alkali concentration (the higher the pH), the higher the kinematic viscosity would be.

Table 1.7. Kinematic Viscosity of Aqueous Solutions Containing H_3PO_4,[34] H_2SO_4,[35] $HClO_4$,[36] KOH,[37] and NaOH[38] at Room Temperature and 1.0 atm Pressure

H_3PO_4 (w/w%)	Kinematic Viscosity (10^2 cm^2 s^{-1})	$HClO_4$ (w/w%)	Dynamic Viscosity (cP)	H_2SO_4 (w/w%)	Kinematic Viscosity (10^2 cm^2 s^{-1})	KOH (w/w%)	Kinematic Viscosity (10^2 cm^2 s^{-1})	NaOH (w/w%)	Kinematic Viscosity (10^2 cm^2 s^{-1})
5	0.99	10	1.006	0	0.94	5	0.98	10	1.27
10	1.1	20	1.040	5	0.98	10	1.1	20	1.72
15	1.2	30	1.144	9.39	1.05	15	1.2	30	9.85
20	1.4	40	1.338	13.42	1.05	20	1.5	40	22.38
25	1.6	50	1.782	17.42	1.20	25	1.8	50	44.08
30	1.9	60	2.73	20.34	1.25	30	2.2		
35	2.2			24.1	1.35	35	2.7		
40	2.6			29.8	1.55	40	3.6		
45	3.1			39.7	1.90	45	5.3		
50	3.7			51.2	2.48	50	7.3		
55	4.5			62.5	3.90				
60	5.6			70.9	5.90				
65	6.8			78.2	9.10				
70	9.2			81.4	10.8				
75	12			83.5	11.2				
80	17			87.5	10.8				
85	23			90.3	10.0				
90	34			94.75	9.6				
95	55			98.3	11.0				
100	100			99.6	13.2				

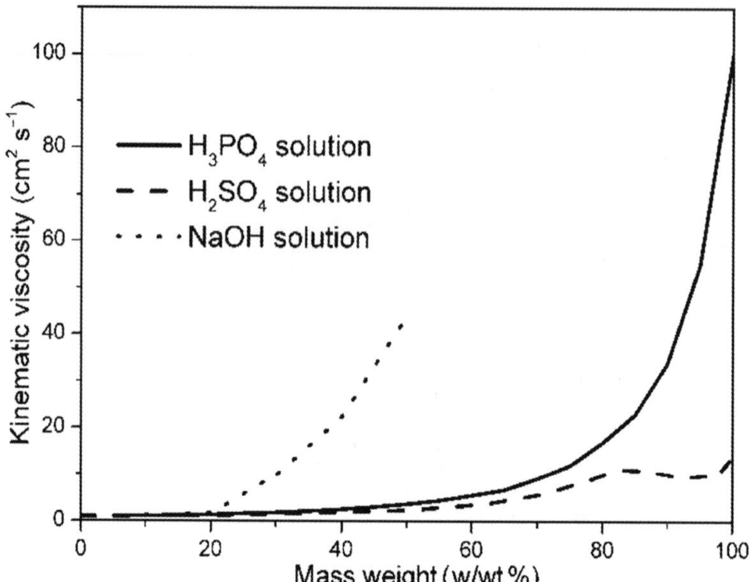

Figure 1.8 Kinetic viscosity as a function of electrolyte concentration at three typical electrolytes. Data from Table 1.7, Refs 34,35,38.

1.5.3. Temperature Effect on Viscosity

Normally, the viscosity of a liquid decreases with increasing temperature, as seen in Table 1.6 for pure liquid water. For quantitative expression of the temperature effect on the viscosity, several models, such as the Eyring model,[39] the exponential model,[40] Arrhenius model,[41] and Williams−Landel−Ferry model,[42] have been proposed and validated using experimental data. The typical equation relating kinematic viscosity (v) of the solution to temperature may be expressed as an Arrhenius form:

$$v = v^{\mathrm{o}} \exp\left(\frac{E_v}{RT}\right) \tag{1.13}$$

where v^{o} is the kinetic viscosity when $T \rightarrow \infty$, and E_v is the active energy of viscous flow. Both v^{o} and E_v can be obtained experimentally by measuring v at different temperatures.

For the usage of electrochemical rotating electrode technique, Table 1.8 lists the data of kinematic viscosity at several typical electrolyte solutions as a function of temperature.

For a clearer observation of the temperature effect on the viscosity, Figure 1.9 shows the plots of kinetic viscosity vs temperature for several typical electrolytes. It can be seen that an increase in temperature leads to the decrease of viscosity of all the five solutions.

Table 1.8. Kinematic Viscosity at Several Typical Electrolyte Solutions as a Function of Temperature (1.0 atm)

	Kinetic Viscosity (Dynamic Viscosity for $HClO_4$) (10^2 cm^2 s^{-1})					
Temperature (K)	Pure Water[33]	H_3PO_4[34] (10 wt%)	$HClO_4$[36] (20 wt%)	H_2SO_4[35] (39.92 wt%)	NaOH[38] (20 wt%)	KOH[37] (20 wt%)
273	1.787	—	1.779	—	—	2.7
283	1.307	—	1.357	3.48	5.24	—
293	1.004	1.2	1.04	2.7	3.94	1.6
303	0.801	0.99	0.867	2.2	2.63	1.3
313	0.658	0.83	0.721	1.85	1.98	1.1
323	0.553	0.71	0.61	1.6	1.58	0.94
333	0.475	0.61	—	1.4	1.34	0.82
343	0.413	0.54	—	1.1	1.06	0.72
353	0.365	0.47	—	—	0.93	0.63
363	0.326	0.42	—	—	0.76	0.57
373	0.290	0.38	—	—	—	0.51

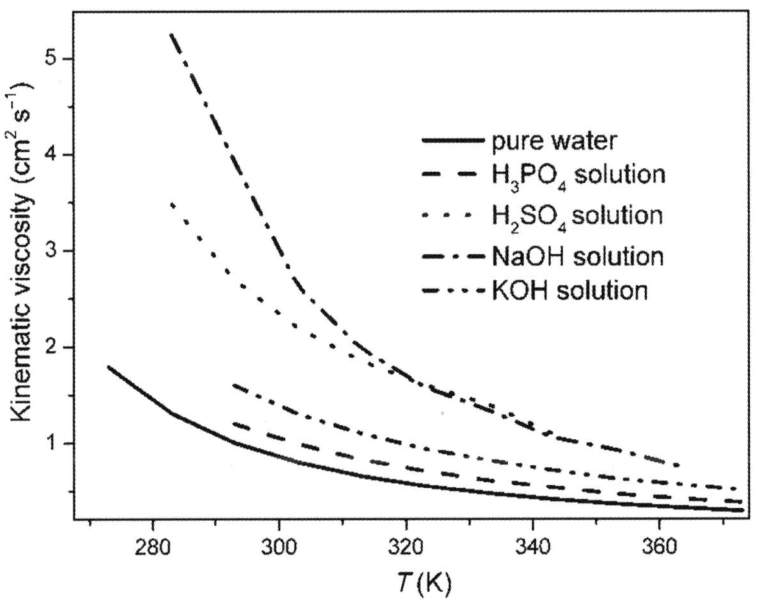

Figure 1.9 Kinetic viscosity as a function of temperature at several typical electrolytes. Data from Table 1.8.

1.5.4. Pressure Effect on Viscosity

In general, the viscosity of an aqueous solution is almost independent of pressure except at very high pressures. While for water, there are two interesting viscosity phenomena at high pressures. Figure 1.10 shows the experimental results for the relative viscosity of water.[43]

From Figure 1.10, it can be seen that the plots of kinematic viscosity vs pressure at temperatures below 308 K have a minimum at 0–1973 atm, and the calculation indicated that the activation energy of viscous flow has a minimum at about 1973 atm. These decreases in viscosity and activation energy with increasing pressure on the low-pressure side of the minima were ascribed to a break of the bulky water structure like hydrogen-bonded tetrahedra. However, when the pressure is increased to a value of higher than 2000 atm, a monotonic increase with further increasing pressure can be observed. The similar trend was observed not only for pure liquid water but also for several electrolyte solutions.[43]

1.6. Oxygen Solubility and Diffusion Coefficient in Nafion® Membranes

In electrochemical measurements of ORR using rotating electrode technique, the surface of working electrode (RDE) is

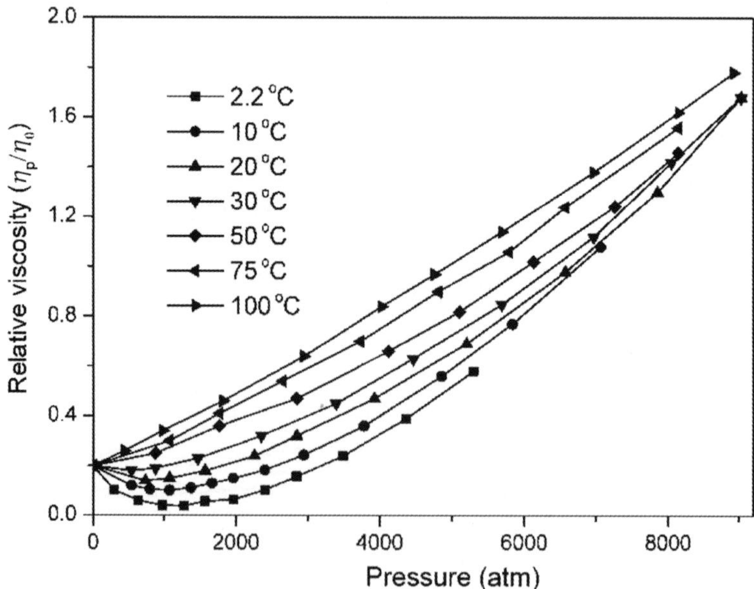

Figure 1.10 Relative viscosity of water as a function of pressure.[43]

normally coated by a catalyst layer. To prevent the catalyst particles from falling off the electrode surface and to make the catalyst layer ionic conductive, a conductive ionomer such as Nafion[®] ionomer is normally applied on the catalyst layer. This layer of Nafion[®] ionomer will have effect on the O_2 diffusion. Therefore, both the O_2 solubility and diffusion coefficient must be known in order to count the ionomer effect on the catalyst ORR current. In this section, the data of O_2 solubility and diffusion coefficient at different temperatures, pressures, and water contents will be given as follows.

1.6.1. Temperature Effect on both the O_2 Solubility and Diffusion Coefficient

In general, both the O_2 solubility and the diffusion coefficient in Nafion[®] membrane or ionomer increase with increasing temperature. The experimental data for C_{O_2} and D_{O_2} on Nafion[®] 112, 211, and 117 membranes are summarized in Table 1.9 from different literature. It can be seen that both the values of C_{O_2} and D_{O_2} are varied with type of the Nafion[®] membranes and the temperature.

1.6.2. Pressure Effect on both the O_2 Solubility and Diffusion Coefficient

The solubility of O_2 in Nafion[®] increases linearly with pressure and follows Henry's law for dilute gases, as indicated by Eqn (1.5). The experimental results on Nafion[®] 117 are shown in Table 1.10.[46] As expected, the solubility of O_2 in Nafion[®] increases linearly with pressure. However, the diffusion coefficient of O_2 in Nafion[®] 117 is almost invariant with O_2 pressure between 2 and 3.5 atm, and then increased with increasing pressure when the pressure is higher than 3.5 atm. This result is consistent with the recent results, which found D_{O_2} was constant within the O_2 pressure between 1 and 3 atm.[45]

1.6.3. Water Content Effect on both the O_2 Solubility and Diffusion Coefficient

In normal, water content in Nafion[®] membrane or ionomer layer has a significant effect on both the O_2 solubility and diffusion coefficient. For a detailed discussion about the relationship among water content, bulk membrane structure, and O_2 solubility/diffusion, the readers may go to Refs 47–49.

Table 1.9. Oxygen Solubility and Diffusion Coefficients in Nafion® 112, 211, and 117 Membranes at Different Temperatures and 1.0 atm

Nafion® Membrane	Temperature (°C)	D_{O_2} (10^6 cm^2 s^{-1})	C_{O_2} (10^6 mol cm^{-3})	References
211[a]	30	1.13	11.6	44
	40	2.29	10.6	
	50	4.20	9.45	
	60	4.02	10.6	
	65	7.90	7.99	
	70	30.0	3.84	
117[b]	25	0.577	15.50	45
	40	1.079	16.91	
	50	1.349	15.26	
	60	1.350	16.54	
	70	1.225	17.19	
	80	1.316	14.15	
112[b]	25	0.1048	102.63	45
	35	0.2049	84.13	
	45	0.1877	113.50	
	55	0.1715	142.72	
	65	0.1895	139.46	
	75	0.2331	115.62	

[a]100% Relative humidity (RH), 30 psi oxygen.
[b]Dry solid content of Nafion® is 25%.

The experimental data of oxygen solubility (C_{O_2}) and diffusion coefficient (D_{O_2}) in Nafion® 211 membrane, as an example, are listed in Table 1.11.[44] Note that the water content in Table 1.11 can be expressed as λ, which is defined as hydration number. For a clearer observation, based on Table 1.11, Figure 1.11 shows both O_2 solubility and diffusion coefficient as a function of water content. It can be seen that oxygen diffusion coefficient is proportional to water content, and oxygen solubility is not very sensitive to the water content.

As a matter of fact, a large variation in the values of O_2 solubility and diffusion coefficient has been reported from different sources. This may be attributed to the difference in the experiment conditions. As seen from Figure 1.11, both these parameters

Table 1.10. Oxygen Solubility and Diffusion Coefficient Data in Nafion® 117 Membrane for O_2 Pressure Ranging from 2 to 5 atm and at a Temperature of 323 K[46]

Pressure (atm)	D_{O_2} (10^6 cm^2 s^{-1})	C_{O_2} (10^6 mol cm^{-3})
2.0	5.24	6.36
2.5	5.46	7.89
3.0	5.48	9.35
3.5	5.47	10.93
4.0	5.71	11.83
4.5	6.71	12.10
5.0	7.07	12.52

Table 1.11. O_2 Solubilities and Diffusion Coefficients in Nafion® 211 Membrane at Different Water Contents, 30 °C and 1.0 atm Pressure[44]

Water Content (λ) (Hydration Number)	C_{O_2} (10^5 mol cm^{-3})	D_{O_2} (10^5 cm^2 s^{-1})
12.4	1.16	0.144
14.6	1.06	0.258
15.8	1.05	0.416
17	0.502	0.876
20	0.402	3.03

are dependent on the water content, suggesting that different experiment conditions such as temperature and humidity could give different membrane water contents, leading to different O_2 solubilities and diffusion coefficients. Table 1.12 lists several values measured at different experiment conditions.

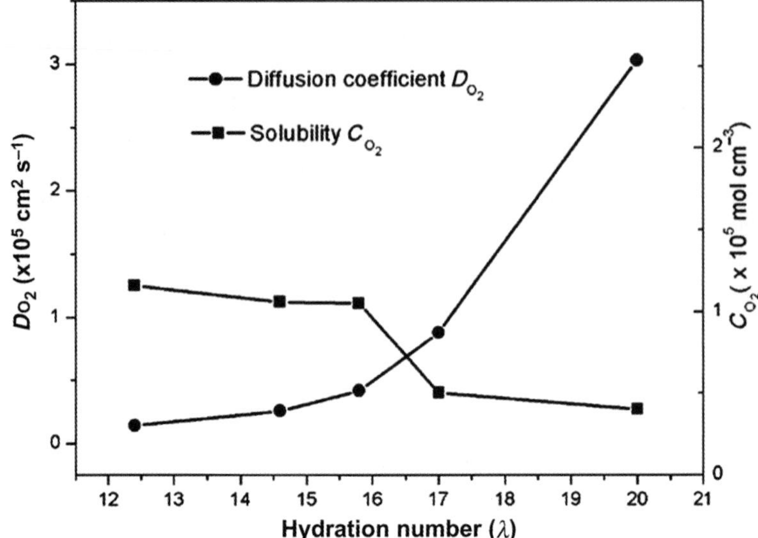

Figure 1.11 Diffusion coefficient, solubility, and permeability of oxygen vs water content for Nafion® 211, measured at 100% RH, 30 psi O_2 pressure.[44]

Table 1.12. O_2 Solubilities and Diffusion Coefficients in Nafion® 117 Membrane at Different Conditions

Experimental Condition	C_{O_2} (10^6 mol cm^{-3})	D_{O_2} (10^6 cm^2 s^{-1})	References
30% RH, 25 °C, 1.0 atm O_2	5.2	0.33	50
56% RH, 25 °C, 1.0 atm O_2	5.1	0.56	50
82% RH, 25 °C, 1.0 atm O_2	4.0	1.1	50
100% RH, 25 °C, 1.0 atm O_2	4.8	2.6	51
100% RH, 30 °C, 3.0 atm O_2	6.0	9.2	52
100% RH, 40 °C, 5.0 atm O_2	5.76	2.88	53
35% RH, 60 °C, 1.0 atm O_2	4.6	1.1	50
56% RH, 60 °C, 1.0 atm O_2	4.1	1.4	50
75% RH, 60 °C, 1.0 atm O_2	3.8	1.7	50
20 °C, 1.0 atm O_2	13	0.7	54
25 °C, 1.0 atm O_2	18.7	0.62	55

1.7. Chapter Summary

In electrochemical measurements and data analysis for ORR using both rotating electrode techniques and theory, the O_2 solubility and diffusion coefficient as well as the viscosity of the aqueous electrolyte solution employed are necessary parameters. In this chapter, the scientific fundamentals of these parameters are given in a detailed level to facilitate the readers to understand the meanings of these parameters. The effects of type/concentration of electrolyte, temperature, and pressure on values of these parameters are also discussed. In order to provide the readers with useful information, the values of O_2 solubility, diffusion coefficient as well as the viscosity of the electrolyte solution are collected from literature, and summarized in several tables. In addition, the values of both the O_2 solubility and diffusion coefficient in Nafion® membranes or ionomers are also listed in the tables. It is our wish that this chapter would be able to serve as a data source for the later chapters of this book, and also the readers could find it useful in their experimental data analysis.

References

1. Faulkner LF, Bard AJ. *Electrochemical methods: fundamentals and applications.* New York: Wiley; 2005.
2. Kearns DR. Physical and chemical properties of singlet molecular oxygen. *Chem Rev* 1971;**71**(4):395–427.
3. Liu FL, Ma LP, Cheng HM. Advanced materials for energy storage. *Adv Mater* 2010;**22**:E1–35.
4. Stanis RJ, Kuo MC, Turner JA, Herring AM. Use of W, Mo, and V substituted heteropolyacids for CO mitigation in PEMFCs. *J Electrochem Soc* 2008;**155**(2):B155–62.
5. Schumpe A, Adler I, Deckwer WD. Solubility of oxygen in electrolyte solutions. *Biotechnol Bioeng* 1978;**20**(1):145–50.
6. Tromans D. Oxygen solubility modeling in inorganic solutions: concentration, temperature and pressure effects. *Hydrometallurgy* 1998;**50**(3):279–96.
7. Narita E, Lawson F, Han KN. Solubility of oxygen in aqueous electrolyte solutions. *Hydrometallurgy* 1983;**10**(1):21–37.
8. Davis RE, Horvath GL, Tobias CW. The solubility and diffusion coefficient of oxygen in potassium hydroxide solutions. *Electrochim Acta* 1967;**12**(3):287–97.
9. Yasunishi A. Solubility of oxygen in aqueous electrolyte solutions. *Kagaku Kogaku Ronbun* 1978;**4**(2):5.
10. Khomutov SG. Solubility of oxygen in electrolyte solutions. *Russ Chem Rev* 1990;**59**(8):21.
11. Benson B, Krause D, Peterson M. The solubility and isotopic fractionation of gases in dilute aqueous solution. I. Oxygen. *J Solution Chem* 1979;**8**(9):655–90.
12. Geng M, Duan Z. Prediction of oxygen solubility in pure water and brines up to high temperatures and pressures. *Geochim Cosmochim Acta* 2010;**74**(19):5631–40.
13. Shoor SK, Walker RD, Gubbins KE. Salting out of nonpolar gases in aqueous potassium hydroxide solutions. *J Phys Chem* 1969;**73**(2):312–7.

14. Bruhn G, Gerlach J, Pawlek F. Untersuchungen über die Löslichkeiten von Salzen und Gasen in Wasser und wäßrigen Lösungen bei Temperaturen oberhalb 100°C. *Z Anorg Allg Chem* 1965;**337**(1−2):68−79.

15. Tan TE, Zhou MH. *Principle of chemical engineering*. 3rd ed. Beijing University of Technology Press; 2006.

16. Wilke CR, Chang P. Correlation of diffusion coefficients in dilute solutions. *AICHE J* 1955;**1**(2):264−70.

17. Hung GW, Dinius RH. Diffusivity of oxygen in electrolyte solutions. *J Chem Eng Data* 1972;**17**(4):449−51.

18. Holtzapple MT, Eubank PT, Matthews MA. A comparison of 3 models for the diffusion of oxygen in electrolyte solutions. *Biotechnol Bioeng* 1989;**34**(7):964−70.

19. Case B. Diffusivity of oxygen in dilute alkaline solution from 0 degree to 65 degrees. *Electrochim Acta* 1973;**18**(4):293−9.

20. Tammann G, Jessen V. Diffusion coefficients of gases in water and their dependence on temperature. *Z Anorg Allg Chem* 1929;**179**(1/3):125−44.

21. Cullen EJ, Davidson JF. Absorption of gases in liquid jets. *Trans Faraday Soc* 1957;**53**(1):113−20.

22. Baird MHI, Davidson JF. Annular jets. 2. Gas absorption. *Chem Eng Sci* 1962;**17**(6):473−80.

23. Jordan J, Bauer WE. Correlations between solvent structure, viscosity and polarographic diffusion coefficients of oxygen. *J Am Chem Soc* 1959;**81**(15):3915−9.

24. Yatskovs A, Fedotov NA. Solubility and diffusion of hydrogen in solutions of potassium hydroxide and phosphoric acid. *Russ J Phys Chem, USSR* 1969;**43**(4):575.

25. Miller CC. The Stokes Einstein law for diffusion in solution. *Proc R Soc Lond Ser A* 1924;**106**(740):724−49.

26. Jin W, Du H, Zheng S, Xu H, Zhang Y. Comparison of the oxygen reduction reaction between NaOH and KOH solutions on a Pt electrode: the electrolyte-dependent effect. *J Phys Chem B* 2010;**114**(19):6542−8.

27. Zhang C, Fan FRF, Bard AJ. Electrochemistry of oxygen in concentrated NaOH solutions: solubility, diffusion coefficients, and superoxide formation. *J Am Chem Soc* 2009;**131**(1):177−81.

28. Klinedinst K, Bett JAS, Macdonald J, Stonehart P. Oxygen solubility and diffusivity in hot concentrated H_3PO_4. *J Electroanal Chem* 1974;**57**(3):281−9.

29. Shimizu N, Kushiro I. Diffusivity of oxygen in jadeite and diopside melts at high pressures. *Geochim Cosmochim Acta* 1984;**48**(6):1295−303.

30. Angell CA, Cheeseman PA, Tamaddon S. Pressure enhancement of ion mobilities in liquid silicates from computer simulation studies to 800-kilobars. *Science* 1982;**218**(4575):885−7.

31. Horne RA, Johnson DS. Viscosity of water under pressure. *J Phys Chem* 1966;**70**(7):2182−90.

32. Schmidt TJ, Jusys Z, Gasteiger HA, Behm RJ, Endruschat U, Boennemann H. On the CO tolerance of novel colloidal PdAu/carbon electrocatalysts. *J Electroanal Chem* 2001;**501**(1−2):132−40.

33. http://www.engineeringtoolbox.com/water-dynamic-kinematic-viscosity-d_596.html.

34. Technical data from innophos phosphoric acid.

35. Rhodes FH, Barbour CB. The viscosities of mixtures of sulfuric acid and water. *Ind Eng Chem* 1923;**15**(8):850−2.

36. Arenz M, Schmidt TJ, Wandelt K, Ross PN, Markovic NM. The oxygen reduction reaction on thin palladium films supported on a Pt(111) electrode. *J Phys Chem B* 2003;**107**(36):9813−9.

37. Technical data from Puchkov and Sargaev.
38. Technical data from DOW. Caustic soda solution handbook.
39. Hirai N, Eyring H. Bulk viscosity of liquids. *J Appl Phys* 1958;**29**(5):810−6.
40. Cheng NS, Law AWK. Exponential formula for computing effective viscosity. *Powder Technol* 2003;**129**(1):156−60.
41. Hrma P. Arrhenius model for high-temperature glass-viscosity with a constant pre-exponential factor. *J Non-Cryst Solids* 2008;**354**(18):1962−8.
42. Out DJP, Los JM. Viscosity of aqueous solutions of univalent electrolytes from 5 to 95°C. *J Solution Chem* 1980;**9**(1):19−35.
43. Bett KE, Cappi JB. Effect of pressure on the viscosity of water. *Nature* 1965;**207**(4997):620−1.
44. Mani A, Holdcroft S. Highly temperature dependent mass-transport parameters for ORR in Nafion (R) 211. *J Electroanal Chem* 2011;**651**(2):211−5.
45. Sethuraman VA, Khan S, Jur JS, Haug AT, Weidner JW. Measuring oxygen, carbon monoxide and hydrogen sulfide diffusion coefficient and solubility in Nafion membranes. *Electrochim Acta* 2009;**54**(27):6850−60.
46. Beattie PD, Basura VI, Holdcroft S. Temperature and pressure dependence of O_2 reduction at Pt vertical bar Nafion((R)) 117 and Pt vertical bar BAM((R)) 407 interfaces. *J Electroanal Chem* 1999;**468**(2):180−92.
47. Yeager HL, Steck A. Cation and water diffusion in Nafion ion-exchange membranes—influence of polymer structure. *J Electrochem Soc* 1981;**128**(9):1880−4.
48. Zawodzinski TA, Davey J, Valerio J, Gottesfeld S. The water-content dependence of electroosmotic drag in proton-conducting polymer electrolytes. *Electrochim Acta* 1995;**40**(3):297−302.
49. Ban S, Huang C, Yuan XZ, Wang H. Molecular simulation of gas adsorption, diffusion, and permeation in hydrated Nafion membranes. *J Phys Chem B* 2011;**115**(39):11352−8.
50. Gode P, Lindbergh G, Sundholm G. In-situ measurements of gas permeability in fuel cell membranes using a cylindrical microelectrode. *J Electroanal Chem* 2002;**518**(2):115−22.
51. Buchi FN, Wakizoe M, Srinivasan S. Microelectrode investigation of oxygen permeation in perfluorinated proton exchange membranes with different equivalent weights. *J Electrochem Soc* 1996;**143**(3):927−32.
52. Basura VI, Beattie PD, Holdcroft S. Solid-state electrochemical oxygen reduction at Pt vertical bar Nafion (R) 117 and Pt vertical bar BAM3G (TM) 407 interfaces. *J Electroanal Chem* 1998;**458**(1−2):1−5.
53. Parthasarathy A, Srinivasan S, Appleby AJ, Martin CR. Temperature-dependence of the electrode-kinetics of oxygen reduction at the platinum Nafion(R) interface—a microelectrode investigation. *J Electrochem Soc* 1992;**139**(9):2530−7.
54. Lehtinen T, Sundholm G, Holmberg S, Sundholm F, Bjornbom P, Bursell M. Electrochemical characterization of PVDF-based proton conducting membranes for fuel cells. *Electrochim Acta* 1998;**43**(12−13):1881−90.
55. Haug AT, White RE. Oxygen diffusion coefficient and solubility in a new proton exchange membrane. *J Electrochem Soc* 2000;**147**(3):980−3.

2

ELECTRODE KINETICS OF ELECTRON-TRANSFER REACTION AND REACTANT TRANSPORT IN ELECTROLYTE SOLUTION

Weiwei Cai[a,b], Xiao Zhao[a,b], Changpeng Liu[b], Wei Xing[a] and Jiujun Zhang[c]

[a]*State Key Laboratory of Electroanalytical Chemistry, Changchun Institute of Applied Chemistry, Changchun, PR China,* [b]*Laboratory of Advanced Power Sources, Changchun Institute of Applied Chemistry, Changchun, PR China,* [c]*Energy, Mining and Environment, National Research Council of Canada, Vancouver, BC, Canada*

CHAPTER OUTLINE

Rotating Electrode Methods and Oxygen Reduction Electrocatalysts. http://dx.doi.org/10.1016/B978-0-444-63278-4.00002-1

2.1. Introduction

The current density measured from the rotating electrode is contributed by both the current densities of electrode electron-transfer reaction and the reactant diffusion. In order to obtain the kinetic parameters of these two processes and their associated reaction mechanisms based on the experiment data, both the theories of electrode electron-transfer reaction and reactant diffusion should be studied and understood. In this chapter, the general theories for electrode kinetics of electron-transfer reaction and reactant diffusion will be given in a detailed level, and we hope these theories will form a solid knowledge for a continuing study in the following chapters of this book.

2.2. Kinetics of Electrode Electron-Transfer Reaction

Electrochemical (or electrode) reaction kinetics is one kind of the chemical reaction kinetics. To obtain a better understanding of the theory of electrode reaction kinetics, understanding the basic knowledge of chemical reaction kinetics is necessary. In this section, the general chemical reaction kinetics will be presented first for facilitating the fundamental understanding of electrode kinetics and mechanism, particularly, for oxygen reduction reaction (ORR).

2.2.1. Fundamental Chemical Reaction Kinetics

Consider two substances, A and B, which are linked by simple unimolecular elementary reactions:

$$A \underset{k_b}{\overset{k_f}{\longleftrightarrow}} B \tag{2-I}$$

Both elementary reactions at two different directions are active at all times, and the rate of the forward process, v_f (mol cm^{-3} s^{-1}), is

$$v_f = k_f C_A \tag{2.1}$$

whereas the rate of the reverse reaction is

$$v_b = k_b C_B \tag{2.2}$$

In Eqns (2.1) and (2.2), C_A and C_B are the concentrations of Substances A and B with a unit of mol cm^{-3} or M, k_f and k_b are the reaction rate constants (unit: s^{-1}) for Reactions (2.1) and (2.2), respectively. The net reaction rate (v_{net}) of A to B is:

$$v_{net} = k_f C_A - k_b C_B \tag{2.3}$$

At $v_{net} = 0$, the net reaction rate is zero; then

$$\frac{k_f}{k_b} = K = \frac{C_B}{C_A} \tag{2.4}$$

where K is the equilibrium constant. Equation (2.4) indicates that the kinetics must collapse to relations of the thermodynamic form when reaction time goes to unlimited, otherwise the kinetic picture cannot be accurate. Therefore, Eqn (2.4) is only valid at the state of equilibrium, or a case predicted by reaction thermodynamics. Due to the limitation of reaction time period, it is not easy for a reaction to reach its equilibrium state unless the reaction rates for both directions are extremely high. Therefore, in the study of chemical kinetics, all cases of equilibrium are assumed based on the error tolerance.

2.2.2. Fundamentals of Electrode Reactions (Bulter–Volmer Equation)[1]

To discuss some basic concepts about the electron-transfer kinetics of electrochemical systems, here we present a simple model reaction as shown in Reaction (4-II).

$$O + n_\alpha e^- \underset{k_b}{\overset{k_f}{\leftrightarrow}} R \tag{2-II}$$

where O and R represent the oxidant and reductant, respectively; n_α is the electron-transfer number; k_f and k_b are the forward and backward reaction rates, respectively. Note that this Reaction (2-II) is an elementary reaction. We use n_α to distinguish the electron-transfer number in a simple elementary reaction from that of overall electron-transfer number (n) in a complex reaction such as ORR. For an electrochemical reaction, its reaction mechanism may consist of several such elementary reactions, among which there should be one such elementary reaction as the reaction rate-determining step. The value of n_α is normally 1, and for some special cases, it could be 2. For introducing the concept of electrode reaction kinetics, we will focus on such an elementary reaction in this chapter.

As shown in Figure 2.1, due to the electron-transfer reaction occuring on the electrode surface, only those oxidant species with a concentration of $C_O(0,t)$ and reductant species with a

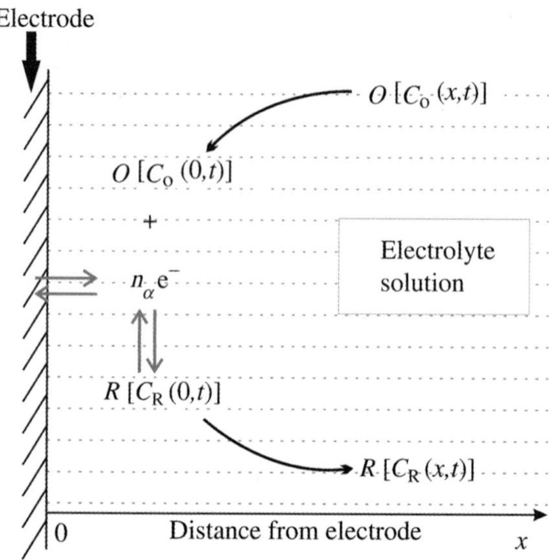

Figure 2.1 Schematic of the electrode electron-transfer and reactant diffusion process in an electrochemical system. $C_O(0,t)$ is the surface concentration of oxidant species, $C_R(0,t)$ is the surface concentration of reductant species, $C_O(x,t)$ is the bulk solution concentration of oxidant, and $C_R(x,t)$ is the bulk solution concentration of reductant species. In these four expressions of concentration, t is the reaction time. (For color version of this figure, the reader is referred to the online version of this book.)

concentration of $C_R(0,t)$ on the electrode surface can participate in the reaction. The oxidant species with a concentration of $C_O(x,t)$ and reductant species with a concentration of $C_R(x,t)$ in the bulk solution have to diffuse to the electrode surface in order to participate the reaction (note that in both these concentration expressions, t is the reaction time). Therefore, there are basically two processes during the reaction: one is the electron-transfer reaction at the electrode/electrode interface, and that other is the reactant diffusion from a bulk solution to the electrode surface. In this section, we will only discuss the former process, that is, the kinetics of electron-transfer process. The reactant diffusion kinetics will be discussed in the latter section of this chapter.

The forward and backward reaction rates (v_f and v_b, respectively) for Reaction (2-II) can be expressed as:

$$v_f = k_f C_O(0, t) \tag{2.5}$$

$$v_b = k_b C_R(0, t) \tag{2.6}$$

when reaction reaches to an equilibrium, that is, $v_f = v_b$; then according to Eqns (2.5) and (2.6), we have equation:

$$k_f C_O(0, t) = k_b C_R(0, t) \tag{2.7}$$

For any chemical or electrochemical reaction, the reaction rate constants in Eqn (2.5) or Eqn (2.6) can be expressed as in Arrhenius forms[2]:

$$k_f = A_f \exp\left(-\frac{W_f}{RT}\right) \tag{2.8}$$

$$k_b = A_b \exp\left(-\frac{W_b}{RT}\right) \tag{2.9}$$

where A_f and A_b are the Arrhenius constants representing the reaction rate constants for forward and backward reactions at unlimited temperature (T), W_f and W_b are the Gibbs free energies of reaction activation for the forward and backward reactions in Reaction (2-II), respectively. Combining Eqns (2.5), (2.6), (2.8) and (2.9), and changing reaction rates into current densities (i_f for forward reaction, and i_b for backward reaction), Eqns (2.7) and (2.8) can be obtained as:

$$i_f = n_\alpha F A_f \exp\left(-\frac{W_f}{RT}\right) C_O(0, t) \tag{2.10}$$

$$i_b = n_\alpha F A_b \exp\left(-\frac{W_b}{RT}\right) C_R(0, t) \tag{2.11}$$

The Gibbs free energy of activation in Eqn (2.10), W_f, can be considered the energy barrier the oxidant must climb in order to become a reductant, or the energy difference between the oxidant and the reaction transition-state, W_b, in Eqn (2.11) can be considered that for a backward reaction, as shown in Figure 2.2.

As shown in Figure 2.2, if the electrode potential is changed from E_1 to E_2, the Gibbs free energy of forward reaction in Reaction (2-II) will be changed from $W_{f,1}$ to $W_{f,2}$, and the backward reaction from $W_{b,1}$ to $W_{b,2}$, respectively. For Reaction (2-II), the electrode potential change can only affect the energy of electrons, resulting in a net change of $n_\alpha F(E_2 - E_1)$ in oxidant energy, where F is the Faraday's constant (98,487 C mol^{-1}). Apparently, this change will affect the energy transition-state's energy, but it is not necessary for the entire energy change, $n_\alpha F(E_2 - E_1)$, going to this transition-state, only a portion may go to the transition-state. In electrochemistry, a parameter called electron-transfer coefficient (α) is normally used to describe this portion. The value of α is between 0 and 1. In an extreme case when $\alpha = 1$, all the energy

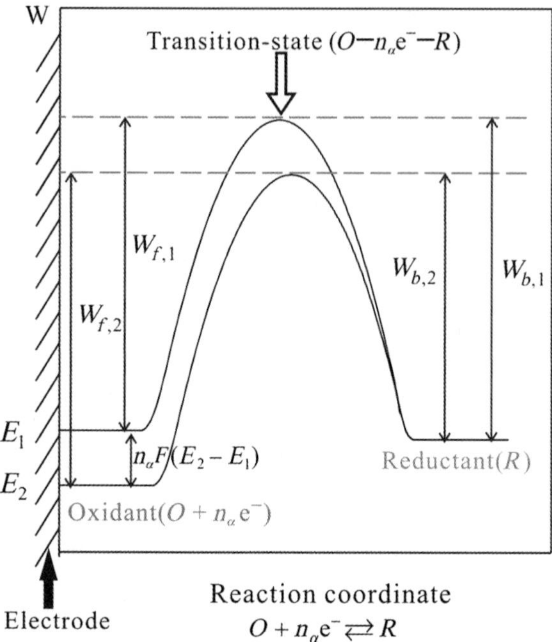

Figure 2.2 Schematic reaction activation energies as a function of reaction coordination. (For color version of this figure, the reader is referred to the online version of this book.)

change will go to the transition-state, the electrode potential change will not affect the reaction activation energy of the forward reaction, but the entire $n_\alpha F(E_2 - E_1)$ will become the extra energy barrier of the backward reaction; and in another extreme case when $\alpha = 0$, the transition-state's energy will not be affected, the entire $n_\alpha F(E_2 - E_1)$ will used to reduce the energy barrier for the forward reaction, and the activation of the backward reaction will not be affected. As a normal case, the energy difference of transition-state induced by the electrode potential change from E_1 to E_2 can be written as $\alpha n_\alpha F(E_2 - E_1)$. Based on Figure 2.2, it can be seen that $W_{f,1} + n_\alpha F(E_2 - E_1) = W_{f,2} + \alpha n_\alpha F(E_2 - E_1)$, and $W_{b,2} + \alpha n_\alpha F(E_2 - E_1) = W_{b,1}$; then the following relationships can be obtained:

$$W_{f,2} = W_{f,1} + (1 - \alpha)n_\alpha F(E_2 - E_1) \tag{2.12}$$

$$W_{b,2} = W_{b,1} - \alpha n_\alpha F(E_2 - E_1) \tag{2.13}$$

If the initial state is an equilibrium state at the standard conditions (25 °C, 1.0 atm) and $C_O(0,t) = C_R(0,t) = 1.0 \ \mathrm{mol \ dm^{-3}}$, E_1 will become the standard equilibrium electrode potential,

E^o, and both $W_{f,1}$ and $W_{b,1}$ will become W_f^o and W_b^o, respectively. To discuss a general case, $W_{f,2}$ and $W_{b,2}$ can be written as W_f to W_b, respectively, Then both Eqns (2.12) and (2.13) will become:

$$W_f = W_f^o + (1 - \alpha)n_\alpha F(E - E^o) \tag{2.14}$$

$$W_b = W_b^o - \alpha n_\alpha F(E - E^o) \tag{2.15}$$

Then the current densities induced by the electrode potential change from E^o to E, as described by Eqns (2.10) and (2.11), can be rewritten as:

$$i_f = n_\alpha F A_f \exp\left(-\frac{W_f^o}{RT}\right)\exp\left(\frac{(1-\alpha)n_\alpha F E^o}{RT}\right)$$

$$\times \exp\left(-\frac{(1-\alpha)n_\alpha F E}{RT}\right)C_O(0,t) \tag{2.17}$$

$$= n_\alpha F K_f^o \exp\left(-\frac{(1-\alpha)n_\alpha F E}{RT}\right)C_O(0,t)$$

$$i_b = n_\alpha F A_b \exp\left(-\frac{W_b^o}{RT}\right)\exp\left(-\frac{\alpha n_\alpha F E^o}{RT}\right)\exp\left(\frac{\alpha n_\alpha F E}{RT}\right)C_R(0,t)$$

$$= n_\alpha F K_b^o \exp\left(\frac{\alpha n_\alpha F E}{RT}\right)C_R(0,t)$$

$$\tag{2.18}$$

where $K_f^o = A_f \exp\left(-\frac{W_f^o}{RT}\right)\exp\left(\frac{(1-\alpha)n_\alpha F E^o}{RT}\right)$ is the Nernst potential-dependent standard reaction rate constant for the forward reaction in Reaction (2-II), and $K_b^o = A_b \exp\left(-\frac{W_b^o}{RT}\right)\exp\left(\frac{\alpha n_\alpha F E^o}{RT}\right)$ is the Nernst potential-dependent standard reaction constant for the backward reaction.

Equations (2.17) and (2.18) can also be alternatively expressed as:

$$i_f = n_\alpha F A_f \exp\left(-\frac{W_f^o}{RT}\right)\exp\left(-\frac{(1-\alpha)n_\alpha F(E - E^o)}{RT}\right)C_O(0,t)$$

$$= n_\alpha F k_f^o \exp\left(-\frac{(1-\alpha)n_\alpha F(E - E^o)}{RT}\right)C_O(0,t)$$

$$\tag{2.19}$$

$$i_b = n_\alpha F A_b \exp\left(-\frac{W_b^o}{RT}\right)\exp\left(\frac{\alpha n_\alpha F(E - E^o)}{RT}\right)C_R(0, t)$$

$$= n_\alpha F k_b^o \exp\left(\frac{\alpha n_\alpha F(E - E^o)}{RT}\right)C_R(0, t) \tag{2.20}$$

where $k_f^o = A_f \exp\left(-\frac{W_f^o}{RT}\right)$ is the standard reaction rate constant for the forward reaction in Reaction (2-II), and $k_b^o = A_b \exp\left(-\frac{W_b^o}{RT}\right)$ is the standard reaction constant for the backward reaction. At standard conditions (25 °C, 1.0 atm, and $C_O(0,t) = C_R(0,t) = 1.0$ mol dm^{-3}), according to Eqn (2.7), k_f^o should be equal to k_b^o, that is:

$$k_f^o = k_b^o = A_f \exp\left(-\frac{W_f^o}{RT}\right) = A_b \exp\left(-\frac{W_b^o}{RT}\right) = k^o \tag{2.21}$$

k^o is called the standard rate constant. Please note that both k_f^o and k_b^o in Eqns (2.20) and (2.21) are different from those of K_f^o and K_b^o in Eqns (2.18) and (2.19).

At the situation ($E \neq E^o$), combining Eqns (2.18) and (2.19), the net current density of electrode Reaction (2-II) can be obtained:

$$i = i_f - i_b = n_\alpha F\left[K_f^o C_O(0, t)\exp\left(-\frac{(1 - \alpha)n_\alpha FE}{RT}\right)\right.$$
$$\left. - K_b^o C_R(0, t)\exp\left(\frac{\alpha n_\alpha FE}{RT}\right)\right] \tag{2.22}$$

If based on Eqns (2.20) and (2.21), the alternative net current density can be written as:

$$i = i_f - i_b = n_\alpha F\left[k_f^o C_O(0, t)\exp\left(-\frac{(1 - \alpha)n_\alpha F(E - E^o)}{RT}\right)\right.$$
$$\left. - k_b^o C_R(0, t)\exp\left(\frac{\alpha n_\alpha F(E - E^o)}{RT}\right)\right]$$
$$= n_\alpha F k^o\left[C_O(0, t)\exp\left(-\frac{(1 - \alpha)n_\alpha F(E - E^o)}{RT}\right)\right.$$
$$\left. - C_R(0, t)\exp\left(\frac{\alpha n_\alpha F(E - E^o)}{RT}\right)\right] \tag{2.23}$$

Equation (2.22) or Eqn (2.23) is one of the most important equations in dealing with electrochemical surface reactions, which is called the Bulter–Volmer equation.

Based on Eqn (2.22) or Eqn (2.23), we can discuss several important concepts of electrochemical kinetics, including the overpotential, the Nernst reversible electrode potential, the exchange current density, the standard reaction rate, the electron-transfer coefficient, and the reversible and irreversible reactions.

1. *Nernst reversible electrode potential.* When the reaction reaches equilibrium, the electrode potential becomes equilibrium or Nernst potential, that is, $E = E^{eq}$, and the net current density will become zero ($i = i_f - i_b = 0$), and Eqn (2.22) or Eqn (2.23) will become Eqn (2.24) if combining with Eqn (2.21):

$$E^{eq} = E^{o} + \frac{RT}{n_\alpha F} \ln\left(\frac{C_O(0,t)}{C_R(0,t)}\right) \tag{2.24}$$

This Eqn (2.24) is the expression of Nernst electrode potential of Reaction (2-II).

2. *Exchange current density and alternative Bulter–Volmer equation.* Based on Eqn (2.22), we can define another important kinetic parameter called the exchange current density (i^o), which is either the forward or backward current density when the forward and the backward reaction rates become equal. In other words, the reaction is at the equilibrium state with the oxidant concentration of $C_O^*(0,t)$ and the reductant concentration of $C_R^*(0,t)$:

$$
\begin{aligned}
i^o &= n_\alpha F K_f^o C_O^*(0,t) \exp\left(-\frac{(1-\alpha)n_\alpha F E^{eq}}{RT}\right) \\
&= n_\alpha F K_b^o C_R^*(0,t) \exp\left(\frac{\alpha n_\alpha F E^{eq}}{RT}\right)
\end{aligned} \tag{2.25}
$$

From Eqn (2.23), the exchange current density can be alternatively expressed as:

$$
\begin{aligned}
i^o &= n_\alpha F k^o C_O^*(0,t) \exp\left(-\frac{(1-\alpha)n_\alpha F(E^{eq}-E^o)}{RT}\right) \\
&= n_\alpha F k^o C_R^*(0,t) \exp\left(\frac{\alpha n_\alpha F(E^{eq}-E^o)}{RT}\right)
\end{aligned} \tag{2.26}
$$

If combining Eqn (2.24) with Eqn (2.26), another expression of exchange current density (i^o) can be obtained:

$$i^o = n_\alpha F k^o \left(C_O^*(0,t)\right)^\alpha \left(C_R^*(0,t)\right)^{1-\alpha} \tag{2.27}$$

Combining Eqns (2.23) and (2.27), an alternative Bulter–Volmer equation can be obtained as:

$$
i = i^o \left(\frac{C_O(0,t)}{C_O^*(0,t)} \exp\left(-\frac{(1-\alpha)n_\alpha F(E - E^{eq})}{RT} \right) - \frac{C_R(0,t)}{C_R^*(0,t)} \exp\left(\frac{\alpha n_\alpha F(E - E^{eq})}{RT} \right) \right)
\tag{2.28}
$$

The term, $E - E^{eq}$, in Eqn (2.28) is called the overpotential, which will be discussed more in a latter paragraph in this section. At a low current density and the reaction rate is totally controlled by the electron-transfer kinetics, $C_O(0,t) \approx C_O^*(0,t)$ and $C_R(0,t) \approx C_R^*(0,t)$, Eqn (2.28) will become Eqn (2.28a):

$$
i = i^o \left(\exp\left(-\frac{(1-\alpha)n_\alpha F(E - E^{eq})}{RT} \right) - \exp\left(\frac{\alpha n_\alpha F(E - E^{eq})}{RT} \right) \right)
\tag{2.28a}
$$

The exchange current density is a very important parameter in determining the reversibility of the electrochemical reaction. Normally, when $i^o \to 0$, the corresponding reaction will be an irreversible reaction, and when $i^o \to \infty$, the corresponding reaction will be the reversible reaction. Figure 2.3 shows the effect of i^o value on the shape of $i \sim (E - E^{eq})$ curves.

3. *Electron-transfer coefficient (α).* As discussed previously, this α is called the electron-transfer coefficient, which is one of the important parameters for the electrode electron-transfer kinetics. For majority of electrochemical reaction systems, the value of this α is in the range of 0.2–0.8, depending on the nature of the studied system. However, in the electrochemical research, if this value is not measured, people normally assume its value to be 0.5.

According to Eqn (2.28), the magnitude of α has a strong effect on the shape of the current–potential curve. Figure 2.4 shows the $i \sim (E - E^{eq})$ curves at three different values of α. It can be seen that it reflects the symmetry of the curves, and only when $\alpha = 0.5$, the curve is a symmetric one.

4. *Overpential and Tafel equation.* In Eqn (2.28), the term of $E - E^{eq}$ is called the overpotential (η), that is, $\eta = E - E^{eq}$, which is used to measure the reversibility of the electrochemical reaction. Overpotential is the driving force of the electrode reaction: the larger the overpotential, the faster the electrode reaction rate would be. From Bulter–Volmer Eqn (2.28a), it can be seen that when this overpotential is negative enough,

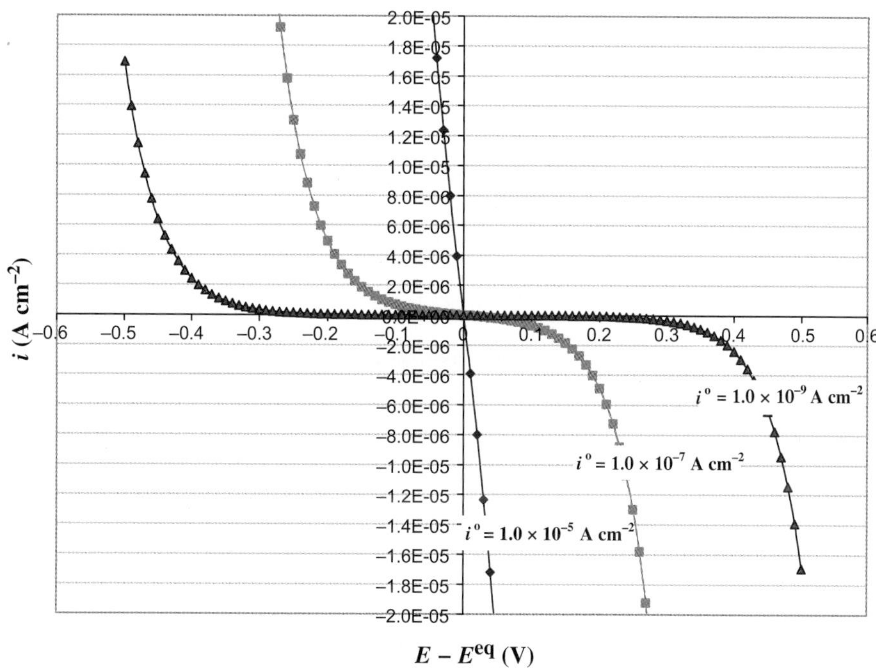

Figure 2.3 Current density–overpotential curves of $O + n_\alpha e^- \underset{k_b}{\overset{k_f}{\leftrightarrow}} R$ reaction at three different exchange current densities ($i = 1.0 \times 10^{-6}$, 1.0×10^{-7}, and 1.0×10^{-9} A cm^{-2}, respectively), calculated according to Eqn (2.28a) using the parameter values of $n_\alpha = 1$, $R = 8.314$ J K^{-1} mol^{-1}, $T = 298$ K, $F = 96{,}487$ C mol^{-1}, and $\alpha = 0.5$. (For color version of this figure, the reader is referred to the online version of this book.)

such as $\eta < -60$ mV, the second term in this equation would be much smaller than the first one. Then, Eqn (2.28) becomes:

$$\eta = E - E^{\mathrm{eq}} = \frac{RT}{(1-\alpha)n_\alpha F}\ln(i^\mathrm{o}) - \frac{RT}{(1-\alpha)n_\alpha F}\ln(i) \qquad (2.29)$$

or

$$E = E^{\mathrm{eq}} + \frac{RT}{(1-\alpha)n_\alpha F}\ln(i^\mathrm{o}) - \frac{RT}{(1-\alpha)n_\alpha F}\ln(i) \qquad (2.30)$$

This Eqn (2.30) is called the Tafel equation for the forward reaction, where both the intercept $\left(E^{\mathrm{eq}} + \frac{RT}{(1-\alpha)n_\alpha F}\ln(i^\mathrm{o})\right)$ and the slope $\left(-\frac{RT}{(1-\alpha)n_\alpha F}\right)$ can be experimentally measured, from which two important kinetic parameters, the electron-transfer coefficient (α) and the exchange current density (i^o), can be obtained if E^o is known.

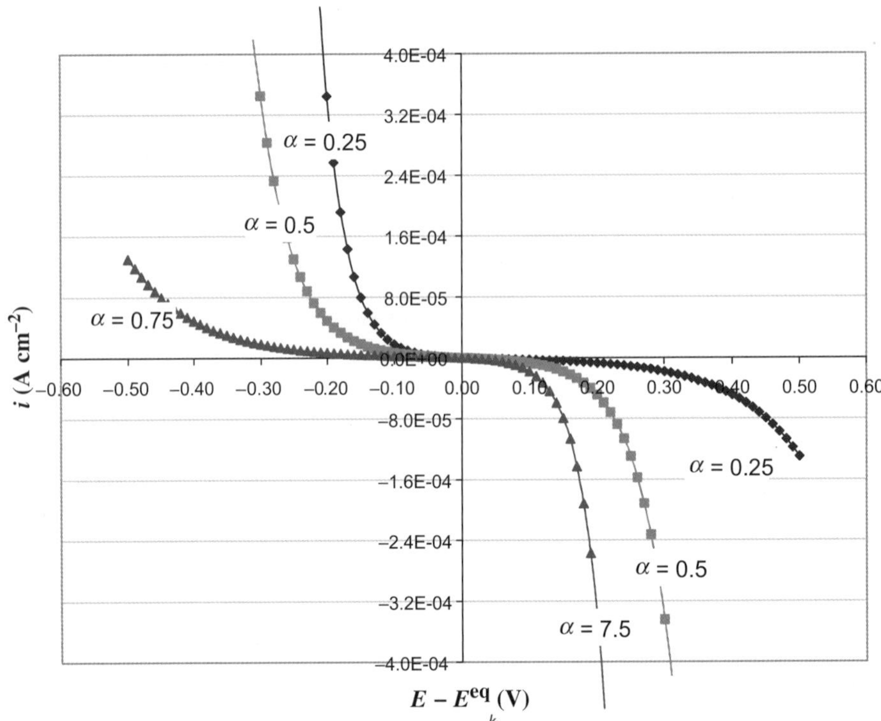

$E - E^{eq}$ (V)

Figure 2.4 Current density–overpotential curves of $O + n_\alpha e^- \overset{k_f}{\underset{k_b}{\leftrightarrow}} R$ reaction at three different electron-transfer coefficients ($\alpha = 0.25$, 0.5, and 0.75, respectively), calculated according Eqn (2.28a) using the parameter values of $n_\alpha = 1$, $R = 8.314$ J K^{-1} mol^{-1}, $T = 298$ K, $F = 96{,}487$ C mol^{-1}, and $i = 1.0 \times 10^{-6}$ A cm^{-2}. (For color version of this figure, the reader is referred to the online version of this book.)

Actually, with the proceeding of the electrode reaction (Reaction (2-II)) from left to right with zero concentration of reductant $(C_R(0,t) = 0)$ at the beginning, the concentrations $C_O(0,t)$ and $C_R(0,t)$ will be changed, that is, $C_O(0,t)$ will be reduced and $C_R(0,t)$ will be increased at the electrode surface. The oxidant (O) will move from the bulk solution to the electrode surface and the reductant (R) will move from the electrode surface to the bulk solution. Therefore, these two reactant transport processes will contribute to the entire electrode reaction, which will be discussed in the following section.

2.3. Kinetics of Reactant Mass Transport Near Electrode Surface[3,4]

Besides the kinetics of electron-transfer reaction discussed above, the process of reactant transport near an electrode surface is the other necessary portion of the entire electrode kinetics. In order to understand the rotating electrode theory and correctly perform the corresponding experiment techniques in evaluating electrocatalyst's activity toward ORR, understanding the kinetics of reactant transport near the electrode surface is necessary.

2.3.1. Three Types of Reactant Transport in Electrolyte (Diffusion, Convection, and Migration)

Reactant transportation in electrolyte near the electrode surface occurs by the following three different modes as illustrated in Figure 2.5.

1. *Diffusion.* As shown in Figure 2.5, diffusion is the spontaneous movement of the species under the influence of concentration gradient (from high concentration regions to low concentration regions). The purpose of spontaneous diffusion is to achieve a minimizing concentration difference within the electrolyte.

For simplification, we chose the direction of reactant diffusion as the perpendicular direction to the electrode surface, as shown in Figure 2.6 (the x direction). We also assume that the reactant in the solution is oxidant with a concentration of $C_O(x,t)$. Figure 2.6 schematically shows two parallel planes near the electrode in the electrolyte solution. These two planes, with a distance of dx, are both parallel to the electrode surface. The reactant diffusion

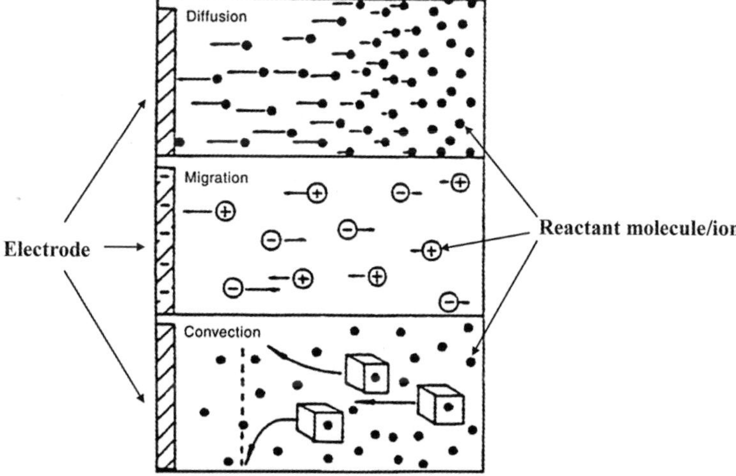

Figure 2.5 Schematics of three modes of mass transport.[5]

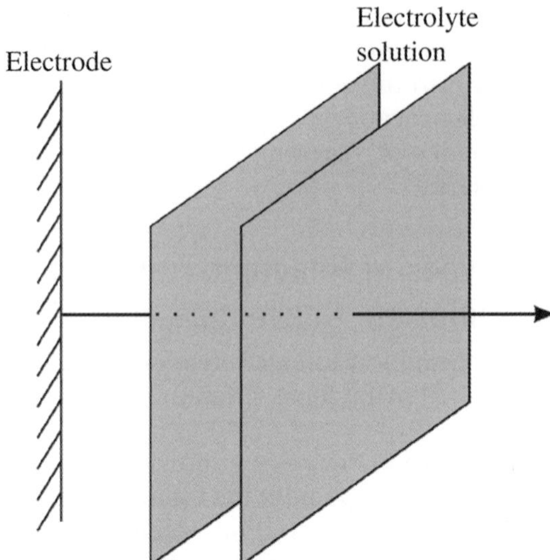

Figure 2.6 Schematic of diffusion planes in the electrolyte solution. (For color version of this figure, the reader is referred to the online version of this book.)

rate ($J_{D,O}$), with unit of the diffused mole number per second at a unit surface area ($mol\ s^{-1}\ cm^{-2}$), can be expressed as:

$$J_{D,O} = -D_O \frac{dC_O(x,t)}{dx} \tag{2.31}$$

where D_O is called the diffusion coefficient ($cm^2\ s^{-1}$)—which has been given a detailed discussion in Chapter 1—$C_O(x,t)$ is the concentration of the reactant ($mol\ dm^{-3}$), dx is the distance the reactant diffused (cm), and $\frac{dC_O(x,t)}{dx}$ is the concentration gradient at the distance x. This Eqn (2.31) is called the Fick's first law.

For Reaction (2-II), the current density ($i_{D,O}$, $A\ cm^{-2}$) produced by the oxidant diffusion from solution toward the electrode surface can be expressed as Eqn (2.32) based on Eqn (2.30):

$$i_{D,O} = n_\alpha F J_{D,O} = -n_\alpha F D_O \frac{dC_O(x,t)}{dx} \tag{2.32}$$

2. *Migration.* As shown in Figure 2.5, migration is the movement of charged ion along an electrical field in the electrolyte. The moving direction is therefore decided by both the directions of the electrical field and the charge state of the particle. The positively charged particles move along with the direction of the electrical field while the negatively charged particles migrate against the direction of the electrical field. The

reactant migration rate ($J_{M,O}$) with unit of the migrated mole number per second at a unit surface area ($mol\ s^{-1}\ cm^{-2}$) can be expressed as:

$$J_{M,O} = u_O^o C_O(x, t) \frac{d\phi(x)}{dx} \tag{2.33}$$

where u_O^o is called the mobility of the oxidant ($cm^2\ s^{-1}\ V^{-1}$) if the oxidant is the charged ion, $\phi(x)$ is the electric field (V) at the distance x in the solution, and $\frac{d\phi(x)}{dx}$ is the gradient of electric field at the distance x. Note that this migration is only for charged ion, for neutral molecule such as dissolved O_2 in the electrolyte solution, $J_{M,O} = 0$. The current density ($i_{M,O}$, $A\ cm^{-2}$) produced by the charged oxidant or reductant migration from solution toward the electrode surface can be expressed as Eqn (2.34) based on Eqn (2.33):

$$i_{M,O} = n_\alpha F J_{M,O} = n_\alpha F |u_O^o| C_O(x, t) \frac{d\phi(x)}{dx} \tag{2.34}$$

where depending on the sign of the charged ion, the u_O^o can have either positive or negative value.

3. *Convection.* As shown in Figure 2.5, convection is the transport of the species near the electrode brought by a gross physical movement. Any stirring, thermal fluctuation, density gradient, or electrode movement such as rotating can cause a convection of the reactant inside the electrolyte solution. Therefore, convention of the reactant has to be considered when dealing with the reactant transport near the electrode surface. The reactant convection rate ($J_{C,o}$) with unit of the transported mole number per second at a unit surface area ($mol\ s^{-1}\ cm^{-2}$) can be expressed as:

$$J_{C,O} = v_O C(x, t) \tag{2.35}$$

where v_O is called the flow rate of the electrolyte solution at x direction ($cm\ s^{-1}$). The current density ($i_{C,O}$, $A\ cm^{-2}$) produced by the solution movement from solution toward the electrode surface can be expressed as Eqn (2.36) based on Eqn (2.35):

$$i_{C,O} = n_\alpha F J_{C,O} = n_\alpha F v_O C_O(x, t) \tag{2.36}$$

Actually, the reactant transport by these three processes (diffusion, migration, and convection) occurs at the same time along the x direction. The total oxidant transport current density

(i, A cm^{-2}) can be obtained by adding those three current densities expressed by Eqns (2.32), (2.34) and (2.36):

$$i = i_{D,O} + i_{M,O} + i_{C,O} = n_\alpha F (J_{D,O} + J_{M,O} + J_{C,O})$$

$$= n_\alpha F \left(- D_O \frac{dC_O(x,t)}{dx} + |u_O^o| C_O(x,t) \frac{d\phi(x)}{dx} + v_O C_O(x,t) \right)$$

$$(2.37)$$

In an electrolyte solution, in the region far from the electrode surface, the dominating reactant transport is the convection process ($J_{C,O}$) because the magnitude of v_O is several order larger than those of D_O and u_O^o. However, in the region very close to the electrode surface, the magnitude of v_O is much smaller than those of both D_O and u_O^o, the dominating current densities should be ($J_{D,O} + J_{M,O}$). In a normal electrochemical measurement, some high concentration of electrolyte is used, for example >0.1 M; the electric field gradient, $\frac{d\phi(x)}{dx}$ in Eqn (2.34), is very small, and therefore the dominating reactant transport is the diffusion process ($J_{D,O}$). Furthermore, this $J_{M,O}$ is only meaningful for the charged ions, and not for those neutral molecules such as dissolved O_2. For the study of oxygen reduction (ORR), this migration process will not be considered, that is, $J_{M,O} = 0$. Therefore, in the following subsections, we will only focus on the reactant diffusion and convection processes.

2.3.2. Nonsteady-State Diffusion Process of Reactant when the Electrolyte Solution at the Static State

In practice, the electrolyte solution is always moving due to the thermal fluctuation or vibration. In order to better understand the diffusion behavior, we assume that the electrolyte solution is at the static state, and the only reactant transport is through diffusion.

Before we go further, it is necessary to introduce another important concept, the Fick's second law. This law deals with the change of reactant concentration with time during the diffusion. It can be deduced using Figure 2.6. At the diffusion Plane 1, the oxidant diffusion rate can be expressed as:

$$J_{D,O,1} = -D_O \left(\frac{dC_O(x,t)}{dx} \right)_{x=x} \tag{2.38}$$

and at Plane 2, the diffusion rate can be expressed as:

$$J_{D,O,2} = -D_O \left(\frac{dC_O(x,t)}{dx} \right)_{x=x+dx}$$

$$= -D_O \left[\left(\frac{dC_O(x,t)}{dx} \right)_{x=x} + \frac{d}{dx} \left(\frac{dC_O(x,t)}{dx} \right) dx \right] \quad (2.39)$$

Then the concentration change between Planes 1 and 2 with time can be expressed as:

$$\left(\frac{dC_O(x,t)}{dt} \right)_D = \frac{J_{D,O,1} - J_{D,O,2}}{dx} = D_O \left(\frac{d^2 C_O(x,t)}{dx^2} \right) \quad (2.40)$$

This Eqn (2.40) is a very important equation dealing with the diffusion-induced change in reactant concentration with time.

As we know, the electrochemical Reaction (2-II) can only occur at the electrode surface ($x = 0$). Then the diffusion current density expressed by Eqn (2.32) can be alternatively expressed as:

$$i_{D,O} = -n_\alpha F D_O \left(\frac{dC_O(0,t)}{dx} \right)_{x=0,t} \quad (2.41)$$

In order to obtain a visual expression of the reactant diffusion behavior with the reaction time, the term of $\left(\frac{dC_O(0,t)}{dx} \right)_{x=0,t}$ in Eqn (2.41) has to be resolved. Several boundary conditions are normally used for this purpose: (1) the diffusion coefficient (D_O) is constant, independent on the reactant concentration; (2) at the beginning of reaction ($t = 0$), the reactant concentration is uniform across the entire electrolyte solution, that is, $C_O(x,0) = C_O^0$; and (3) at any time, the reactant concentration at the unlimited distance is not changed with the reaction process, that is, $C_O(\infty, t) = C_O^0$. Based on these three boundary conditions, we can discuss two typical cases as follows:

1. *Diffusion process at a constant current density.* If the electrochemical reaction is carried out at a constant current density (I_O^0), Eqn (2.41) can be rewritten as:

$$\left(\frac{dC_O(0,t)}{dx} \right)_{x=0,t} = \frac{I_O^0}{n_\alpha F D_O} \quad (2.42)$$

Using Eqn (2.40) and the three boundary conditions, the reactant concentration ($C_O(\infty, t) = C_O^0$) at any time and distance can be resolved as:

$$C_O(x,t) = C_O^0 + \frac{I_O^0 x}{n_\alpha F D_j} \left[\text{erfc} \left(\frac{x}{2\sqrt{D_O t}} \right) - 2\sqrt{\frac{D_O t}{\pi}} \exp \left(-\frac{x^2}{4 D_O t} \right) \right]$$

$$(2.43)$$

where erfc function can be expressed as:

$$\mathrm{erfc}\left(\frac{x}{2\sqrt{D_O t}}\right) = 1 - \mathrm{erf}\left(\frac{x}{2\sqrt{D_O t}}\right) = 1 - \frac{2}{\sqrt{\pi}} \int_0^{\frac{x}{2\sqrt{D_O t}}} e^{-t^2}\,dt \quad (2.44)$$

To visualize the concentration distribution near the electrode surface, Figure 2.7 shows several curves of $C_O(x,t)$ vs x at several different time periods, which are calculated according to Eqn (2.43). It can be seen that the reactant concentration is decreased with increasing reaction time due to the reaction consumption. The larger decrease in concentration can be seen at the region more close to the electrode surface. At the electrode surface, the concentration can be expressed as Eqn (2.45) by setting $x = 0$ in Eqn (2.43):

$$C_O(0, t) = C_O^o - \frac{2I_j^o}{n_\alpha F}\sqrt{\frac{t}{\pi D_O}} \quad (2.45)$$

This equation gives the trend of surface concentration of the reactant with increasing reaction time. Actually, the controlled

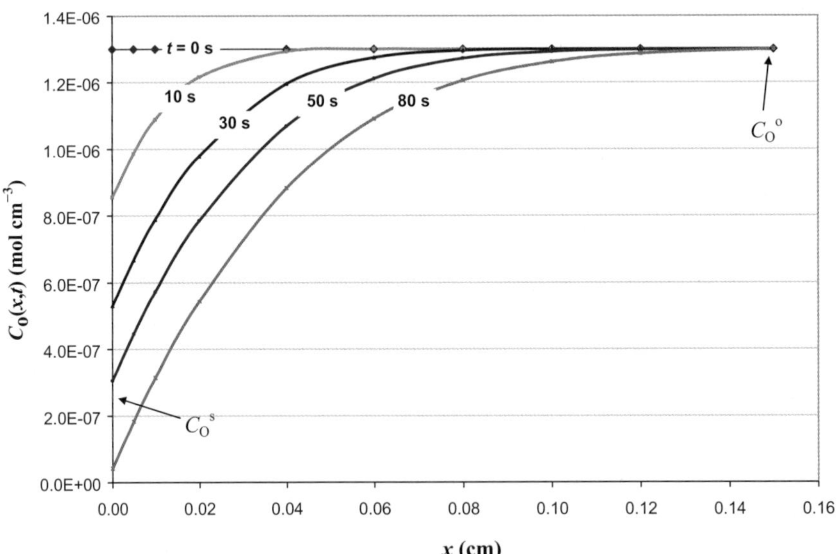

Figure 2.7 Distribution of the reactant concentration near the electrode surface at a constant current density (I_0^o) for the electrode reaction at different reaction time periods as marked beside each curve, calculated using Eqn (2.44). $I_0^o = 5.0 \times 10^{-5}$ A cm^{-2}; $n_\alpha = 1.0$; $F = 96{,}487$ C mol^{-1}; $D_0 = 1.9 \times 10^{-5}$ cm^2 s^{-2}; $C_0^o = 1.3 \times 10^{-6}$ mol cm^{-3}; and $\tau = 89$ s. (For color version of this figure, the reader is referred to the online version of this book.)

current density (I_O^0) can only be sustained before the time when the surface concentration reaches zero, after that the diffusion process cannot supply enough reactant for the desired constant current density. This time period is called the transition time (τ). According to Eqn (2.45), when $C_O(0,t) = 0$, the transition time can be expressed as:

$$\tau = \pi D_O \left(\frac{n_\alpha F C_O^0}{2I_O^0}\right)^2 \tag{2.46}$$

Combining Eqns (2.45) and (2.46), another alternative equation for surface concentration of reactant can be obtained:

$$C_O(0,t) = C_O^0 \left(1 - \left(\frac{t}{\tau}\right)^{1/2}\right) \tag{2.47}$$

Equation (2.47) can be used to obtain the electrode potential if we could assume that Reaction (2-II) is a totally reversible reaction, and the reductant is insoluble ($C_R(0,t) = 1$). Substituting the oxidant's surface concentration $C_O(0,t)$ into the Nernst Eqn (2.24), the following electrode potential can be obtained:

$$E^{eq} = E^o + \frac{RT}{n_\alpha F} \ln \left(\frac{\tau_O^{1/2} - t^{1/2}}{t^{1/2}}\right) \tag{2.48}$$

where τ_O is the transition time for oxidant. This equation can be used to obtain the electron-transfer number based on the experimental linear relationship between E^{eq} and $\frac{\tau_O^{1/2} - t^{1/2}}{t^{1/2}}$.

2. *Diffusion process at a constant electrode potential.* Assuming that Reaction (2-II) is a totally reversible reaction, and the reductant is insoluble ($C_R(0,t) = 1$). According to the Nernst Eqn (2.24), the oxidant's surface concentration should be constant if the electrode potential is held as a constant. In this case, $C_O(0, t) = C_O^s =$ constant (C_O^s is the reactant concentration at electrode surface). Using the other three conditions as: (1) the diffusion coefficient (D_O) is constant, independent on the reactant concentration; (2) at the beginning of reaction ($t = 0$), the reactant concentration is uniform across the entire electrolyte solution, that is, $C_O(x,0) = C_O^0$; and (3) at any time, the reactant concentration at unlimited distance is not changed with reaction process, that is, $C_O(\infty, t) = C_O^0$, Eqn (2.40) can be resolved to give the expression of $C_O(x,t)$:

$$C_O(x,t) = C_O^s + (C_O^0 - C_O^s)\mathrm{erf}\left(\frac{x}{2\sqrt{D_O t}}\right) \tag{2.49}$$

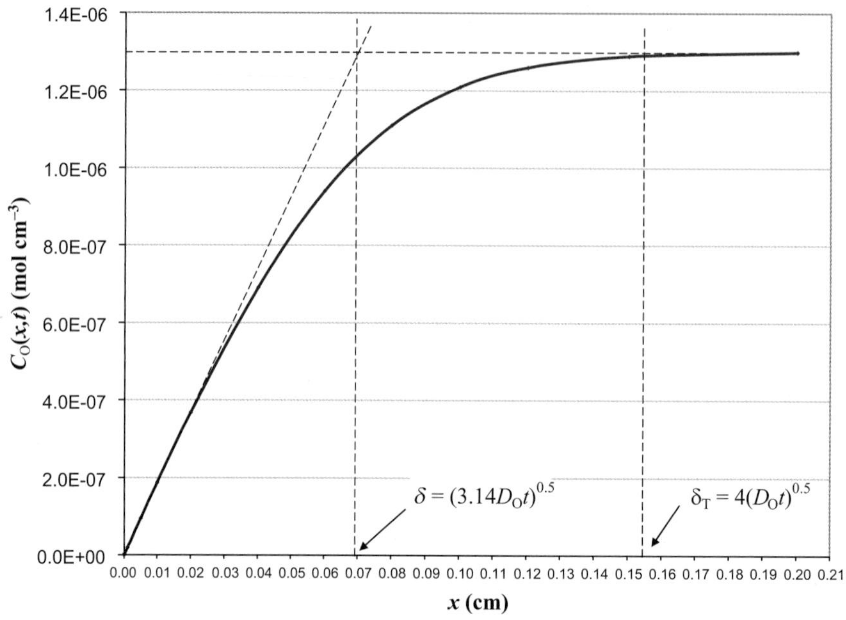

Figure 2.8 Distribution of the reactant concentration near the electrode surface at a constant surface concentration ($C_O^s = 0$) for the electrode reaction at steady state, calculated based on Eqn (2.50). $D_O = 1.9 \times 10^{-5}$ cm^2 s^{-2}; and $C_O^o = 1.3 \times 10^{-6}$ mol cm^{-3}. (For color version of this figure, the reader is referred to the online version of this book.)

At an extremely situation when $C_O^s \rightarrow 0$ at any time, Eqn (2.49) will becomes:

$$C_O(x, t) = C_O^o \,\mathrm{erf}\left(\frac{x}{2\sqrt{D_Ot}}\right) \tag{2.50}$$

In this equation, when $\frac{x}{2\sqrt{D_Ot}} \geq 2$ (or $x \geq 4\sqrt{D_Ot}$) $\mathrm{erf}\left(\frac{x}{2\sqrt{D_Ot}}\right) \approx 1$, then $\frac{C_O(x,t)}{C_O^o} \approx 1$. Here the $\delta_T = 4\sqrt{D_Ot}$ is defined as the total thickness of diffusion layer, as shown in Figure 2.8, which is the distance when $C_O(x,t)$ reaches C_O^o. Another useful definition is called the effective thickness of diffusion layer (δ), which is the bulk reactant concentration (C_O^o) divided by the reactant concentration gradient at the electrode surface $\left(\left(\frac{\mathrm{d}C_O(0,t)}{\mathrm{d}x}\right)_{x=0}\right)$:

$$\delta = \frac{C_O^o}{\left(\frac{\mathrm{d}C_O(0,t)}{\mathrm{d}x}\right)_{x=0}} = \sqrt{\pi D_Ot} \tag{2.51}$$

where $\delta = \sqrt{\pi D_Ot}$ is calculated based on Eqn (2.50), which can also be obtained from Figure 2.8. It can be seen that this effective thickness is a function of time, suggesting that the diffusion layer

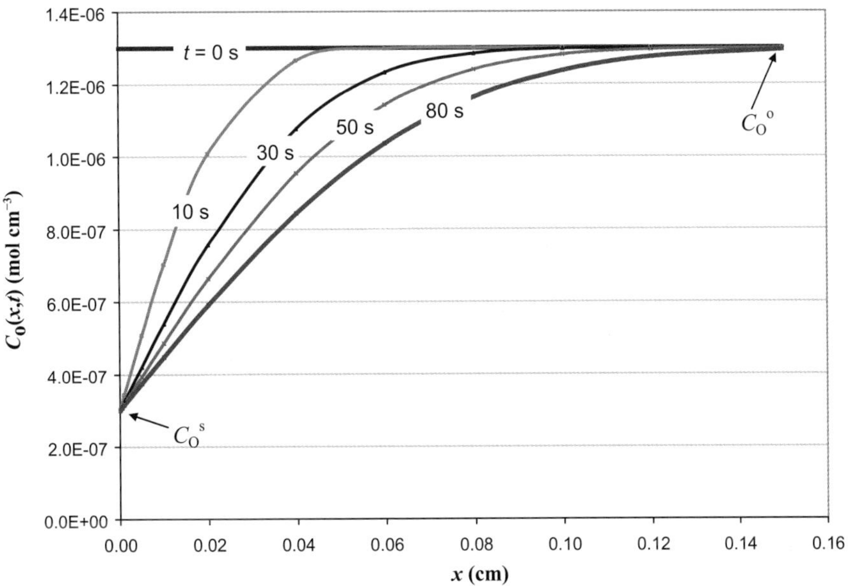

Figure 2.9 Distribution of the reactant concentration near the electrode surface at a constant surface concentration (C_O^s) for the electrode reaction at different reaction time periods as marked beside each curves, calculated using Eqn (2.49). $C_O^s = 3.0 \times 10^{-6}$ mol cm^{-3}; $D_O = 1.9 \times 10^{-5}$ cm^2 s^{-2}; and $C_O^o = 1.3 \times 10^{-6}$ mol cm^{-3}. (For color version of this figure, the reader is referred to the online version of this book.)

thickness will become thicker with increasing the electrode reaction time. However, this diffusion layer thickness is only dependent on the diffusion coefficient of the reactant and the reaction time period, but independent on the reactant concentration. For example, if the diffusion coefficient of dissolved O_2 in the electrolyte solution is taken as 1.9×10^{-5} cm^2 s^{-1}, the effective diffusion-layer thicknesses at different reaction time periods can be calculated as 7.7×10^{-3} ($t = 1$ s), 2.4×10^{-2} ($t = 10$ s), 7.7×10^{-2} ($t = 100$ s), and 2.4×10^{-1} ($t = 1000$ s) cm, respectively.

Regarding the nonsteady-state current density ($i_{D,O}$) induced by the diffusion process at any time shown in Figure 2.9, it can be expressed as Eqn (2.52) after a modification from Eqn (2.32):

$$i_{D,O} = n_\alpha F D_O \left(\frac{dC_O(0,t)}{dt} \right)_{(x=0),t} \tag{2.52}$$

Combining Eqn (2.49) with Eqn (2.52), the current density can be rewritten as:

$$i_{D,O} = n_\alpha F (C_O^o - C_O^s) \sqrt{\frac{D_O}{\pi t}} = n_\alpha F (C_O^o - C_O^s) \left(\frac{D_O}{\pi} \right)^{1/2} t^{-1/2} \tag{2.53}$$

When the reactant surface concentration reaches zero ($C_O^s = 0$) (the case that the reaction rate is totally controlled by the diffusion rate), Eqn (2.53) will become:

$$i_{D,O} = n_\alpha FC_j^o \left(\frac{D_O}{\pi}\right)^{1/2} t^{-1/2} \tag{2.54}$$

Based on this equation, the experiment plot of $i_{D,O}$ vs $t^{-1/2}$ allows the measurement of either n_α or D_O if the reactant concentration is known.

From Figures 2.7 and 2.9 and Eqn (2.51), it can be seen that the diffusion layer thickness is increased with prolonging the reaction time period, suggesting that it is impossible to sustain a steady-state reaction rate if only based on the reactant diffusion transport. Therefore, in order to maintaining a sustain reaction rate (or current density), some forced convection measures such as stirring and electrode rotating have to be used. For example, rotating electrode can create a steady-state convection near the electrode surface and stop the continuing expansion of the diffusion layer with the reaction time, leading to a fixed thickness when the rotating rate is fixed. Therefore, the combination between the reactant diffusion and solution convection must be considered in understanding the transport kinetics, which will be discussed in the following subsection.

2.3.3. Steady-State Diffusion–Convection Process of Reactant

In order to make some sense about the combination of diffusion and convection process and their resulting stead-state reactant transport near the electrode surface, an ideal situation is presented here. Figure 2.10(A) shows the experiment device for separating the diffusion from the convection regions. A container connected to a capillary with a length of l, at the end of which a planner working electrode is attached. This container contains an electrolyte solution containing only an oxidant Reaction (2-II) with concentration of C_O^o, a counter electrode, and a stirrer at a vigorous stirring during the current passes through the working and counter electrode. Due to the diameter of the capillary is very small, there should not be any convection inside it. Furthermore, due to the electrolyte's concentration inside the capillary is high, the electric mobility could also be negligible, so that there is only diffusion transport inside the capillary. Due to the stirring, the concentration of oxidant inside the container should be uniformly distribution—that is, the concentration at the mouth of

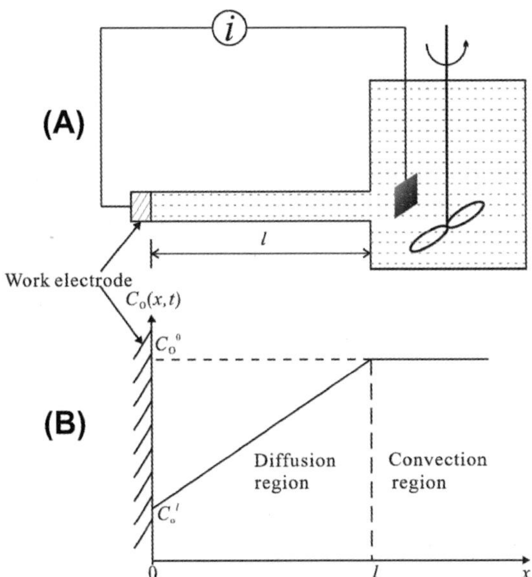

Figure 2.10 (A) Experiment device for measuring the reactant concentration distribution in separating the diffusion from the convection regions; (B) Reactant distribution near the electrode surface. (For color version of this figure, the reader is referred to the online version of this book.)

capillary is constant at C_O^o. The oxidant concentration distribution near the electrode surface can be expressed as in Figure 2.10(B). It can be seen that the diffusion layer thickness is limited at the distance of l, beyond which the region is the convection one.

According to Eqn (2.32), the diffusion current density ($i_{D,O}$) can be expressed as:

$$i_{D,O} = n_\alpha FD_O \frac{dC_O(x,t)}{dx} = n_\alpha FD_O \frac{C_O(l,t) - C_O(0,t)}{l}$$

$$= n_\alpha FD_O \frac{C_O^o - C_O^s}{l} \tag{2.55}$$

when C_O^s reaches zero, the value of $i_{D,O}$ will become the maximum ($I_{D,O}$):

$$I_{D,O} = n_\alpha FD_O \frac{C_O^o}{l} \tag{2.56}$$

This maximum current density is normally called the diffusion-limiting current density. It can be seen that both Eqns (2.55) and (2.56) represent the steady-state situation created by the convection process outside the region of diffusion layer. Due to this solution convection, the diffusion layer thickness cannot expend with time anymore. While if there is no such a solution convection process, the diffusion layer will expend and become thicker with prolonging the reaction time, as shown in Figures 2.7 and 2.8.

Unfortunately, the case presented in Figure 2.10 is ideal, not necessary to reflect the real situation. For a practical electrode, both diffusion and convection processes coexist. Even inside the diffusion layer, there is some degree of convection—that is, the solution within the diffusion layer is not static. Therefore, the diffusion layer thickness should be determined by both the diffusion and convection processes. Fortunately, using mathematical modeling, the reactant concentration distribution profile near the electrode surface has been found to be similar to that shown in Figure 2.9, from which the effective (or equivalent) diffusion-layer thickness can also be defined in the same way as Eqn (2.51). For a detailed expression about this diffusion layer thickness induced by both diffusion and convection process, we will give more discussion in Chapter 5.

In a similar way for Eqns (2.55) and (2.56), the current density for the steady-state diffusion–convection process of oxidant in Reaction (2-II) ($i_{D,O}$) can be expressed as:

$$i_{D,O} = n_\alpha F D_O \frac{C_O^o - C_O^s}{\delta_O} \tag{2.57}$$

where δ_O is the effective diffusion layer thickness of the oxidant. When C_O^s reaches to zero, the value of $i_{D,O}$ will become the maximum ($I_{D,O}$):

$$I_{D,O} = n_\alpha F D_O \frac{C_O^o}{\delta_O} \tag{2.58}$$

where the maximum current density ($I_{dl,O}$) is also called the diffusion limiting current density. For reductant oxidation in Reaction (2-II), the similar expression as Eqn (2.58) can also obtained:

$$i_{D,R} = n_\alpha F D_R \frac{C_R^o - C_R^s}{\delta_R} \tag{2.59}$$

$$I_{D,R} = n_\alpha F D_R \frac{C_R^o}{\delta_R} \tag{2.60}$$

where $I_{D,R}$ is the diffusion limiting current density of the reductant, C_R^s is the surface concentration of reductant, and δ_R is the effective diffusion layer thickness of the reductant.

2.4. Effect of Reactant Transport on the Electrode Kinetics of Electron-Transfer Reaction

Reactant transport can not only affect the kinetics of the electron transfer but also affect the thermodynamics of the electrode reaction. In this section, we will discuss these two effects separately.

2.4.1. Effect of Reactant Transport on the Kinetics of Electron-Transfer Process

To investigate the effect of reactant transport on the electron-transfer current density, we can express Eqn (2.28) as Eqn (2.61):

$$i = i^{\mathrm{o}}\left(\frac{C_O^s}{C_O^o}\exp\left(-\frac{(1-\alpha)n_\alpha F(E - E^{\mathrm{eq}})}{RT} \right) - \frac{C_R^s}{C_R^o}\exp\left(\frac{\alpha n_\alpha F(E - E^{\mathrm{eq}})}{RT} \right) \right) \tag{2.61}$$

This equation describes the situation where both oxidant (O) and reductant (R) in Reaction (2-II) are coexisting in the electrolyte solution with the initial concentrations of C_O^o and C_R^o, respectively. As discussed previously, when the overpotential $(E - E^{\mathrm{eq}})$ is negative enough so that the second term in the right-hand side of Eqn (2.61) can be negligible compared to the first term, Eqn (2.61) will become:

$$i = i^{\mathrm{o}}\frac{C_O^s}{C_O^o}\exp\left(-\frac{(1-\alpha)n_\alpha F(E - E^{\mathrm{eq}})}{RT} \right) \tag{2.62}$$

When the overpotential $(E - E^{\mathrm{eq}})$ is positive enough so that the first term in the right-hand side of Eqn (2.61) can be negligible compared to the first term, Eqn (2.61) will become:

$$i = i^{\mathrm{o}}\frac{C_R^s}{C_R^o}\exp\left(\frac{\alpha n_\alpha F(E - E^{\mathrm{eq}})}{RT} \right) \tag{2.63}$$

As described by Eqns (2.57) and (2.58) for oxidant reduction and Eqns (2.59) and (2.60) for the reductant oxidation, the ratios of $\frac{C_O^s}{C_O^o}$ and $\frac{C_R^s}{C_R^o}$ can be expressed as Eqns (2.64) and (2.65), respectively:

$$\frac{C_O^s}{C_O^o} = 1 - \frac{i_{D,O}}{I_{D,O}} \tag{2.64}$$

$$\frac{C_R^s}{C_R^o} = 1 - \frac{i_{D,R}}{I_{D,R}} \tag{2.65}$$

If $i_{D,O}$ and $i_{D,R}$ can be simply expressed as i, and Eqns (2.64) and (2.65) are substituted into Eqns (2.62) and (2.63), separately, two new expressions can be obtained:

$$i = i^o\left(1 - \frac{i}{I_{D,O}}\right)\exp\left(-\frac{(1-\alpha)n_\alpha F(E - E^{eq})}{RT}\right) \tag{2.66}$$

$$i = -i^o\left(1 - \frac{i}{I_{D,R}}\right)\exp\left(\frac{\alpha n_\alpha F(E - E^{eq})}{RT}\right) \tag{2.67}$$

where the terms of $\left(1 - \frac{i}{I_{D,O}}\right)$ and $\left(1 - \frac{i}{I_{D,R}}\right)$ represent the effect of reactant diffusion–convection transport on the electron-transfer kinetics. It can be seen that when $i \ll I_{D,O}$ or $i \ll I_{D,R}$, the reactant transport effect can be negligible, and when i has the similar magnitude to that of $I_{D,O}$ or $I_{D,R}$, the reactant transport effect will be significant.

To give some sense about the reactant transport effect on the current density, Figure 2.11 shows the current–overpotential

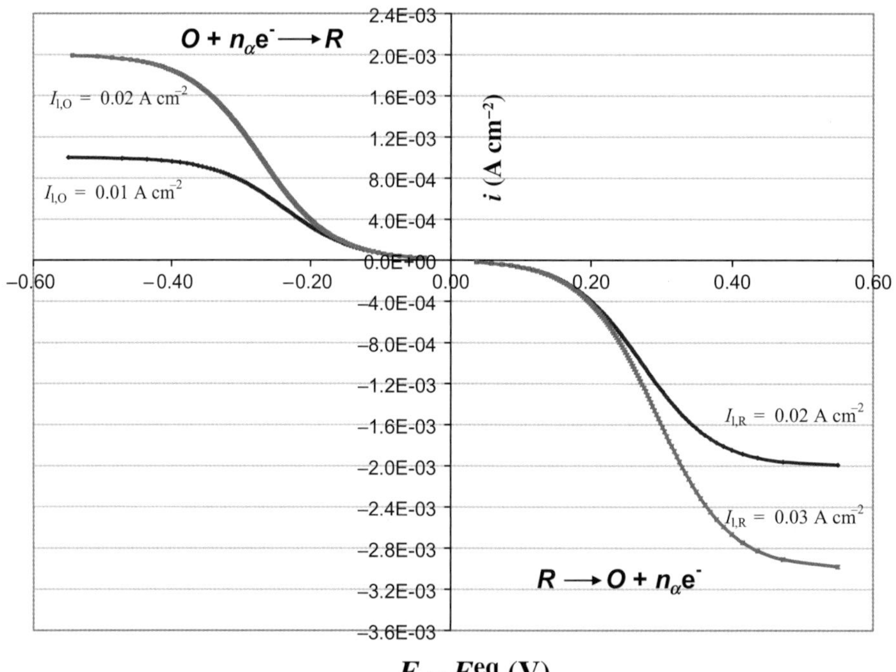

Figure 2.11 Current–potential curves calculated using Eqns (2.66) and (2.67). The diffusion limiting current densities are as those marked in the figure; $i^o = 1.0 \times 10^{-5}$ A cm⁻²; $n_\alpha = 1.0$; $\alpha = 0.5$; $T = 298$ K; $R = 8.314$ J mol⁻¹ K⁻¹; $F = 96,487$ C mol⁻¹. (For color version of this figure, the reader is referred to the online version of this book.)

curves at two sets of different diffusion limiting current density, as marked beside each curves. Note that when performing the calculation for reductant oxidation using Eqn (2.67), both the values of i and $I_{D,R}$ are taken as negative values.

Figure 2.11 shows that due to the reactant transport limitation, the electrode current densities are all limited below or equal to the diffusion limiting current densities.

Note that if the current density is only determined by the electron-transfer process at the electrode surface without any effect from the reactant transport, according to Eqn (2.66) or Eqn (2.67), this current density can be expressed as $i_{k,O} = i^o \exp\left(-\frac{(1-\alpha)n_\alpha F(E-E^{eq})}{RT}\right)$ for oxidant reduction or $i_{k,R} = -i^o \exp\left(\frac{\alpha n_\alpha F(E-E^{eq})}{RT}\right)$ for reductant oxidation. Substituting this i_k into Eqn (2.66) or Eqn (2.67), the following two Equations can be obtained for oxidant reduction and reductant oxidation, respectively:

$$\frac{1}{i} = \frac{1}{i_{k,O}} + \frac{1}{I_{D,O}} \tag{2.68}$$

$$\frac{1}{i} = \frac{1}{i_{k,R}} + \frac{1}{I_{D,R}} \tag{2.69}$$

These two equations have the form predicted by the Koutecky–Levich theory, which will be discussed more in Chapters 5 and 6.

Equations (2.66) and (2.67) can also be expressed as the Tafel form:

$$E = E^{eq} + \frac{RT}{(1-\alpha)n_\alpha F}\ln(i^o) - \frac{RT}{(1-\alpha)n_\alpha F}\ln\left(\frac{iI_{D,O}}{I_{D,O}-i}\right) \tag{2.70}$$

$$E = E^{eq} - \frac{RT}{\alpha n_\alpha F}\ln(i^o) + \frac{RT}{\alpha n_\alpha F}\ln\left(\frac{iI_{D,R}}{i-I_{D,R}}\right) \tag{2.71}$$

Using these equations, the experiment data of current density at various electrode potentials can be analyzed, from which the kinetic parameters of electron-transfer reaction such as the electron-transfer coefficient and the exchange current density can be obtained.

2.4.2. Effect of Reactant Transport on the Thermodynamics of Electrode Reaction

Regarding the effect of reactant transport on the thermodynamics of electrode reaction, we have to consider the reversible electrochemical system, where both the forward and the backward reaction rates in Reaction (2-II) are very fast (or the exchange current density is very large) so that the reaction is always at its equilibrium state, and then the electrode potential (E) can be described by the following Nernst equation:

$$E = E^\circ + \frac{RT}{n_\alpha F} \ln\left(\frac{C_O^s}{C_R^s}\right) \tag{2.72}$$

According to this equation, we can discuss two cases: one is the case when the reductant in Reaction (2-II) is insoluble (such as the metal ions' deposition) or solvent (such as O_2 reduction reaction in aqueous solution to produce water), and the other case is both oxidant and reductant are soluble.

1. *Reductant is insoluble or the solvent.* In this case, the activity of reductant is always equal to 1 (or $C_R^s = 1$), Eqn (2.72) can be rewritten as:

$$E = E^\circ + \frac{RT}{n_\alpha F} \ln\left(C_O^s\right) \tag{2.73}$$

Substituting Eqn (2.64) into Eqn (2.73), Eqn (2.74) can be obtained:

$$E = E^\circ + \frac{RT}{n_\alpha F} \ln\left(C_O^0\right) + \frac{RT}{n_\alpha F} \ln\left(\frac{I_{D,O} - i}{I_{D,O}}\right) \tag{2.74}$$

When $i = \frac{1}{2} I_{D,O}$, the electrode potential defined by Eqn (2.74) will become the half-wave potential: $E_{1/2} = E^\circ + \frac{RT}{n_\alpha F} \ln\left(\frac{1}{2} C_O^0\right)$, and then Eqn (2.74) can be rewritten as:

$$E = E_{1/2} - \frac{RT}{n_\alpha F} \ln\left(\frac{1}{2}\right) + \frac{RT}{n_\alpha F} \ln\left(\frac{I_{D,O} - i}{I_{D,O}}\right) \tag{2.75}$$

Note that Eqn (2.74) is for the case of reversible reaction where Nernst potential is applicable to its electrode potential.

2. *Both the oxidant and reductant are soluble.* When the electrochemical reaction proceeds according to Reaction (2-II) from left to right, the rate of R production at the electrode surface should equal to that diffused out if the reaction is at a steady state. Then C_R^s can be expressed as:

$$C_R^s = C_R^0 + \frac{i\delta_R}{n_\alpha F D_R} \tag{2.76}$$

If at the beginning of the reaction, there is no reductant in the solution ($C_R^0 = 0$), Eqn (2.76) will become:

$$C_R^s = \frac{i\delta_R}{n_\alpha F D_R} \tag{2.77}$$

According to Eqns (2.57) and (2.58), the surface concentration of oxidant can be expressed as:

$$C_O^s = C_O^o + \frac{i\delta_O}{n_\alpha FD_O} = \frac{\delta_O(I_{D,O} - i)}{n_\alpha FD_O} \qquad (2.78)$$

Substituting Eqns (2.77) and (2.78) into Eqn (2.72), the Nernst electrode potential can be expressed as:

$$E = E^o + \frac{RT}{n_\alpha F}\ln\left(\frac{\delta_O D_R}{\delta_R D_O}\right) + \frac{RT}{n_\alpha F}\ln\left(\frac{i}{I_{D,O} - i}\right) \qquad (2.79)$$

When $i = \frac{1}{2}I_{D,O}$, the electrode potential defined by Eqn (2.79) will become the half-wave potential: $E_{1/2} = E^o + \frac{RT}{n_\alpha F}\ln\left(\frac{\delta_O D_R}{\delta_R D_O}\right)$, then Eqn (2.79) can be rewritten as:

$$E = E_{1/2} + \frac{RT}{n_\alpha F}\ln\left(\frac{i}{I_{D,O} - i}\right) \qquad (2.80)$$

This half-wave potential is very useful in evaluating the reactant transport-affected electrochemical reaction. Note again that Eqn (2.80) is for the case of reversible reactions. As a rough estimation, this half-wave potential may be useful for those pseudoreversible reactions. However, one should be careful when using this half-wave potential to evaluation the irreversible electrochemical reactions.

2.5. Kinetics of Reactant Transport Near and within Porous Matrix Electrode Layer[6]

In the above sections, we have presented the electrode kinetics of electron-transfer reaction and reactant transport on planar electrode. However, for practical application, the electrode is normally the porous electrode matrix layer rather than a planner electrode surface because of the inherent advantage of large interfacial area per unit volume. For example, the fuel cell catalyst layers are composed of conductive carbon particles on which the catalyst particles with several nanometers of diameter are attached. On the catalyst particles, some proton or hydroxide ion-conductive ionomer are attached to form a solid electrolyte, which is uniformly distributed within the whole matrix layer. Due to the electrode layer being immersed into the electrolyte solution, this kind of electrode layer is called the "flooded electrode layer".

The detailed description of the mass transfer in this flooded porous electrode layer is not straightforward. The behavior of the porous electrode is more complicated than that of planar electrodes because of the intimate contact between electrode particles and electrolyte. With regard to describing the reactant transport near and within porous electrode layer, there are various types of model. In the earlier models, the pores in flooded porous electrode layer were treated as straight cylinders arranged perpendicular to the external face of the electrode. It is considered that the electrode particle (or catalyst particle) phases and reactants are evenly distributed throughout the volume.[7,8] Further development is the flooded agglomerate models described by Giner and Hunter[9] and Iczkowski and Cutlip.[10] They treated the porous electrode as a collection of flooded catalyst-containing agglomerates. In the agglomerate model described by Durand et al.,[11] the catalyst particle phase was treated as a network of nanoparticles and the reactants were considered to be nonuniformly distributed on a surface or in a volume. For detailed information about these models, please study the literature indicated.

Here we are intending to give a sense as to how to treat the reactant transports within the electrode layer roughly as well as what is its effect on the electron-transfer kinetics. Figure 2.12(A) shows the schematic of such an electrochemical electrode system consisted of the current collector, matrix electrode layer, and the electrolyte.

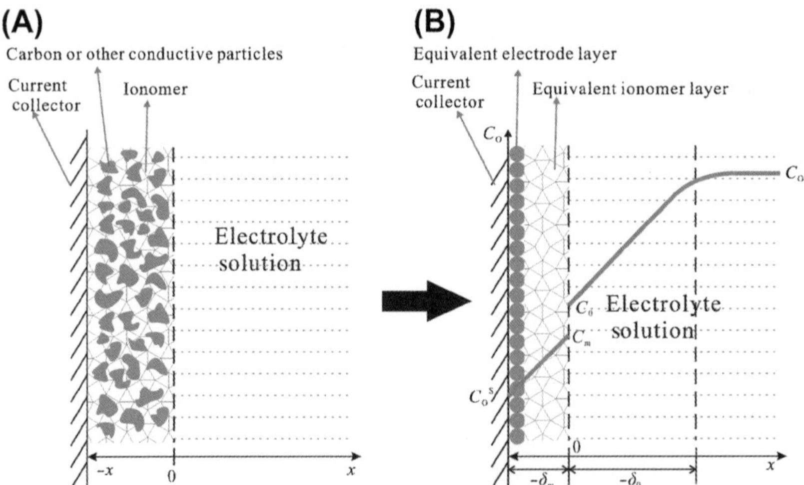

Figure 2.12 (A) Schematic of the electrode/electrolyte interface of the porous matrix layer, and (B) equivalent electrode/electrolyte interfaces of the porous matrix layer and the oxidant distribution within the interfaces. (For color version of this figure, the reader is referred to the online version of this book.)

For simplification of the treatment, we may adopt the treatment described by Gough and Leypoldt[11,12] and treat the electrode particles as a very thin equivalent electrode layer covered by an equivalent solid electrolyte membrane with a thickness of δ_m, as shown in Figure 2.12(B). At the steady state of the reaction, the concentration of oxidant on the thin electrode layer surface is defined as C_O^s, the concentration at the interface of the equivalent electrode membrane/electrolyte solution at the membrane side is defined as C_m, and that at the solution side is defined as C_O, respectively. The oxidant diffusion coefficient in the electrolyte solution is defined as D_O and in the membrane as D_m, respectively. Note that the zero point of the x-axis is at the interface of the equivalent electrode membrane/electrolyte solution. In such an electrode configuration, the obtained steady-state diffusion current density was obtained as Eqn (2.81)[6,13–15]:

$$\frac{1}{i} = \frac{1}{i_{k,O}} + \frac{1}{I_{dl,O}} + \frac{1}{I_{m,O}}$$

(2.81)

where $\;i_{K,O} = i^o \exp\left(-\frac{(1-\alpha)n_\alpha F(E-E^{eq})}{RT}\right),\;\; I_{dl,O} = n_\alpha FD_O\frac{C_O^o}{\delta_O},\;$ and $I_{m,O} = \frac{n_\alpha FC_m D_m \delta_m}{\delta_m}$. Substituting these three expressions into Eqn (2.81), we can obtain:

$$\frac{1}{i} = \frac{1}{i^o \exp\left(-\dfrac{(1-\alpha)n_\alpha F(E-E^{eq})}{RT}\right)} + \frac{1}{n_\alpha FD_O\dfrac{C_O^o}{\delta_O}} + \frac{1}{\dfrac{n_\alpha FC_m D_m \delta_m}{\delta_m}}$$

$$= \frac{1}{i^o}\exp\left(\frac{(1-\alpha)n_\alpha F(E-E^{eq})}{RT}\right) + \frac{\delta_O}{n_\alpha FD_O C_O^o} + \frac{\delta_m}{n_\alpha FC_m D_m}$$

(2.82)

In Eqn (2.82), C_m is normally treated as the solubility of the oxidant in the ionomer membrane. The diffusion layer thickness δ_O can be obtained using a rotating disk electrode technique, which will be given in a very detailed discussion in Chapter 5. The equivalent thickness of the ionomer membrane can be calculated according to the amount of ionomer applied in the electrode layer using the following equation:

$$\delta_m = \frac{W_m}{\rho_m A}$$

(2.83)

where W_m is the weight of the ionomer loaded in the electrode matrix layer (g), ρ_m is the density of the ionomer (g cm^{-3}), and A is the electrode geometric surface area (cm^2).

2.6. Chapter Summary

In investigating an ORR, the fundamental understanding of the kinetics of electrode reaction $(O + n_\alpha e^- \underset{k_b}{\overset{k_f}{\leftrightarrow}} R)$ is necessary. In order to facilitate this understanding and preparing the basic knowledge for rotating electrode theory, in this chapter both the electron-transfer and reactant transport theories at the interface of electrode/electrolyte are presented. Regarding the electron-transfer kinetics, the commonly used equations such as Bulter–Volmer equation and Tafel equation are derived for describing the current–potential relationship. Using these two equations and based on the experiment data, the kinetic parameters such as exchange current density and electrotransfer coefficient can be obtained. Regarding the reactant transport, three transportation modes such as diffusion, migration, and convection are described. A focusing discussion is given to the reactant diffusion near the electrode surface using both Fick's first and second laws. The effects of reactant diffusion/convection on the electrode electron-transfer kinetics are also discussed in this chapter. In addition, based on the approach in literature, the kinetics of reactant transport near and within porous matrix electrode layer and its effect on the electron-transfer process are also presented using a simple equivalent electrode/electrolyte interface.

It is our belief that these fundamentals of electrode kinetics are necessary in understanding the ORR kinetics. It is also our intention to give such a knowledge and information to the reader for their continuing journey to the rest part of this book.

References

1. Bard AJ, Faulkner LF. *Electrochemical methods – fundamentals and applications*. New York Wiley; 2005.
2. Gardiner Jr WC. *Rate and mechanism of chemical reactions*. New York: Benjamin; 1969.
3. Zha Q. *Introduction to electrode kinetics*. Beijing: Chinese Science Press; 2005.
4. Gagotsky VS. *Fundamentals of electrochemistry*. New Jersey: Wiley-Interscience; 2006.
5. Maloy JT. Factors affecting the shape of current-potential curves. *J Chem Educ* 1983;**60**(4):285.
6. Vielstich W, Lamm A, Gasteiger HA. Rotating thin film method for supported catalysis. In: *Handbook of fuel cells-fundamentals, technology and applications*. New York: John Wiley & Sons; 2003. pp. 316–33.
7. Bultel Y, Ozil P, Durand R. Modelling the mode of operation of PEMFC electrodes at the particle level: influence of ohmic drop within the active layer on electrode performance. *J Appl Electrochem* 1998;**28**(3):269–76.

8. Antoine O, Bultel Y, Durand R, Ozil P. Electrocatalysis, diffusion and ohmic drop in PEMFC: particle size and spatial discrete distribution effects. *Electrochim Acta* 1998;**43**(24):3681–91.
9. Giner J, Hunter C. Mechanism of operation of teflon-bonded gas diffusion electrode — a mathematical model. *J Electrochem Soc* 1969;**116**(8):1124.
10. Iczkowski RP, Cutlip MB. Voltage losses in fuel-cell cathodes. *J Electrochem Soc* 1980;**127**(7):1433–40.
11. Gough DA, Leypoldt JK. Membrane-covered, rotated disc electrode. *Anal Chem* 1979;**51**(3):439–44.
12. Gough DA, Leypoldt JK. Rotated, membrane-covered oxygen electrode. *Anal Chem* 1980;**52**:1126–30.
13. Gottesfeld S, Raistrick ID, Srinivasan S. Oxygen reduction kinetics on a platinum RDE coated with a recast Nafion film. *J Electrochem Soc* 1987;**134**(6):1455–62.
14. Lawson DR, Whiteley LD, Martin CR, Szentirmay MN, Song JI. Oxygen reduction at Nafion film-coated platinum electrodes: transport and kinetics. *J Electrochem Soc* 1988;**135**(9):2247–53.
15. Watanabe M, Igarashi H, Yosioka K. An experimental prediction of the preparation condition of Nafion-coated catalyst layers for PEFCs. *Electrochim Acta* 1995;**40**(3):329–34.

3

ELECTROCATALYSTS AND CATALYST LAYERS FOR OXYGEN REDUCTION REACTION

Ligang Feng[a], Xiujuan Sun[a], Shikui Yao[a], Changpeng Liu[c], Wei Xing[a] and Jiujun Zhang[b]

[a]*State Key Laboratory of Electroanalytical Chemistry, Changchun Institute of Applied Chemistry, Changchun, PR China,* [b]*Energy, Mining and Environment, National Research Council of Canada, Vancouver, BC, Canada,* [c]*Laboratory of Advanced Power Sources, Changchun Institute of Applied Chemistry, Changchun, PR China*

CHAPTER OUTLINE

Rotating Electrode Methods and Oxygen Reduction Electrocatalysts. http://dx.doi.org/10.1016/B978-0-444-63278-4.00003-3

3.1. Introduction

In this book, the targeted research systems using the rotating electrode techniques are the electrochemically catalyzed oxygen reduction reactions (ORRs). Therefore, some background and research progress in the ORR catalysts are necessary, which will be the major content in this chapter.

In polymer electrolyte membrane (PEM) fuel cells and metal-air batteries, the most challenge reaction is the catalytic ORR due to its slow kinetics when compared to that of anode fuel or metal electrooxidation. Therefore, developing ORR electrocatalysts and their associated catalyst layers (CLs) are the major efforts in overcoming both the technology and commercialization challenges. Thus, exploring new catalysts, improving catalyst activity and stability/durability, and reducing catalyst cost are currently the major tasks in fuel cell and metal-air battery technology and commercialization.

At the current state of technology, the most practical ORR catalysts are Pt-based materials. However, for Pt-based catalysts, there are several drawbacks, such as high cost, insufficient availability, sensitivity to contaminants, and Pt dissolution. In the effort to search for alternative low-cost non-Pt catalysts, several others, including supported platinum group metal (PGM) types such as Pd-, Ru-, and Ir-based catalysts, bimetallic alloy catalysts, transition metal macrocycles, and transition metal chalcogenides have been explored. It seems that non-noble metal catalysts would therefore appear to be a possible solution for the sustainable commercialization of PEM fuel cells and metal-air batteries. Unfortunately, these approaches are still in the research stage, because both the catalyst activities and stabilities are still too low to be practical in comparison with Pt-based catalysts. Although reducing Pt loading in a catalyst or CL using alloying and carbon supports has shown some progress, all efforts have thus far been offset by rising prices of Pt materials.

In order to give some basic knowledge and concepts, this chapter will review both the electrocatalysts and CLs and their applications for ORR in terms of their types, properties, catalytic activity/stability, as well as their research progress in the past several decades. Furthermore, the synthesis method and characterization methods for electrocatalysts, and the fabrication procedures for CLs, are also reviewed. It is believed that the technical breakthroughs in terms of both ORR activity and stability of the electrocatalysts and CLs should be heavily relied on the new and innovative material synthesis methods or procedures. It is our attention, in this chapter, to present the background and knowledge as well as the research progress about the ORR catalysts, which are the targeted systems for the evaluation using rotating electrode technique, which may benefit the readers for their continuing study throughout this entire book.

3.2. Concepts of Catalytic Activity and Stability

For electrocatalysis of an ORR, two of the most important criteria are catalyst's activity and stability. The activity measures how fast the electrochemical reaction can be speeded up by the catalyst. The faster the catalyzed reaction, the higher the activity of that catalyst would be. For example, a high active ORR electrocatalyst used at a fuel cell cathode would give a high performance of the fuel cell, that is, the higher the activity of the electrocatalyst, the higher the fuel cell performance would be. The catalyst stability means how long this catalyst can last for the catalyzed reaction before its activity degrades to the required minimum level. Of course, the more stable or higher stability, the better the catalyst would be.

In order to get a better understanding about the activity and stability of the ORR electrocatalysts, the following sections will give some fundamentals about the catalysis.

3.2.1. Catalyst and Catalytic Reaction

Catalyst is a substance that can change the rate or kinetics of a chemical reaction. Unlike other reagents that participate in the chemical reaction, a catalyst is not consumed by the reaction itself. Basically, there are two kinds of catalyst, one is called the homogeneous catalysts, and the other called the heterogeneous catalysts. The homogeneous catalyst is in the same phase as the reactants of the chemical reaction, while the heterogeneous catalyst has a different phase from that of the reactants. For example, in some chemical reaction systems, the catalyst is a form of solid particles, and the reactants are the form of the mixed liquid gases. In this sense, an ORR electrocatalyst is a heterogeneous catalyst because it is fixed on the electrode surface to catalyze the reduction reaction of the dissolved oxygen in solution phase.

A catalyst has several features that may need to be mentioned here: (1) catalyst only participates in the catalytic reaction but not be consumed throughout the reaction process; (2) catalyst can only speed up the catalytic reaction's rate or change the reaction's kinetics but not change the reaction's thermodynamics or reaction's equilibrium; (3) a chemical reaction catalyzed by a catalyst has a lower rate-limiting free energy of activation than the corresponding uncatalyzed reaction, resulting in higher reaction rate at the same temperature, as shown in Figure 3.1[1]; (4) a catalyst has its selectivity, it can only catalyze a specific chemical reaction; and (5) catalyst can be easily poised by a very small amount of contaminant in the reaction system, causing activity degradation or failure.

3.2.2. Catalytic Activity

Normally, catalytic activity of a catalyst can be defined by its turnover frequency (TOF). The TOF in a catalytic reaction can be defined as the mole change of product with reaction time $\left(\xi = \frac{1}{\nu_{pdt}} \frac{dm_{pdt}}{dt} \right)$, or the production rate of product or the reaction

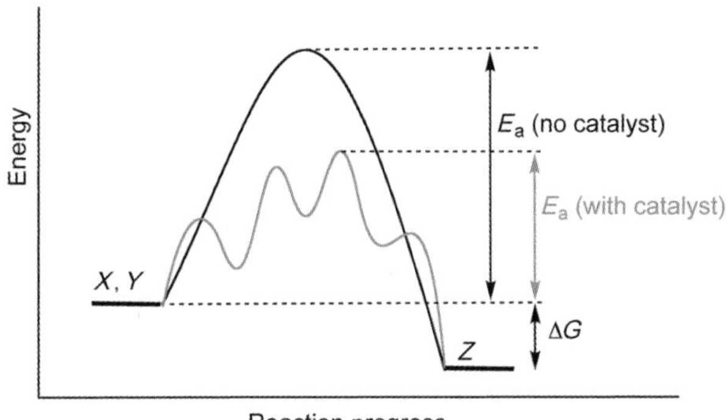

Figure 3.1 Generic potential energy diagram showing the effect of a catalyst in a hypothetical exothermic chemical reaction X + Y to give Z. The presence of the catalyst opens a different reaction pathway (shown in red) with lower activation energy. The final result and the overall thermodynamics are the same.[1] (For interpretation of the references to color in this figure legend, the reader is referred to the online version of this book.)

rate of the catalytic reaction divided by the mole of catalyst used (m_{cat}):

$$\text{TOF} = \frac{1}{m_{cat}}\xi = \frac{1}{m_{cat}}\frac{1}{\nu_{pdt}}\frac{\mathrm{d}m_{pdt}}{\mathrm{d}t} \qquad (3.1)$$

where ν_{pdt} is the stoichiometry of the product in the catalytic reaction, and ξ is the reaction rate of the catalytic reaction with a unit of mol s^{-1}. Obviously, this TOF has a unit of s^{-1}. This Eqn (3.1) indicates that the faster the catalytic reaction rate, the higher the catalyst activity. However, in practice, the catalysts may be composite materials, and it is not easy to exactly know the structure or mole quantity of the active component in the catalyst, and several alternative definitions have been appeared in literature. Their usage is strongly dependent on the individual application areas. The first alternative expression for catalytic activity, called specific rate of catalytic reaction (r_S), can be expressed as:

$$r_S = \frac{1}{w_{cat}}\xi \qquad (3.2)$$

where w_{cat} is the weight or mass of the catalyst (g). Obviously, this specific rate of catalytic reaction, r_S, has a unit of mol s^{-1} g^{-1}. The second alternative expression, called rate of catalytic reaction per unit volume of catalyst (r_V), can be written as:

$$r_V = \frac{1}{v_{cat}}\xi \qquad (3.3)$$

where v_{cat} is the volume of the catalyst (cm^3). Apparently, this rate of catalytic reaction per unit volume of catalyst, r_V, has a unit of mol s^{-1} cm^{-3}. The third alternative expression, called rate of catalytic reaction per unit surface area of catalyst, (r_A), can be written as:

$$r_A = \frac{1}{A_{cat}}\xi \tag{3.4}$$

where A_{cat} is the surface area of the catalyst (cm^2). Apparently, this rate of catalytic reaction per unit surface area of catalyst, r_A, has a unit of mol s^{-1} cm^{-2}.

3.2.3. Electrocatalytic Activity—Catalytic TOF and Current Density

Regarding the electrocatalyst, the similar concepts about the catalytic activity can be defined in the similar ways as Eqns (3.1)−(3.4). Actually, electrocatalysts are a specific form of catalysts that function at electrode surfaces or may be the electrode surface itself. An electrocatalyst can be heterogeneous such as a platinum surface or nanoparticles, or homogeneous like a coordination complex or enzyme. However, in this book, we are only focused on the heterogeneous electrocatalysts. The role of electrocatalyst is to assist in transferring electrons between the electrode catalytic active sites and reactants, and/or facilitates an intermediate chemical transformation. One important difference between catalytic chemical reaction and electrocatalytic reaction is that the electrode potential of the electrocatalyst can also assist in the reaction. By changing the potential of the electrocatalyst, which is attached onto the electrode surface, the electrocatalytic activity can be enhanced or depressed significantly.

The electrocatalytic activity of an electrocatalyst can also be described using the same concept of TOF as shown in Eqn (3.1). However, the electrocatalytic reaction involves the electron transfer from the catalyst to the reactant, the reaction rate or TOF is better being expressed as the electron transfer rate at one catalytic active site of the catalyst:

$$(TOF)_{ecat} = \frac{1}{n_{cas}} \frac{1}{\nu_e} \frac{dn_e}{dt} \tag{3.5}$$

where $(TOF)_{ecat}$ is the TOF or electrocatalytic rate of the electrocatalyst (e site^{-1} s^{-1}), n_{cas} is the concentration of electrocatalytic active site (site per square centimeter of electrocatalyst surface, site cm^{-2} or site per cubic centimeter of electrocatalyst volume, site cm^{-3}), ν_e is the stoichiometry of the electron in the electrocatalytic reaction, n_e is the concentration of electron (e cm^{-2} or e cm^{-3}), and t is the reaction time (s). In general, different electrocatalysts have different values of $(TOF)_{ecat}$. The higher the value of $(TOF)_{ecat}$, the more active the electrocatalyst would be. Figure 3.2 shows several typical electrocatalysts used in PEM fuel cell oxygen reduction reaction.[2]

Figure 3.2 shows that different kinds of electrocatalysts can give different TOFs (or different catalytic activities) toward an ORR. These nanostructured electrocatalysts include Pt, Pt alloy, and non-noble metal (Fe-based) complexes among which the core–shell structured Pt-based catalyst gives the largest TOF for an ORR. Unfortunately, the structure retention would be the challenge when these core–shell structured catalysts are used in fuel cell environments.

For electrocatalytic reaction, the activity of an electrocatalyst can also be expressed as the reaction current density:

$$i_{ecat} = q_e n_{cas}(TOF)_{ecat} \tag{3.6}$$

where i_{ecat} is the electrocatalytic current density (mA cm^{-2} or mA cm^{-3}), q_e is the electron charge ($=1.602 \times 10^{-19}$ C e^{-1}), and both n_{cas} and $(TOF)_{ecat}$ have the same meanings and units as those in Eqn (3.5). Equation (3.6) indicates that in order to obtain high electrocatalytic activity, increasing both catalyst's active site concentration or density and TOF are necessary.

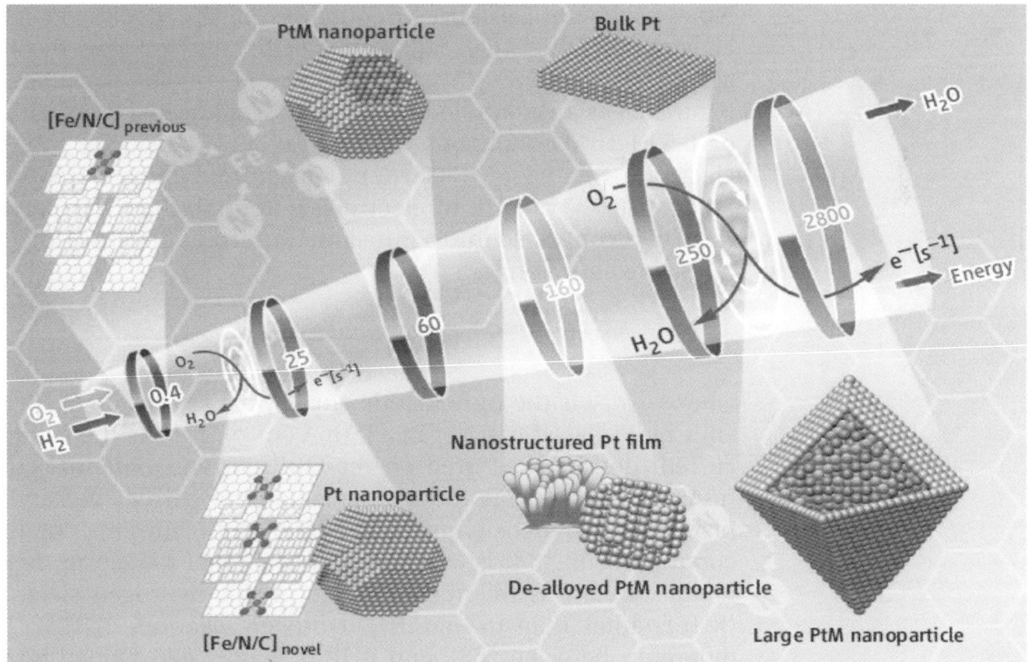

Figure 3.2 Turnover frequencies of several typical electrocatalysts for PEM fuel cell oxygen reduction reaction.[2] (For color version of this figure, the reader is referred to the online version of this book.)

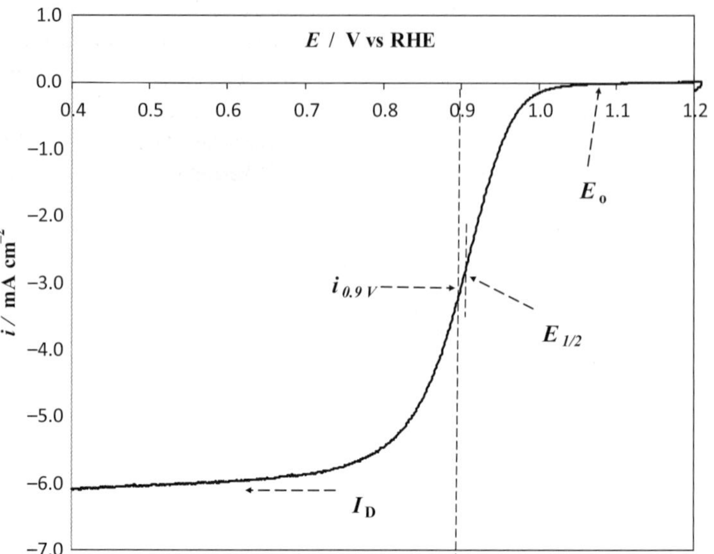

Figure 3.3 Current–potential curve recorded on a rotating disk glassy carbon electrode coated with 20wt %Pt/C catalyst, measured in O_2-saturated 0.1 M $HClO_4$ at 30 °C. Potential scan rate: 5 mV s^{-1}, electrode rotating rate: 1600 rpm. Catalyst loading: 0.048 mg$_{Pt}$ cm^{-2}. Ryan Baker, Jiujun Zhang, Unpublished data.

For an ORR activity of a Pt-based catalyst in a CL, two catalytic activities are also used frequently: one is called the Pt-mass activity, and the other called Pt-specific activity. This Pt-mass activity is measured using the current–potential curve (in Figure 3.3 as an example, recorded in O_2-saturated 0.1 M $HClO_4$ or 0.5 M H_2SO_4 aqueous solution at 30 °C) using a rotating electrode rotated at 1600 rpm. The current density ($i_{0.9\,V}$) is taken at 0.9 V vs RHE of electrode potential, then this current density is converted into a pure kinetic current density using the following equation:

$$i_{k,0.9V} = \frac{i_{0.9V} I_d}{I_D - i_{0.9V}} \tag{3.7}$$

where $i_{k,0.9\,V}$ is the pure kinetic current density of the catalyzed ORR by Pt-based catalyst in CL (mA cm^{-2}), $i_{0.9\,V}$ is the catalytic current density measured on the ORR current–potential curve (mA cm^{-2}), and I_D is the diffusion limiting current density (or plateau current density) measured on the ORR current–potential curve (mA cm^{-2}). It is worthwhile to note that the electrode potential of 0.9 V is referred to the reversible hydrogen electrode (RHE) rather than the normal hydrogen electrode (NHE). The difference between NHE and RHE is mainly determined by the electrolyte used in the reference electrode chamber. The potential of NHE is always at 0.000 V at any temperature, while the electrode potential of RHE is dependent on the concentration (mainly the PH) of electrolyte used. For example, in 0.1 M $HClO_4$ solution, the electrode potential difference between NHE and RHE is about 60 mV.

Based on Eqn (3.7), the Pt-mass activity $(i_{ma,Pt})$ can be calculated using the following equation:

$$i_{ma,Pt} = \frac{i_{k,0.9V}}{m_{Pt}} = \frac{1}{m_{Pt}} \frac{i_{0.9V} I_D}{I_D - i_{0.9V}} \tag{3.8}$$

where m_{Pt} is the Pt loading in the CL with a unit of $mg\,cm^{-2}$. Obviously the Pt-mass activity $i_{ma,Pt}$ has a unit of $mA\,mg^{-1}_{Pt}$. For example, the Department of Energy of USA (DOE) defines the target of a fuel cell ORR catalyst activity according to the performance of Pt-based catalysts in the fuel cell testing. It says that by the year 2020 the ORR activity of a Pt-based catalyst must reach a mass activity 440 mA mg^{-1}_{Pt} at an iR-free cell voltage of 0.9 V (440 mA mg^{-1}_{Pt}@0.9 $V_{iR\text{-}free}$) below 80 °C.[3]

The Pt-specific activity $(i_{S,Pt})$ is defined as:

$$i_{S,Pt} = \frac{i_{ma,Pt}}{EPSA} \tag{3.9}$$

where EPSA is the electrochemical Pt surface area $(cm^2\,cm^{-2}\,mg^{-1}_{Pt})$, which is measured on cyclic voltammogram recorded using the electrode coated with Pt-CL (for a detailed description, please see a latter section of this chapter). Therefore, this Pt-specific activity has a unit of $mA\,cm^{-2}$. Note that the EPSA data are only meaningful for Pt catalysts. For Pt-alloy catalysts, the measured EPSA data may not be useful. As a result, if the EPSA data are in uncertainty, the Pt-specific activity expressed by Eqn (3.9) is also less meaningful.

It is worthwhile to note that Eqns (3.7)−(3.9) are useful in both acidic and basic electrolyte solutions. This is because the measured electrode potential refers to the RHE rather than NHE.

For non-noble metal ORR catalysts, the definition of catalytic activity is different from that of Pt-based catalysts. For a non-noble metal catalyst, the similar preparation procedure for CL and ORR measurement steps using rotating disk electrode technique to that for Pt-based catalyst have been widely used in literature.[2] However, due to both the ORR onset and half-wave potentials catalyzed by non-noble metal catalysts are much lower than those of Pt-based catalysts, it is difficult or impossible to observed ORR current density at 0.9 V vs RHE. A current density at other lower potentials may be used to define the catalyst activity for the purpose of comparison. In this case, Eqn (3.7) may still usable except the electrode potential is not 0.9 V, instead of the other chosen potential. Due to the ORR active site of non-noble metal catalyst consisting of both the metal center and the ligand, only using metal loading to calculate the ORR mass activity like that expressed by Eqn (3.8) is not appropriate anymore. Therefore, expressing ORR catalytic activity for a non-noble metal catalyst with the unit of mA per cubic centimeter of CL (mA cm^{-3}) may be more appropriate. For example, the DOE defines the target of fuel cell ORR catalyst activity according to the performance of the non-noble metal catalyst in the fuel cell testing. It says that by the year 2020 the ORR activity of a non-noble metal catalyst must reach a volumetric current density of 300 A cm^{-3} at an iR-free cell voltage of 0.8 V below 80 °C.[3] In the fuel cell measurement for catalyst volumetric current density, the studied non-noble metal catalyst is used to make the cathode CL, and a commercial Pt/C is used to make the anode CL for the construction of membrane electrolyte assembly (MEA), which is then assembled into an acidic or basic H_2/O_2 PEM fuel cell for testing at 80 °C. Then the measured current density at IR-free 0.8 V ($i_{0.8\ V}$, mA cm^{-2}) is taken. This current density is divided by the thickness of the cathode CL (l_{CCL}, cm) to obtain the volumetric ORR catalytic activity ($i_{V,\ 0.8\ V}$, mA cm^{-3}) of the non-noble metal catalyst used:

$$i_{V,0.8\ V} = \frac{i_{0.8\ V}}{l_{CCL}} \tag{3.10}$$

Note that this CL thickness can be measured on the cross section of the MEA using scanning electron microscope (SEM).

3.2.4. Electrocatalytic Activity—Onset Potential and Half-wave Potential

For catalytic ORR measured by a rotating electrode technique, other two measures for catalytic activity are also commonly used: the onset potential (E_{onset}) and the half-wave potential ($E_{1/2}$), as shown in Figure 3.3. Onset potential is defined as the potential at which the catalyzed ORR current starts to appear. The more positive the onset potential, the more active the electrocatalyst would be. There are no simple theoretical expressions about E_{onset} and $E_{1/2}$. Fortunately, they can be easily obtained from the current–potential curves recorded by rotating electrode techniques. However, there exists some arbitrary in determining this onset potential point from the current–potential curve shown in Figure 3.3. The more accurate way may be the half-wave potential. The more positive the half-wave potential, the more active the electrocatalyst would be. Note that this half-wave potential is contributed by both the electrode kinetics of catalyst and the O_2 diffusion processes in the bulk solution and within the CL, suggesting that this half-wave potential may not be accurate in representing the catalytic activity of the catalyst.

To give some sense about the catalytic activities of electrocatalysts, Table 3.1 lists some typical ORR electrocatalysts and their catalytic activities.

3.2.5. Stability of Electrocatalysts

There is still no consistent definition about the catalyst's stability in literature. The stability of electrocatalyst may be defined as the percentage loss of electrocatalytic activity (or electrocatalytic current density) after a certain period of electrocatalytic reaction:

$$\Delta i_{ecat}\% = \frac{(i_{ecat})_{BOL} - (i_{ecat})_{EOL}}{(i_{ecat})_{BOL}} \times 100 \qquad (3.11)$$

where $\Delta i_{ecat}\%$ is the percentage loss of electrocatalytic current density or mass activity ($mA\ cm^{-2}$ or $mA\ cm^{-3}$), $(i_{ecat})_{BOL}$ is the catalytic current density at the beginning of lifetime (BOL) test, and $(i_{ecat})_{EOL}$ is the current density at the end of lifetime (EOL) test. From Eqn (3.7), it can be seen that the smaller the value of $\Delta i_{ecat}\%$, the more stable the electrocatalyst would be. For an electrocatalytic ORR, the kinetic current densities of recorded BOL and EOL at a fixed electrode potential can be used to evaluate the electrocatalyst's stability. Actually, there are many different methods and conditions for the catalyst stability measurements, and the stability results from different sources may not be comparable. It is suggested that to obtain a meaningful stability result, the experiment design and test should be carried out according to the targeted applications. For fuel cell applications, the DOE has set up the stability target by the year 2020, which says that the ORR mass activity loss is less than 40% after the EOL test.[3]

3.2.6. Composition and Structure of ORR Electrocatalysts

In general, both the catalyst's catalytic activity and stability are strongly dependent on the catalyst material's chemical and physical properties, composition, morphology, and structure. For ORR catalysis, in order to improve both the activity and stability of electrocatalysts, various kinds of materials including noble metals, non-noble metals, metal oxides, chalcogenides, metal

Table 3.1. Catalytic Activities of Some Typical Electrocatalysts for Oxygen Reduction Reaction, Measured Using Rotating Disk Electrode Technique

Catalyst	Electrocatalytic Current Density, and Test Condition	ORR Onset Potential	ORR Half-wave Potential	References
20 wt% Pt/C (E-TEK)	120 mA mg$^{-1}_{Pt}$ at 0.9 V vs RHE in 0.5 M H_2SO_4, 2000 rpm, scan rate: 5 mV s^{-1}	1.0 V vs RHE	0.9 V vs RHE	4
20 wt% $Pt_{0.55}Pd_{0.45}$/ $C_{67\%}(Nb_{0.07}Ti_{0.93}O_2)_{33\%}$	167 mA mg$^{-1}_{Pt}$ at 0.9 V vs RHE in 0.1 M $HClO_4$ at 30 °C, 1600 rpm, scan rate: 5 mV s^{-1}	1.05 V vs RHE	0.9 V vs RHE	5
45 wt% $Pt_{71}Ni_{29}$/C	167 mA mg$^{-1}_{Pt}$ at 0.9 V vs RHE in 0.5 M H_2SO_4 at ambient conditions, 2000 rpm, scan rate 5 mV s^{-1}	1.1 V vs RHE	0.89 V vs RHE	4,6
20 wt% Pd_2Co/C	~2.0 mA cm^{-2} at 0.60 V vs RHE in 0.5 M H_2SO_4 at 25 °C, 1600 rpm, 5 mV s^{-1}	0.95 V vs RHE	0.66 V vs RHE	7
$Mo_{4.2}Ru_{1.8}Se_8$	~1.0 mA cm^{-2} at 0.5 V vs NHE, 0.5 M H_2SO_4 at 25 °C, 1600 rpm, scan rate: 5 mV s^{-1}	~0.76 V vs NHE	0.50 V vs RHE	8
W-Co-Se	~0.8 mA cm^{-2} at 0.55 V vs NHE in 0.5 M H_2SO_4 at 25 °C. Catalyst loading: 270 μg cm^{-2}, 1600 rpm, scan rate: 5 mV s^{-1}	0.755 V vs NHE	~0.55 V vs NHE	9
42.0 wt% Co-N_x/C	~3.0 mA cm^{-2} at 0.6 V vs NHE in 0.5 M H_2SO_4 at 25 °C, 2500 rpm, scan rate: 5 mV s^{-1}	~0.85 V vs NHE	~0.68 V vs NHE	10
PANI-Fe-C and PANI-FeCo-C	~1.9 mA cm^{-2} at 0.80 V vs NHE in 0.5 M H_2SO_4 at 25 °C, 900 rpm, 5 mV s^{-1}	~0.93 V vs NHE	~0.81 V vs NHE	11

complexes, metal carbides, metal nitrides, and so on, and different morphologies including sphere, tubing, fibre, alloy, core–shell, and so on, have been developed and explored as ORR electrocatalysts.

Another important approach to improve the catalyst's activity and stability is to use supporting strategy in achieving high reaction surface area, more reaction active site, and uniformity of the activity site distribution. At the current state of technology, almost all ORR electrocatalysts are supported catalysts. The support materials are normally carbon nanoparticles that have high surface area and good affinity to catalyst particles. Figure 3.4 shows several typical schematic structures of the supported ORR electrocatalysts. This kind of catalyst morphologies have been validated using some sophisticated instruments such as Transmission Electron Microscopy (TEM). Several typical examples are shown in Figure 3.5.

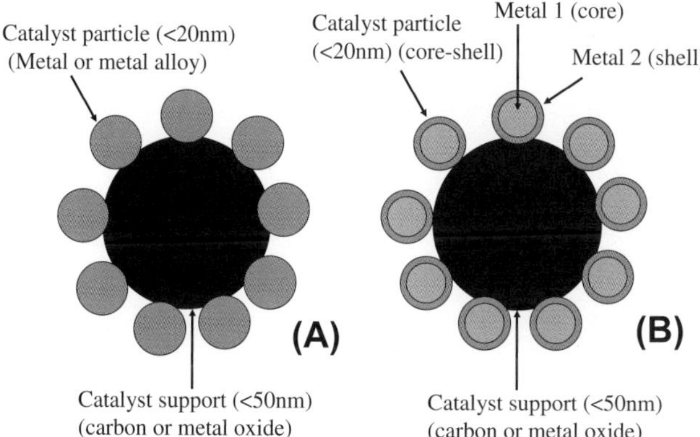

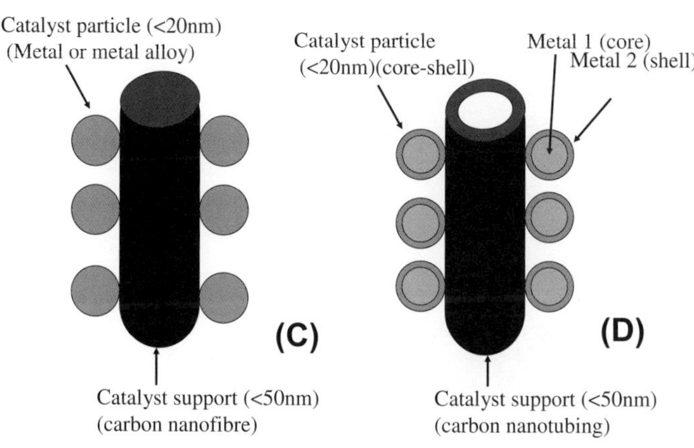

Figure 3.4 Schematics of ORR electrocatalyst's morphologies. (A) Metal catalyst such as Pt and metal alloy catalyst such as PtPd supported on a conductive material such as carbon or metal oxide; (B) core–shell catalyst such as Au@Pt supported on conductive material such as carbon or metal oxide; (C) metal catalyst such as Pt and metal alloy catalyst such as PtPd supported on a nanofibre such as carbon or metal-oxide nanofibre; and (D) core–shell catalyst such as Au@Pt supported on conductive nanotubings such as carbon nanotubings. (For color version of this figure, the reader is referred to the online version of this book.)

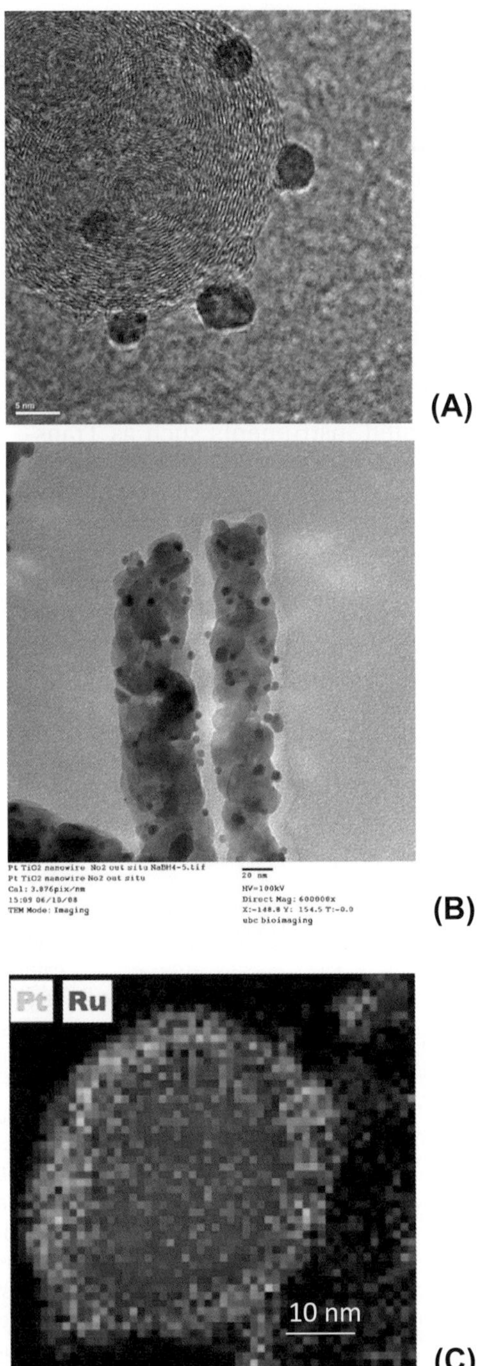

Figure 3.5 TEM imagines for several typical electrocatalysts. (A) PtRu alloy supported on carbon particles (PtRu/C)[12]; (B) Pt supported on TiO_2 nanofibers (Pt/TiO_2) ("TiO2-supported Pt electrocatalysts for oxygen reduction reaction", Robert Hui, and Jiujun Zhang, unpublished work); and (C) Ti_4O_7-supported Ru@Pt core—shell catalyst.[13] (For color version of this figure, the reader is referred to the online version of this book.)

3.2.7. Requirements for ORR Electrocatalyst

An ORR electrocatalyst should meet the following requirements: (1) high catalytic activity, (2) high electrical conductivity, (3) high electrochemical stability (not being oxidized at high electrode potentials), (4) high chemical stability (not being oxidized by proton and oxygen and not soluble in acidic or basic aqueous solution), (5) favorite structure optimum composition, favorite morphology, high specific surface area (high SSA), small particle size, high porosity, and uniform distribution of catalyst particles on the support, (6) strong interaction between the catalyst particle and the support surface, and (7) high catalytic stability. It is impossible for an electrocatalyst to meet all these requirements at the same time. There are some trade-off and tolerance, depending on the type of the applications and the applications' conditions. The following subsections will give some detailed discussion about these requirements.

3.2.7.1. High Catalytic Activity

High catalytic activity is the first and essential requirement for an ORR electrocatalyst. In general, all ORR electrocatalysts reported in literature have some catalytic activity, more or less. The most practical and also commercially available catalysts are carbon-supported Pt catalysts (Pt/C), which have the Pt-mass activity ($i_{ma,Pt}$, defined by Eqn (3.8)) values of about $100-130$ mA mg$^{-1}_{Pt}$. However, for carbon-supported Pt-alloy catalysts including core–shell structure ones, as shown in Figure 3.2, have much higher Pt-mass activity such as $300-800$ mA mg$^{-1}_{Pt}$. For example, Pt_3Ni (111) skin model catalyst could give an ORR activity as high as 530 mA mg$^{-1}_{Pt}$.[14] The ternary Pt-alloy catalysts such as PtCoMn and PtNiFe have demonstrated 8X and 10X the activity of Pt/C.[15] Unfortunately, these catalysts' structure retention and high solubility of the alloyed metals are the limitations for their practical usage in PEM fuel cell operation at this moment. For non-noble metal catalysts, due to their ORR onset potentials or half-wave potentials or their fuel cell performance are much lower than those Pt-based catalysts, there is no catalytic ORR current density that could be observed at 0.9 V vs RHE or 0.9 V cell voltage in some cases. Therefore, the voltage point measuring catalytic activity is set at 0.8 V, and the catalyst activity is expressed as a volumetric catalytic activity instead of a mass activity. The best activity for FeCo-based non-noble metal catalysts are in the range of $130-165$ A cm^{-3} as reported by Dodelet' group at INRS[2] and Zelenay's group at LANL.[16] Regarding the performance of non-noble metal catalyst-based fuel cell, the best maximum power densities could be reached to 400–500 mW cm^{-2}.

3.2.7.2. High Electrical Conductivity

The CLs, as a layer of electrode, must be electrical conductive for a fast and smooth pass of the electrical current. If the CL has high electrical resistance, the internal resistance of the electrochemical cell such as fuel cell will have a high internal resistance, leading to large *iR* drop, resulting in a power loss. To have a high conductive CL, the catalyst particles inside the layer are better to be electrical conductive. On the other hand, in an ORR process catalyzed by supported ORR catalyst such as Pt/C, the adsorbed O_2 on Pt particle will get four electrons from the catalyst particle. These electrons come from the cathode current collector, traveling through both the catalyst carbon support and Pt particle then reaching to O_2. Therefore, both catalyst particle and its support must be electrically conductive in order to have a fast and smooth electron transfer pathway. Generally, the higher the electrical conductivity of the catalyst, the better the catalyst activity would be. As an example, Figure 3.6 shows the correlation between the catalyst support and the ORR catalytic activity: a monotonic increase in ORR catalytic activity with increasing the catalyst support's conductivity.

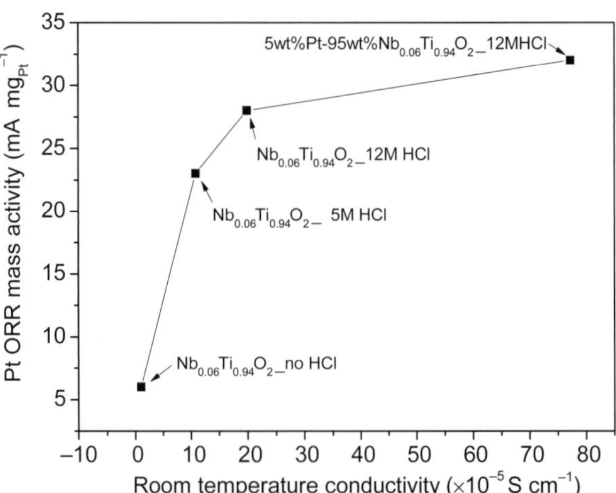

Figure 3.6 Relationship between the supports' room temperature conductivity and their corresponding catalysts' Pt-ORR mass activity. The conductivity data were taken from the supports and the Pt-ORR mass activity data were taken from their supported 20 wt% of Pt catalysts.[17]

Generally, different catalysts have different electrical conductivities. For all carbon-supported metal materials, their conductivities are high enough for ORR applications and meeting the conductivity requirement in constructing fuel cell cathode CL. This is because both the carbon support material and the metal/metal alloy are high electrically conductive ones. However, for some catalysts, such as metal oxides-supported catalysts, the conductivities of such support materials are low, resulting in insufficient conductivities of their supported catalysts. To increase the conductivity of the support, the common way is to make composite support materials. For example, TiO_2 has a very low conductivity. If using carbon particles to make a carbon-TiO_2 composite material, the conductivity of such a composite material will be close to that of carbon materials.

3.2.7.3. High Electrochemical Stability

The ORR catalysts are used in either acidic or basic electrolyte solutions, and when they are used in either acidic or alkaline fuel cells in the presence of high oxidizing oxygen, electrochemical stabilities of both the catalysts and their support materials are very important for their practical applications. In the presence of O_2, the potential of electrode coated with a catalyst (or the catalyst's potential) will be higher than 1.2 V vs RHE. This potential is higher than the oxidation potentials of almost all metal and carbon materials, or in other words, almost all metal and carbon materials are thermodynamically unstable in the sense of electrochemistry. However, due to the slow kinetics of the oxidation process, or the oxidation product is less soluble, the carbon and Pt-based materials can still be used as ORR catalysts for fuel cells even at acidic or basic environment and high temperatures such as 70−80 °C.

For Pt catalyst, the surface cyclic voltammogram shows that the electrooxidation starts to occur when the electrode potential is more positive than 0.6 V vs RHE to form Pt oxides such as PtO. In the presence of O_2, the electrode potential is more positive than 1.0 V, at which Pt will be definitely electrochemically oxidized to form PtO. In an acidic environment, the formed PtO will be dissolved according to Reaction (3-I):

$$PtO + 2H^+ \rightarrow Pt^{2+} + H_2O \qquad (3\text{-}I)$$

With a long period operation, Pt will be gradually dissolved although its dissolution rate is slow, causing degradation of the catalyst's activity. The higher the electrode potential, the faster the electrooxidation rate would be. In a fuel cell, the degradation mechanism of carbon-supported Pt catalyst seems more complicated, as schematically shown in Figure 3.7(A). Due to the local potential distribution on the carbon support is not uniform; the slow dissolved Pt^{2+} ion could be redeposited back to form large-sized Pt particle (called the agglomeration), reducing the catalytic reaction surface, leading to an decrease in an ORR activity. For Pt-alloy catalysts, the electrooxidation then dissolution of the alloying metals can easily happen, leading to morphology change, resulting in reduced catalytic activity.

Not only the Pt catalyst is required to be electrochemical stable, but also the carbon support should be also stable. Unfortunately, due to the thermodynamic potential of carbon electrooxidation ($E^o = 0.207$ V vs RHE) is much more negative than the electrode potential when O_2 presents, the carbon oxidation will happen according to Reaction (3-II):

$$C + 2H_2O \rightarrow CO_2 + 4H^+ + 4e^- \qquad (3\text{-}II)$$

The oxidation mechanism is schematically shown in Figure 3.7(B). It can be seen that after the carbon support is lost due to oxidation, the Pt particle will fall off, becoming electrically isolated particle.

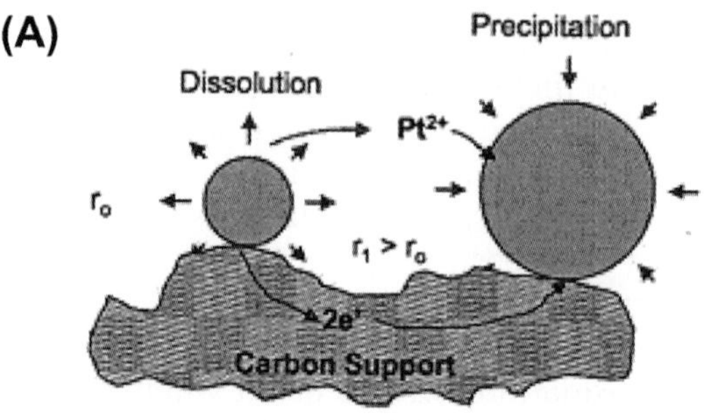

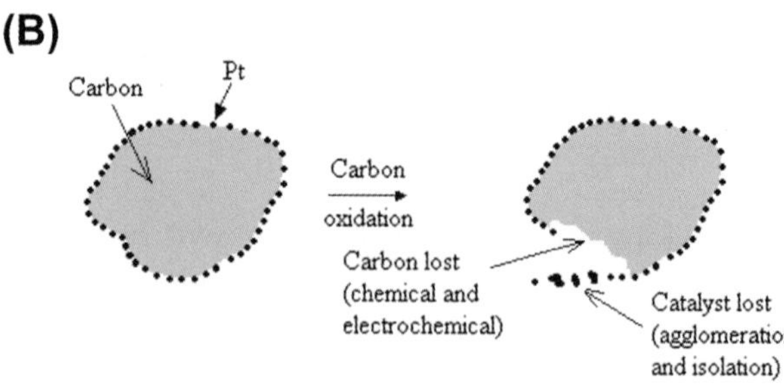

Figure 3.7 Schematic processes of (A) Pt dissolution and (B) carbon corrosion.[18]

Some non-carbon supports are also electrochemically unstable. For example, when TiC was used as a catalyst support to form Pt/TiC and Pt$_3$Pd/TiC ORR catalysts, its electrooxidation at potential higher than 0.8 V vs RHE was found.[19] When TiC@TiO$_2$ core–shell composite was used for the support, the electrochemical stabilities were significantly improved.

3.2.7.4. High Chemical Stability

A chemically stable ORR electrocatalyst should not be oxidized or corroded by O$_2$ or proton. However, some catalyst components such as the alloy metals such as Fe, Co, Ni, Mn, Cu, etc., in Pt alloy catalysts could be chemically oxidized by either proton or O$_2$ to leach out. Some metal oxide supports and non-noble metal catalysts could also dissolve in acidic environment. Therefore, developing ORR catalysts and support materials with chemical stability is necessary.

3.2.7.5. Favorite Structure

A catalyst powder should have a favorite structure in order to be a high active and stable ORR electrocatalyst. Here the catalyst structure contains catalyst's composition, morphology (particle shape), SSA, particle size, porosity, and particle distribution.

Regarding the catalyst's composition, there should be an optimum composition that could give the best ORR catalytic activity. For example, a Pt-alloy catalyst (Pt$_X$M$_y$) (M = Ni, Co, Fe, Cu, etc., $X + Y = 1$), different ratios of X/Y could give different ORR catalytic activities. Through experiment and theoretical modeling, the optimum X/Y ratio could be obtained. For supported catalyst such as Pt/C, the weight percentage (wt%) of Pt catalyst with respect to carbon support could also affect its ORR activity. This is because if the Pt wt% is too low, the concentration of ORR reaction site in the CL could be diluted, leading to low catalytic current density; if the Pt wt% is too high, the Pt utilization in the CL could be reduced. At current state of technology, the commercially available Pt/C catalysts are normally having the Pt wt% of 20–50.

Regarding the catalyst morphology, as discussed previously, there are varieties of catalyst shapes such as sphere, nanorod, nanofiber, nanotubing, core–shell, etc. For catalyst support, there are also different shapes similar to those of catalyst. Obviously, different combinations between different shape of catalyst and support could result in different ORR catalytic activities and stabilities. Normally, Pt-alloy catalysts with core–shell morphology could give the best ORR activities but their stabilities are questionable. The high ORR activity of Pt-alloy catalysts is mainly due to the favorite interaction and energy matching between the adsorbed O_2 and the catalyst site.

The catalyst's SSA is a property of the catalyst powder, which is the total surface area per gram of powder ($m^2 g^{-1}$). The SSA can be measured by N_2-adsorption using the BET (Brunauer-Emmett-Teller, three inventors' last names) isotherm, which has the advantage of measuring the surface of fine structures and deep texture on the particles. The magnitude of SSA is strongly dependent on the catalyst particle size/distribution, and porosity. Normally, the smaller the particle size and the narrower the particle distribution, the larger the SSA would be; and the more porous the particle, the larger SSA would be. For a supported ORR electrocatalyst, its SSA is contributed by both catalyst particle and the support particle, but mainly dominated by the SSA of the later. To obtain an active ORR catalyst, it is always a desire to make its SSA larger. This can be done by selecting support material with a high SSA such as active carbon, and depositing small sized/uniformly distributed catalyst particle on the support. For example, carbon particle has a typical SSA of 254 $m^2 g^{-1}$, which is much larger than other support materials. This is one of the reasons why carbon particles are the most common support material for catalyst preparation.

3.2.7.6. Strong Interaction between the Catalyst Particle and the Support Surface

There are two interactions between the catalyst particle and the support surface: one is the physical interaction, and the other is the electronic interaction. Both of these two interactions have strong effects on the ORR catalyst's activity and stability. Physical interaction refers the attachment between the catalyst particle and the support surface. For Pt/C catalyst as an example, if the attachment of Pt on the carbons surface is not strong enough, Pt particle will easily fall off from the carbon surface, becoming isolated Pt particle during the operation. It is believed that such a metal-support interaction could favor the anchoring of Pt nanoparticles on a support surface, promoting uniform catalyst dispersion, and preventing them from migration, coalescence, aggregation, and loss. The electronic interaction between support and Pt could favor the electron transfer from Pt to adsorbed oxygen atoms, resulting in a high ORR activity.

3.2.7.7. High Catalytic Stability

Catalyst's stability can be affected by many factors such as electrochemical stability, chemical stability, structure composition, morphology, SSA, particle size, porosity, distribution of catalyst particles on the support, interaction between the catalyst particle and the support surface, the type of applications as well as the operation conditions. Therefore, the optimization among these affecting factors has to be carried out in order to achieve practically usable ORR catalysts.

3.3. Current Research Effort in ORR Electrocatalysis

In ORR, the electrocatalysts employed can be classified into two categories: one is the noble metal catalysts, and the other is the non-noble metal catalysts. In this section, we will review the current major effort in improving an ORR activity/stability.

3.3.1. Noble Metal-based Electrocatalysts

At the current stage of technology, carbon-supported Pt and Pt-based alloy catalysts are the most active and stable catalysts for an ORR, which have been used for fuel cell cathodes. The major research effort for Pt and Pt-based alloy catalysts is to optimize (1) the size and dispersion of nanoparticles, (2) interaction between Pt catalyst and supporting materials, and (3) Pt-alloying strategy.

3.3.1.1. Catalyst Particle Size Optimization and Dispersion Uniformity

For Pt catalysts, generally to say, the smaller the Pt particles and the larger SSA the particles possess, the better the electrocatalytical performance would be. However, whether or not there exists a critical size that can give the maximum Pt-mass activity is still no clear answer yet, although extensive research efforts have been done on this subject.[20] Calculation using density functional theory (DFT) is often used for providing insight into the effect of nanoparticle size on the electrocatalytical activity.[21]

To achieve small catalyst particle size, and prevent particles from aggregating during the synthesis seems very important. For example, in synthesis of Pt-based catalysts, long-chain polyelectrolyte and Au are usually used to stabilize the platinum nanoparticles.[22–26] The long-chain polyelectrolyte chains were found to be able to fine and stabilize these nanoparticles from aggregating through the electrostatic interaction between Pt and long-chain polyelectrolytes, leading to a high active catalyst. Besides the polyelectrolytes, metal oxides could also help form desired dispersion and prevent the catalyst particles from aggregation during the synthesis.[27–36]

3.3.1.2. Interaction between Pt Catalysts and Supporting Materials

It is believed that some doped carbon nanotubes can provide the Pt catalyst with strong interaction. For example, highly dispersed Pt nanoparticles supported on surface of thiolation functional carbon nanotubes, such as S-doped CNTs (SH-CNTs), $CNTs-C_6H_6-CH_2-SH$, $CNTs-CONH-(CH_2)_2-SH$, and $CNTs-CONH-C_6H_6-SH$, were explored.[37,38] Furthermore, when some metal oxides such as $NbTiO_2$ and metal carbides were explored as the catalyst supports, strong interaction between catalyst and the support was also observed.[5,39]

3.3.1.3. Pt-Alloying Strategy

It has been recognized that Pt-alloy catalysts have higher ORR activity than those of pure Pt catalysts. Pt alloys such as $Pt-Pd$,[40,41] $Pt-Ru$,[40,42] $Pt-Au$,[20,43–45] $Pt-Co$,[26,46–49] $Pt-Cr$ and $Pt-Ni$,[14,50] $Pt-Ti$,[51] $Pt-Mn$,[48] $Pt-Fe$,[47,48,52] and $Pt-Sn$,[53] and so on demonstrated some improved catalytic ORR activity when compared to pure Pt because of the synergistic role of the transition metals. It was reported that the catalytic activity of Pt_3M (M = V, Ti, Co, Fe, Ni) could be significantly improved with strong resistance to poisonous substances.[54,55] Recently, Stamenkovic et al.[56] demonstrated that the extended single crystal surfaces of $Pt_3Ni(111)$ exhibited an enhanced ORR activity that was 10-fold higher than Pt(111) and 90-fold higher than the current state-of-the-art Pt/C catalysts. Such a remarkable activity was claimed to attribute to the weaker OH adsorption arising from the decrease of the d-band center on the Pt skin formed by surface segregation. In addition, a monolayer of Pt on a Pd surface showed about five times of higher Pt mass-specific activity than that of Pt/C.[57]

However, the mechanism of the alloying metal's roles is still not fully understood. Several explanations have been put forward[58]: *(1) Nearest-neighbor distance theory.* It is believed that the distance of the nearest-neighbor atoms on the catalyst surface should play a major role in the ORR. In the ORR process, the mechanism could be depicted as that the oxygen molecule adsorbed on the alloy catalyst surface underwent an electron transfer to form the intermediation $M-HO_{2(ads)}$, followed by a step of O–O bond break.[59] The bonding energy of the intermediation $M-HO_{2(ads)}$ was affected most by the nearest-neighbor distance between two sites at which the O–O bond rupture is occurring. This distance should play a critical role in the overall reaction rate and an optimum distance should exist. At larger than optimum distance, dissociation would have to occur prior to adsorption, and at smaller than optimum spacing, repulsive forces would retard dual site adsorption. However, for pure Pt, the atom distance of Pt–Pt is not the optimum spacing. The bringing of foreign atom (alloying atom) would decrease the distance, and then reach the optimum spacing. In other words, foreign atoms could facilitate the oxygen dissociation and adsorption on the alloy surface, which favors the ORR; *(2) Surface roughening effect.* Due to the depletion of the second metal in the alloy metal, the catalyst surface becomes roughened, then increasing the SSA of Pt, and leading to the enhanced ORR activity[60]; and *(3) Electronic structure theory.* Toda et al.[61] discovered a significant enhancement of electrocatalytic activity of Pt by alloying with Fe, and experimentally confirmed that at Pt–Fe bulk alloy's surface consisted of a pure Pt skin-layer (<1 nm in thickness). This Pt skin-layer's electronic structure could be modified by the bulk alloy. The enhancement of ORR catalytic activity could be well explained by the 5d-vacancy of the surface, but not by Pt interatomic distance or roughening of the surfaces. Yeager et al.[62] considered that the rate decisive step for the ORR on a Pt-catalyst surface in acid electrolytes was the adsorption step of an O_2 molecule, as side-on or bridged type accompanied with an electron transfer. This adsorption type involved a lateral interaction of the π-orbitals of the O_2 with empty dz^2-orbitals of a surface Pt atom or with empty dxz- and dyz-orbitals of dual Pt atoms, respectively, with back bonding from the partially filled orbitals of the Pt to the π^*-orbitals of the O_2. The increased d-vacancy of the Pt in the electrode surface, brought about by alloying, might bring about a strong metal–oxygen interaction. Such a strong interaction could cause an increase of O_2 (adsorbed), a weakening of the O–O bond and an increase in its length, resulting in fast bond scission and/or a new bond formation between the O atom and H^+ in the electrolyte.

However, as discussed previously, despite that the ORR activity of the alloy catalysts are higher than Pt and they show excellent tolerance to methanol, the stability of these catalysts still needs to be improved.[20,24,63–65]

3.3.2. Non-noble Metal-based ORR Electrocatalysts

High cost and limited reserve of the noble metal have seriously restricted their commercial applications in PEM fuel cells. Reducing Pt usage and using non-noble metal catalyst to replace

Pt-based catalysts are two necessary approaches at the current technology stage. It is believed that non-noble metal electrocatalysts is probably the sustainable solution for PEM fuel cell commercialization. In the past several decades, various non-noble metal catalysts for ORR have been explored, including non-pyrolyzed and pyrolyzed transition metal nitrogen-containing complexes, transition metal chalcogenides, conductive polymer-based catalysts, metal oxides/carbides/nitrides/oxynitrides/carbonitrides, and enzymatic compounds. The major effort in non-noble metal electrocatalysts for ORR is to increase both the catalytic activity and stability.

Among all kinds of non-noble metal catalysts, pyrolyzed Co- and Fe-porphyrins, phthalocyanines, and tetraazannulenes have been identified as the best kinds of ORR electrocatalysts.[66–73] The key step of the catalyst synthesis of this kind of catalysts is the pyrolysis step. This step was demonstrated to increase the concentration of available ORR active sites. However, long-term stability still remains the problem. One hypothesis is that the hydrogen peroxide produced during the ORR process can attack the catalysts and damage their active sites, resulting in performance degradation. Normally, pyrolysis can destroy the structure of the ligand and form surface $Fe-N_X$ or $Co-N_X$ species, which are active toward the ORR. However, the structures of such active sites are still not well understood, although tremendous analysis has been done using sophisticated instruments. Generally to say, the electrocatalytic activity and stability of pyrolyzed Fe- and Co- macrocyclic complexes can be strongly affected by many factors such as the type of central transition metal ion, the ligand structure, the type of support, the method of synthesis, the pyrolysis temperature, etc.

3.3.2.1. Effect of Central Transition Metal

Generally speaking, the central metal ion of a macrocycle seems to play a decisive role in the ORR mechanism. Most Fe complexes can reduce oxygen directly to water through a 4-electron transfer pathway, while most Co complexes give peroxide as the main product through a 2-electron transfer pathway. For N_4 chelates (phthalocyanines) as catalysts, the activity order of the catalyzed ORR follows the pattern of $Fe > Co > Ni > Cu$, Mn.[68] It has also been reported that O_2 reduction efficiency could also be improved by using two different metal centers, as proposed by Chu et al.[74] They found that the heat-treated mixture of Fe- and Co-tetraphenylporphyrins gave a much higher catalytic ORR activity than the heat-treated single Fe-tetraphenylporphyrin or Co-tetraphenylporphyrin. This was attributed to the formation of a face-to-face structure with two kinds of metal active sites. The interaction between two different transition metals and oxygen molecules promotes O—O bond breakage, resulting in an easier 4-electron transfer to form H_2O.

3.3.2.2. Effect of Complex Ligand

Ligands also play an important role in the catalyzed ORR. They not only serve as part of the active site, but also keep the metal in a stable form on the electrode surface. Biloul et al.[75] tested and compared the stability of 800 °C pyrolyzed Co- tetramethoxyphenylporphyrin (TpOCH$_3$PPCo) and its derivatives Co- trifluoromethylphenylporphyrin (CF$_3$PPCo) and Cotetraazaannulene (CoTAA) as ORR catalysts at a current density of 100 mA cm^{-2} with 300 h of lifetime under fuel cell conditions. The stability order was observed as: CoTAA $>>$ TpOCH3PPCo $>$ TpOCF$_3$PPCo. To improve the stability of these complexes, an attempt was made to enlarge the organic part of the molecule and create a stronger attachment of the catalyst onto the support base.[76]

3.3.2.3. Effect of Heat Treatment Conditions

It has been demonstrated that a pyrolysis step is necessary and critical in improving both the activity and the stability of Fe- and Co-N_X ORR electrocatalysts. van Veen et al.[77] discussed four models in an effort to explain this pyrolysis effect: (1) improving the dispersion of the supported chelate; (2) catalyzing the formation of a special type of carbon, which is actually the active phase; (3) generating the M—N_X species; and (4) promoting a reaction between chelate and subjacent carbon in such a way as to modify the electronic structure of the central metal ion with retention of its N_4 coordinated environment. Actually, the active sites should be M—N_4 or M—N_2^- units, depending on the heat treatment temperature (see Figure 3.8).

The observed nature of the catalytic sites was dependent on the heat treatment temperature. The heat treatment process could be described as the following steps[71]: (1) 220—300 °C: dehydration; (2) 300—400 °C: CO and CO_2 are released due to decarboxylation; (3) 400 °C: completion of the polymerization reaction; (4) 500—600 °C: Co-phthalocyanine polymer is stable up to this temperature range; (5) $\sim$700 °C: pyrolysis of the polymer to produce fragments containing Co bound to C and N on the catalyst surface. Some of the fragments may be involved in the Co-N_4 moiety; (6) above 700 °C: only inorganic Co (Coo and Co(II)) is found on the support; and (7) 1000 °C: Co can still be detected, but no sign of the N-macrocycle was observed.

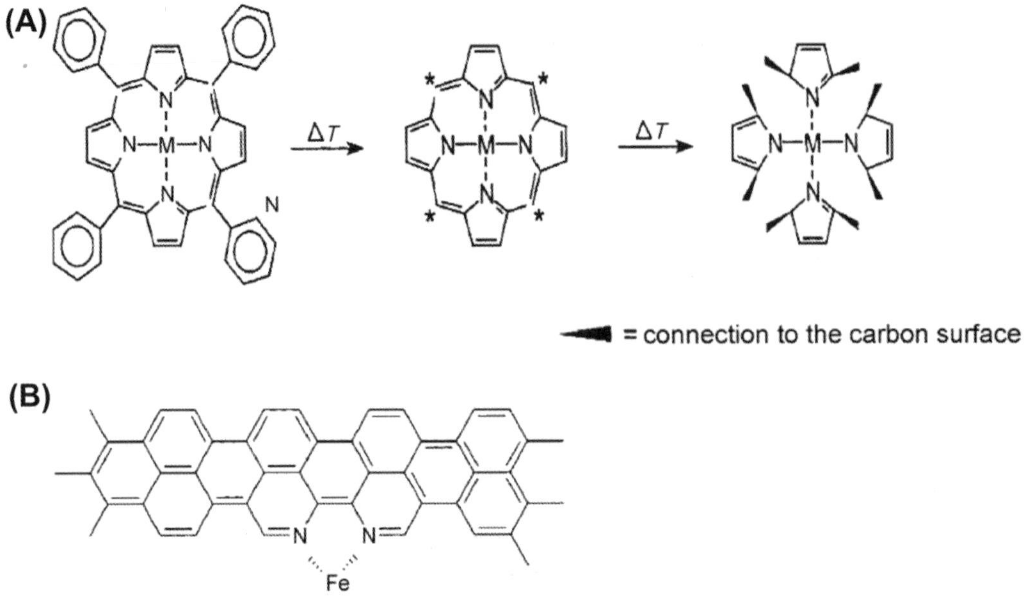

Figure 3.8 (A) Visualization of the reaction of porphyrin with carbon during the pyrolysis; (B) proposed moiety of the FeN$_2$/C catalytic site.[78,79]

In summary, Transition Metal Macrocylic Complexes have shown comparable activity to traditional Pt-based catalysts for an ORR. However, the poor stability and complicacy of preparation are major obstacles before wide application. Further, the active site structures are still a subject of controversy, which hinders efforts to control active site formation.

3.4. Electrocatalysts Synthesis and Characterization

As recognized, the controllable synthesis for ORR electrocatalysts with high electrocatalytic activity has still remained as the challenge. It has been demonstrated that the synthesis methods or procedures have significant influence on the composition, structure, and other properties of the catalysts, which would therefore affect their catalytic activity and stability. In order to achieve high ORR activity and stability of the electrocatalysts, many synthesis methods have been explored.

The currently employed synthesis methods are based on chemical, electrochemical as well as physical principles, including low-temperature chemical precipitation, colloidal, sol–gel, impregnation, microemulsions, electrochemical, spray pyrolysis, vapor deposition, and high-energy ball milling. For catalyst characterization, many sophisticated instrument methods have been used. For analyzing the composition and phase of catalyst, X-ray Diffraction (XRD) and Electron Diffraction (EDS) spectroscopies, X-ray Fluorescence (XRF), X-ray Emission (XRE), and Proton-induced X-ray Emission (PIXE) are frequently employed; for measuring physical surface area and electrochemical active surface area, BET and electrochemical hydrogen adsorption/desorption methods are employed; for catalyst's morphology and active components, scanning electron microscopy (SEM) and TEM are used; for structure and crystallography of surface and small active component particles, electron spectroscopy for chemical analysis (ESCA), X-ray photoelectron spectroscopy (XPS), UV-induced photoelectron spectroscopy (UVPS), and energy dispersive spectroscopy (EDS) are employed; for analysis of the thermal stability of catalysts, differential thermal analysis (DTA), differential scanning calorimetry (DSC) and thermogravimetric analysis (TGA) are often used; and other structural techniques such as FTIR, UV–VIS, and TPD/TPR are also used for characterizing the bulk and surface of electrocatalysts.

In this chapter, we are not going to describe too-detailed information about the catalyst synthesis methods and characterization techniques, which may be beyond the scope of this book. The interested readers may please go to literature for getting those detailed information and knowledge.[80] In the following subsections, we will give some brief description about the synthesis methods and characterization techniques and their applications in ORR electrocatalysts.

3.4.1. Synthesis Methods for ORR Electrocatalysts

3.4.1.1. Impregnation-Reduction Method

In a variety of preparation methods, impregnation-reduction method is one of the mostly used methods in preparing Pt-based catalysts without the help of protective reagents and thermal treatments. Compared to other methods, impregnation-reduction method is relatively simple and direct, easier to conduct as well. This method only includes two steps: impregnation and reduction. The first step is that the support such as carbon or other inert material is dissolved in a solvent such as water, ethanol, isopropanol, or their mixture to form a support containing mixed solution. A mount of metal precursor are impregnated into this support solution, and then the pH of the mixture is adjusted to a desired value. Then the excess reducing agent is introduced (sodium borohydride, formaldehyde, hydrazine, hydrogen, etc.) into this mixture under constant stirring at a desired temperature for several hours. When the reduction is finished, the product is filtered and then dried; finally, the catalysts can be obtained. It has been demonstrated that the properties of the support materials have significant influence on the particle size and distribution of the catalysts. Furthermore, the reducing agent, the metal precursor, solvent, pH value, and the reaction temperature are also very important in determining the catalyst's properties.

3.4.1.2. Galvanic Displacement Method

Galvanic displacement method is also often used for synthesizing catalysts. By this method, low Pt-content electrocatalysts can be obtained. For example, a carbon-supported core–shell structured electrocatalyst with bimetallic IrNi as the core and platinum monolayer as the shell has been successfully synthesized using this method.[81] In this synthesis, IrNi core supported on carbon was first synthesized by a chemical reduction and thermal annealing method and a Ni core and Ir shell structure could be formed finally. The other advantage of this method is that the Ni can be completely encased by Ir shell, which will protect Ni dissolve in acid medium. Secondly, IrNi@Pt$_{ML}$/C core–shell electrocatalyst was prepared by depositing a Pt monolayer on the IrNi substrate by galvanic displacement of a Cu monolayer formed by under potential deposition (UPD).

3.4.1.3. Electrochemical Deposition

Electrochemical deposition method is one of the most useful approaches to prepare nanostructured ORR electrocatalysts. Using this method, the structures, shapes and sizes of the electrocatalysts can be controlled on the surface of conducting materials by altering the conditions of electrochemical deposition. Furthermore, in the electrochemical deposition method, there is another good point, no capping reagent or surfactant or other dispersion agent is involved in the electrodeposition method. Therefore, the synthesis procedures can be greatly simplified. For example, bimetallic PtAu/graphene nanostructure catalysts were successfully synthesized by electrodeposition of Pt–Au nanostructures on the surface of graphene.[44] By changing the molar ratio of Pt and Au, the catalytical activity is altered.

3.4.1.4. Thermal Decomposition Method

$Pd_{100-x}Mo_x/C$ nanoparticle catalysts were synthesized by a simultaneous thermal decomposition with palladium acetylacetonate, platinum acetylacetonate, and molybdenum carbonyl in o-xylene in the presence of Vulcan XC-72R carbon. At first, the metal precursor and carbon support were mixed together and refluxed and finally dried, followed by a heat treatment up to 900 °C in H_2 atmosphere.[82]

3.4.1.5. Template-based Method

In this method, hard template, such as mesoporous silica soft template such as surfactant, directing agent or capping agent are used. Using mesoporous silica, catalysts with large BET surface area and high pore volumes can be obtained. RuSe/C electrocatalysts with highly porous were successfully synthesized via a template-based method, by the impregnation of a hard template (silica gel) with the mixture of $RuCl_3 \cdot xH_2O$, SeO_2 and sucrose.[83] However, the hard templates need to remove after the reaction finished, which would increase the cost of the process. In the initial step, the reactant mixture was infiltrated into the template porosity. After this step, the composite material was heat treated in different gas atmospheres at a high temperature. Finally, the obtained composite was then treated with acid at room temperature in order to remove the template. Guo et al.[84] synthesized raspberry-like hierarchical AuPt hollow spheres using TiO_2 as the template. After removing TiO_2, the AuPt hollow structure can be kept. Compared with normal Pt-based catalysts, the obtained AuPt hollow structure have several advantages: (1) by changing the size of TiO_2 precursor spheres, AuPt hollow structure with different size can also be obtained; (2) the activity site formed after the reaction; (3) the formed nanochannels make the transport of reactant and product easily meaning that the hollow spheres can be used as the reactor, which would probably lead to high electrocatalytic activity. Yeo et al.[43] reported a novel synthesis of Pt nanodendrites by Au-seed-mediated growth inside hollow silica nanospheres, as shown in Figure 3.9.

3.4.1.6. Microwave and Ultrasound—Assisted Process

The advantages of this method are simple and save reaction time. Usually, it only need several minutes for metal precursor to reduce and the formed metal seed can further growth quickly. Furthermore, the microwave-heating can provide uniform hot environment through which the size and the morphology of the metal seed can be well controlled.[85] Pt-encapslulated Pd—Co

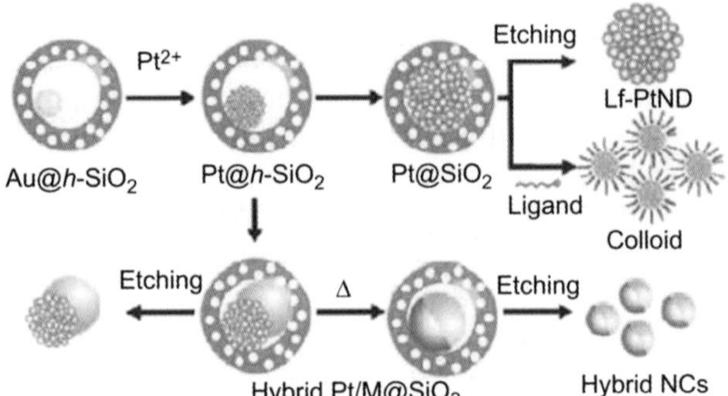

Figure 3.9 Synthesis of platinum-based nano-dendrites and hybrid nanocrystals.[43] (For color version of this figure, the reader is referred to the online version of this book.)

nanoalloy electrocatalysts for ORR have been synthesized by the microwave method at 300 °C with the power of 600 W. Pt—WC/C prepared by the direct reduction of a platinum salt precursor combined with intermittent microwave heating 5s on and 5s pause for six times. Jang et al.[86] reported that Pt_3Co nanoparticles with Pt-enriched shells on a carbon support could be prepared by a one-step ultrasound polyol process using $Pt(acac)_2$ (acac = acetylacetonate) and $Co(acac)_2$. With ultrasound, the conversion of $Co(acac)_2$ facilitated and the conversion of $Pt(acac)_2$ into nanoparticles were retards, thus forming the Co-core and Pt-shell catalysts. In sonication and microwave-assisted reactions, the high-frequency oscillation of ultrasound and microwave could generate extremely high energy conditions. In this synthesis, a solution of $Pt(acac)_2$ and $Co(acac)_2$ in EG with a porous carbon support added was irradiated with ultrasound by a high-intensity ultrasonic probe at 150 °C for 3 h under Ar.

3.4.1.7. Pyrolysis Procedure

Pyrolysis procedure has been recognized as an effective step in improving catalyst's ORR activity, particularly for non-noble metal catalysts.[87] For example, in the synthesis of non-noble metal N_4-chelate catalysts for ORR, a pyrolysis at temperature range of 600—900 °C can significantly enhance both the ORR activity and stability of the catalysts.[88] However, since the N_4-chelate structure would be largely destroyed at high pyrolysis temperatures, it might not be necessary to use expensive metal macrocyclic complexes as precursors for active catalyst preparation.[89] It is expected that if a mixture of basic components, i.e., transition metal sources such as Fe or Co, nitrogen donors, and carbon supports, could be treated at a high temperature, catalytically active sites such as metal-N_4 or $FeN_2C_4^+$ might be formed.[90] Several research groups have reported their work on this pyrolysis route with various transition metal, nitrogen, and carbon-containing species as precursors. This approach creates a very attractive strategy for most popular transition metal precursors that have been employed are Fe or Co inorganic salts such as sulfates, acetates, hydroxides, chlorides, and cyanides, and Fe complexes such as ferrocene.[91–94] Inorganic precursors might be more cost-effective for commercialization.

Lalande et al.[94] proposed a multistep pyrolysis method for ORR electrocatalyst preparation. Wei et al.[95] carried out some comparisons between one step and multistep pyrolysis. In their experiments, catalysts containing carbon, nitrogen, and cobalt were prepared. In the multistep pyrolysis method, the first pyrolysis step was to pyrolyze the carbon support (Vulcan XC-72R) with a nitrogen precursor (acetonitrile) in a flowing argon atmosphere, and then to introduce the cobalt precursor (cobalt sulfate) into the reactor for the second step. The second step was to pyrolyze the mixture of the product and the introduced cobalt precursor in a flowing argon atmosphere. It was also found that the pyrolysis sequence had a strong impact on the catalyst activity. For example, the onset electrode potential of ORR catalyzed by catalysts produced in different pyrolysis sequences was about 790–800 mV (vs RHE)—similar to that obtained with a cobalt-centered macrocycle catalyst synthesized by one-step pyrolysis at 1000 °C.[70,96]

Bogdanoff et al.[97] found that the activity of pyrolyzed carbon-supported CoTMPP could be limited by the morphology of the products. Thus, they introduced a new preparation strategy, which led to an in-situ formed graphite-like carbon matrix with uniform catalytic centers. In their method, metal oxalates were added into the reaction as foaming agents during the pyrolysis. The purpose of adding metal oxalate-foaming agents was to suppress the particle aggregation and increase the surface area of the macrocycle catalysts (up to $800 \, m^2 \, g^{-1}$). In this way, the catalytic activity could be improved by an increase in the catalyst.

3.4.1.8. Modified Sol–Gel Method

More recently, Ye et al.[98] proposed a new sol–gel method combined with a supercritical drying technique for the synthesis of nanocomposite electrocatalysts for ORR. They claimed that their method had a structure-preserving ability and favored a nanoscale mixing of the constituents, and then could produce a remarkably homogeneous solution. The chemical, physical, and morphological properties of catalysts produced using this sol–gel method were claimed to be significantly isotropic. Compared to other synthesis methods, this new sol–gel method had some unique and competitive features. The first step of this synthesis method was to mix an inorganic salt (e.g., $Fe(NO_3)_3$ or $CoCl_2$) or an organometallic compound (e.g., Co(III) acetylacetonate) with polyacrylonitrile (PAN) in N,N-dimethylformamide (DMF)/water to form a well-mixed solution. The mixed solution was degassed when heated to approximately 120 °C. When the solution was cooled, a polymer gel containing the metallic compound was obtained by thermally induced phase separation. The polymer gel was then pretreated at 220 °C, followed by a pyrolysis step at 900 °C under an argon atmosphere. In this way, a non-noble metal catalyst was obtained that showed a strong catalytic activity and relatively high stability for the ORR in acid conditions.

3.4.1.9. Template-Assisted Ultrasonic Spray Pyrolysis Method

A new pyrolysis technique, called the template-assisted ultrasonic spray pyrolysis (TAUSP), for synthesizing highly active macrocyclic complexes has been reported, which could effectively create favorable morphology and increase the catalyst surface area. For example, Liu et al.[10] recently developed this TAUSP technique to synthesize CoTMPP/C catalysts. The TAUSP method is a continuous, one-step, scalable method to prepare unaggregated spherical and uniform particles with controllable particle size. The carbon-supported CoTMPP particles synthesized by this method have a high surface area of $834 \ m^2 \ g^{-1}$. In a rotating ring-disk electrode (RRDE) measurement, the ORR catalyzed by this CoTMPP displayed double the activity of a catalyst prepared by the conventional heat-treated method. The results demonstrated that the favorable morphology of a CoTMPP/C catalyst created by ultrasonic spray pyrolysis has a significant effect on catalyst activity, even under fuel-cell operating conditions.

3.4.1.10. Polymer-Assisted Procedure

Zelenay et al.[99] explored Co-polypyrrole (CoPPy) material as a PEM fuel-cell cathode catalyst. The composite CoPPy catalyst, even without a heat treatment, could generate a power density of $\sim 0.15 \ W \ cm^{-2}$ in a H_2–O_2 fuel cell and displayed no signs of performance degradation for more than 100 h. Their results showed that heteroatomic polymers can be used not only to stabilize the non-noble metals in a PEM fuel cell environment but also to generate active sites for the ORR. Study of the interaction between the catalyst and oxygen also demonstrated that CoPPy could form stable end-on, side-on, and bridged oxygen adducts. Furthermore, the side-on and bridged oxygen adducts were more stable than the end-on adduct. Since side-on and bridged oxygen adducts could elongate O—O bond lengths, they might lead to a 4-electron reduction product.

3.4.1.11. Ball-Milling Method

As the name suggests, the ball milling method consists of balls and a mill chamber. Therefore, a ball mill contains a stainless steel container and many small iron, hardened steel, silicon carbide, or tungsten carbide balls are made to rotate inside a mill. The powder of a material is taken inside the steel container. This powder will be made into nanosize using the ball-milling technique. A magnet is placed outside the container to provide the pulling force to the material and this magnetic force increases the milling energy when milling container or chamber rotates the metal balls. Ball milling is a mechanical process and thus all the structural and chemical changes are produced by mechanical energy.[100] Baek et al.[101] recently proposed that edge-selectively functionalized graphene nanoplatelets (EFGnPs) as metal-free electrocatalysts for ORR can be large-scaled prepared by ball-milling method. The EFGnPs were obtained simply by dry ball-milling graphite in the presence of hydrogen, carbon dioxide, sulfur trioxide, or carbon dioxide/sulfur trioxide mixture. The resultant sulfonic acid- (SGnP) and carboxylic acid/sulfonic acid- (CSGnP) functionalized GnPs were found to show a superior ORR performance to commercially available platinum-based electrocatalyst in an alkaline electrolyte. It was also found that the edge polar nature of the newly prepared EFGnPs without heteroatom doping into their basal plane played an important role in regulating the ORR efficiency.

3.4.2. Characterization of ORR Electrocatalysts

3.4.2.1. Physical Measurement

X-ray diffraction (XRD), SEM and TEM, X-ray absorption spectroscopy (XAS), X-photoelectron microscopy (XPS), Raman Spectrum are the most commonly used techniques to characterize the composition and crystal phase of electrocatalysts. Here, we give only several examples to illustrate the usage of these tools.

XRD pattern can tell us the composition of the product; each element has its own unique peak in the spectrum. Figure 3.10 reveal that the phase of Platinum crystals is the face-centered cubic phase. From XRD pattern, the average particle size can also be obtained using the Scherrer equation (Eqn (3.8)) based on

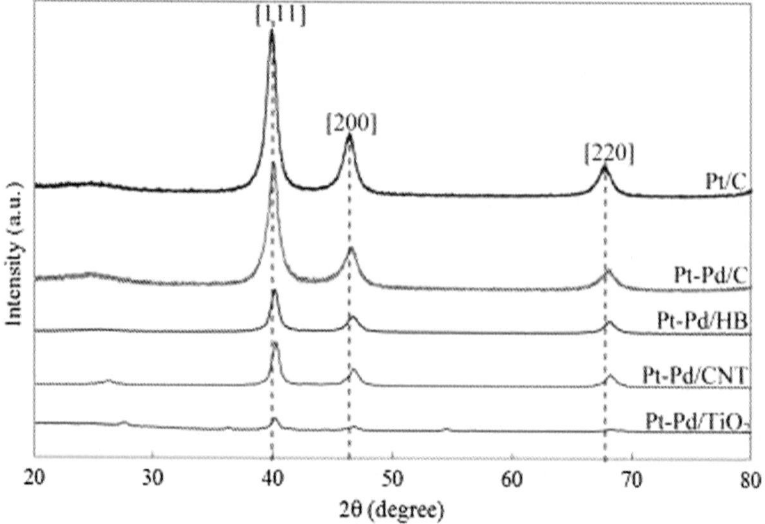

Figure 3.10 Representative XRD patterns of as-prepared supported Pt-Pd electrocatalysts and the commercial Pt/C (20 wt% ETEK).[43] (For interpretation of the references to color in this figure legend, the reader is referred to the online version of this book.)

the peak width in the XRD pattern. By comparing the position and the intensity of the peaks, we can acquire the transformation or changing of the products.

$$L = \frac{0.9\lambda_{K\alpha l}}{B_{(2\theta)}\cos\theta_{max}} \tag{3.12}$$

where $\lambda_{K\alpha l}$ is the incident wavelength (1.54,056 Å) and $B_{(2\theta)}$ is the width of half peak in radians.

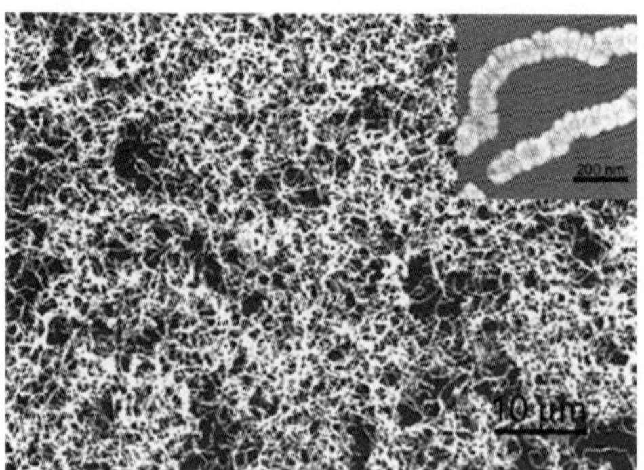

Figure 3.11 SEM image of AuPt bimetallic hollow tube-like 1-D nanomaterials. The inset is a high-magnification SEM image.[102]

Transmission electron microscopy (TEM) in image and scanning transmission or electron microscopy (STEM) modes, and energy dispersive X-ray spectroscopy (EDX) attached to the TEM and SEM are commonly used for particle size and distribution characterization. For example, the low- and high-magnification SEM images in Figure 3.11 indicate the large-scale tube-like 1-D nanomaterials, which have several microns in length, have been fabricated. The shape can be clearly seen from the SEM images and the high magnification indicates the diameter and the composition of the catalysts.[102]

Using both TEM and HRTEM, the morphology and crystalline can be obtained, as shown in Figure 3.12. Other tools, including infrared spectroscopy (FT-IR), X-ray absorption spectroscopy (XAS) and UV–vis absorption spectroscopy are also generally used. Atomic force microscopic (AFM) is sometimes used when

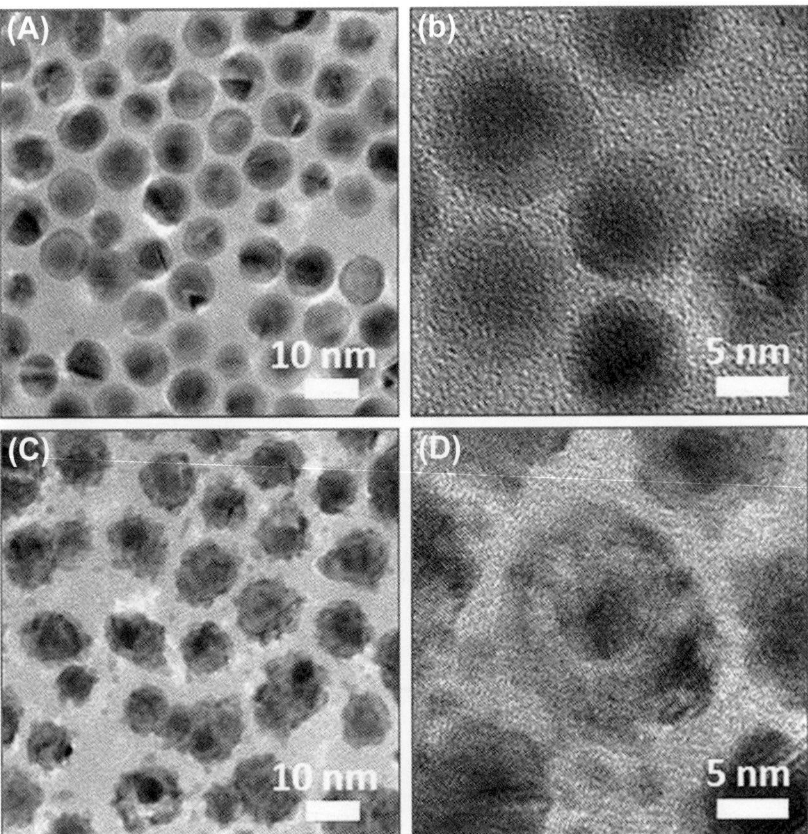

Figure 3.12 TEM (a, c) and HRTEM (b, d) images of as-synthesized nanoparticles.[44]

the support is a layer material, such as graphene, graphene oxide, and reduced graphene, it can tell the thickness of graphene sheets.[44,103]

Furthermore, high-angle annular dark-field scanning TEM (HAADF-STEM) analysis can clearly reveal the distribution of different elements in the catalysts with different colors. Usually, the core and shell of the catalysts can be distinguished from the dark shell and bright core, as shown in Figure 3.13.[20]

In Figure 3.13, the AuPd nanoparticles supported on carbon in the HAADF-STEM image enable a brighter core to be distinguished from the darker shell. It can tell us that the AuPd nanoparticles are composed of a shell and a core. Due to the mechanism of HAAD-STEM, we can clearly tell that the shell containing a lighter element (Pd) and the core containing a heavier element Au.[20]

X-ray photoelectron spectroscopy (XPS) is a useful and powerful tool that can give the configuration information of the catalyst.[104] In the XPS spectra, different element or the same element with different valencies has distinguished peaks. Figure 3.14 shows the XPS spectra of the Pd/C and Pd–WO$_3$/C catalysts. It can be seen that obvious W and Pd peaks appears in the XPS spectra. The W(4f$_{7/2}$) and W(4f$_{5/2}$) peaks for the Pd–WO$_3$/C catalyst located at 34.9 eV and 37.1 eV, respectively. For Pd(3d) of Pd/C catalyst, there are three peaks located at 284.7, 334.9, and 340.2 eV, and for Pd–WO$_3$/C catalyst, they are located at 284.5, 335.7, and 340.6 eV, respectively.

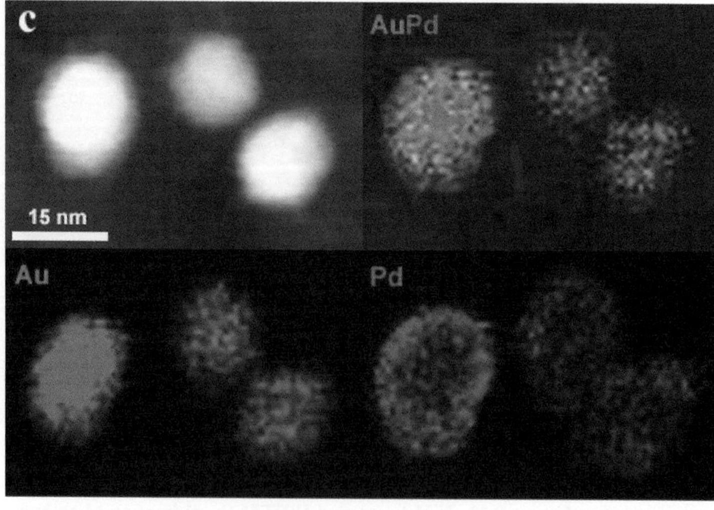

Figure 3.13 Representative HAADF-STEM image (top left panel) of the AuPd(1:0.61)/C and its corresponding Pd (red) and Au (green) elemental mapping image.[20] (For interpretation of the references to color in this figure legend, the reader is referred to the online version of this book.)

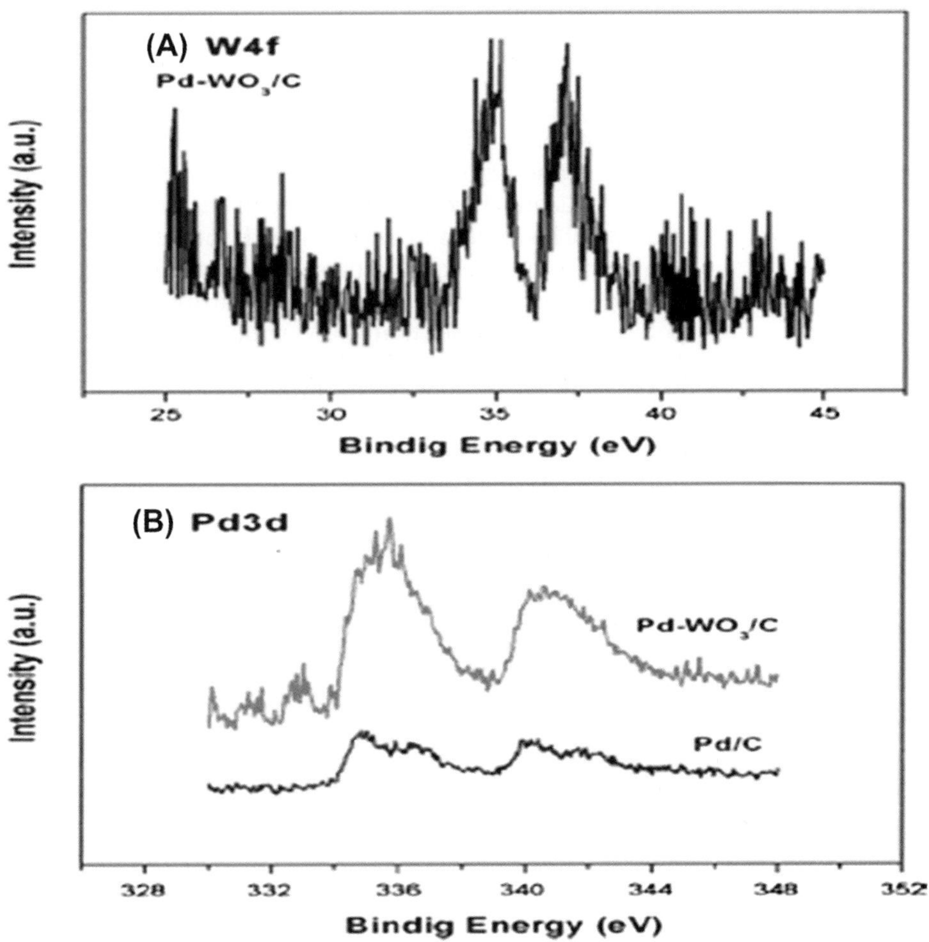

Figure 3.14 XPS spectra of the Pd/C and Pd—WO₃/C catalysts: (A) W(4f) region (only for the Pd—WO₃/C catalyst) and (B) Pd(3d) regions.[104] (For color version of this figure, the reader is referred to the online version of this book.)

3.4.2.2. Electrochemical Characterization

Regarding the electrochemical characterization of the ORR activity, a glassy carbon (GC) electrode or glassy carbon or gold rotating disk electrode (RDE), rotating disk-ring electrode (RRDE) are used for measuring the ORR activity and stability.[105]

Cyclic voltammetry (CV) of the catalysts is normally conducted in either N_2- or Ar- or O_2-saturated electrolyte solution (often in 0.5 M H_2SO_4, 0.1 M $HClO_4$, or KOH).[20] As shown in Figure 3.15 obtained in N_2-saturated 0.1 M $HClO_4$ solution, the normalized CVs exhibit three distinctive regions: a hydrogen adsorption/desorption region in low potential, a double layer capacitive region, and a metal oxide formation/reduction region in a high potential range. By comparing the area of the double layer of catalysts with different composition, we can conclude that they have different electrochemical surface area. In particular, capacitive background current densities between the hydrogen and oxide regions are observed to increase from Au/C to AuPd/C and further increase with increasing Pd:Au ratio in AuPd/C. The hydrogen adsorption/desorption region also gives some information about these catalysts. In Figure 3.15, the hydrogen adsorption/desorption, occurring at a more negative potential range rather than 0.0 V, which is not observed in Au/C. Note that the charge under the hydrogen adsorption/desorption peaks may be partially contributed by the hydrogen migrated into the Pd lattice. Moreover, with the increase of the ratio of Pd:Au, the hydrogen adsorption/desorption region becomes larger, which indicates that more hydrogen migrated into the Pd lattice.

For AuPd/C catalysts, with higher Pd:Au ratio, the integrated charge amounts for the surface Pd oxide reduction (i.e., area under the cathodic peak at 0.39 V) becomes larger than those with a lower Pd:Au ratio. This suggests an increased Pd surface area,

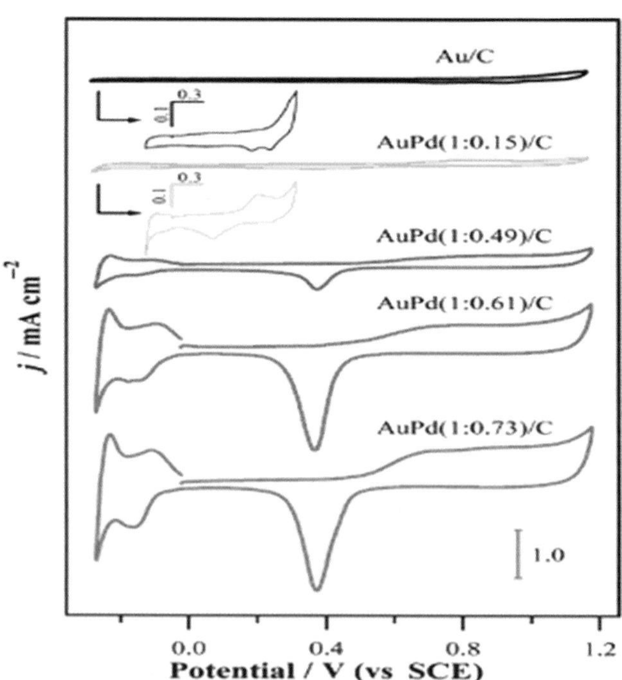

Figure 3.15 Cyclic voltammograms obtained with AuPd/C modified GC electrodes in Ar-saturated 0.1 M HClO$_4$ solution at a scanning rate of 10 mV s^{-1}.[20] (For color version of this figure, the reader is referred to the online version of this book.)

which is in agreement with the observation of capacitive current densities.

When a Pt monolayer was deposited on the surface of bulk Pd(111) single crystals, the ORR activity, characterized using a rotating ring disk electrode in a 0.1 M HClO$_4$ solution, was shown in Figure 3.16.[106] For the disk, the limiting current increases with the increasing of the rotation rate while the onset potential is kept unchanged. Using the Koutecky–Levich plots discussed in Chapter 5, the transferred electron during ORR can be calculated. The current–potential curves also revealed that a very positive onset potential for the ORR (0.95–1.0 V vs NHE), as well as a half-wave potential of 0.838 V. In addition, the ring current was insignificant, indicating that the ORR catalyzed by this catalyst was an almost completed four-electron process on the disk electrode from O$_2$ to H$_2$O.

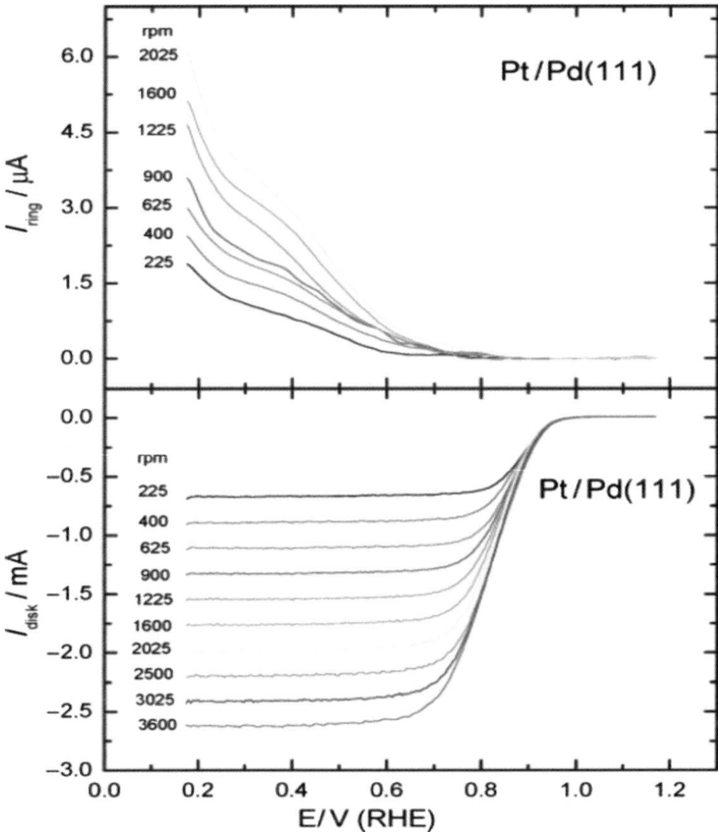

Figure 3.16 Polarization curves for the ORR on a Pt monolayer on a Pt(111) electrode surface in 0.1 M HClO$_4$ solution (sweep rate 20 mV s^{-1}, ring potential 1.27 V vs RHE, ring and disk areas are 0.126 and 0.283 cm^2, respectively).[106] (For color version of this figure, the reader is referred to the online version of this book.)

For detailed electrochemical characterization of the ORR electrocatalysts, Chapter 7 of this book will give more discussion.

3.5. Catalyst Layers, Fabrication, and Characterization

For application of ORR electrocatalyst in PEM fuel cells, the catalyst must be used to prepare the CL, which is then bonded into the MEA for fuel cell testing. The composition, structure, and preparation procedure all have strong effects on the fuel cell performance. For example, even an ORR catalyst has high ORR activity, if the CL and its MEA preparation are not optimized, it is not necessary that fuel cell would have a high performance. Therefore, the optimum preparation of CL and its MEA are very important in fully realizing the catalyst activity.

3.5.1. Introduction to ORR Catalysts Layer

Fuel cell CLs are the key components in the entire fuel cell device because the reactions such as hydrogen—oxidation reaction (HOR) at anode and the ORR at cathode occur inside the CLs. Particularly, in order to carry out the ORR, the catalyst particles inside the cathode CL must be in contact with each other for electrical conductivity and also in contact with protonic conducting (in acidic PEM fuel cells), or hydroxide conducting (in alkaline PEM fuel cells) ionomer for ionic conductivity. In addition, there must be some channels within the CL for transporting the reactants and the products. In other words, the catalyst particles must be in close contact with each other, with the electrolyte, and also with the adjacent diffusion medium (DM). Moreover, the reactants gas (O_2) and the produced water travel mainly through the voids, so the CL must be porous enough to allow gas to diffuse to the reaction sites and liquid water to wick out.

In the preparation of MEA, there are two options. One is to coat the CL onto the DM such as carbon paper, or carbon cloth, the other is to coat the CL onto the PEM, as shown in Figure 3.17. The CL coated on the diffusion medium is called Catalyst-coated Diffusion Medium (CDM), and the CL coated on the membrane is called Catalyst Coated Membrane (CCM). Using a hot-press process, by sandwiching PEM between two CDMs, or CCM between two DMs, an MEA can be fabricated for fuel cell testing.

Regarding the composition of a CL, it is usually consisted of the electrocatalyst, ionomer such as Nafion® and a binder such as

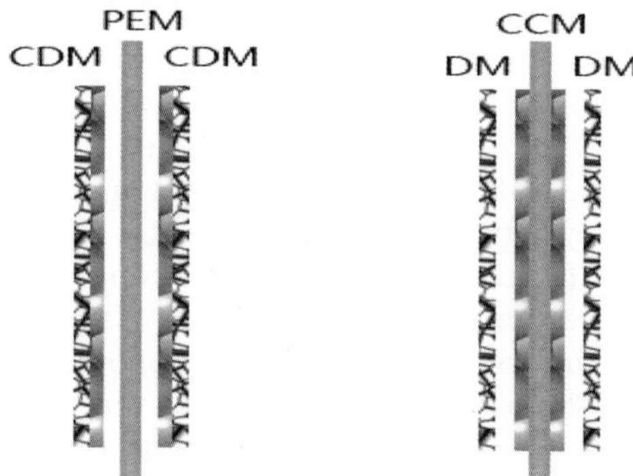

Figure 3.17 Two different options for making MEA. (For color version of this figure, the reader is referred to the online version of this book.)

PTFE. Normally, there exists an optimum composition of these three components. Increasing the ionomer content in the CL can improve the proton migration but reduce void space, thus worsening gaseous oxygen transport; Increasing the catalyst loading can enhance the rate of the electrochemical reaction but increase the cost and the transport resistance, and increasing the PTFE can provide more pore in the CL due to the hydrophobic properties but increase the electrical resistance. Thus, achieving a compromise between electrocatalyst loading, ionomer and PTFE contents poses a challenge for the improvement of the CL performance. Some work around the composition of the CL has been developed[107–115] and they suggested that the CL should be as dense and thin as possible in order to increase catalyst utilization.

The interaction between the ionomer and catalyst is crucial for the electrochemical performance of the CL in fuel cells. If the catalyst nanoparticles that are isolated from the ionomer network, they will become electrochemically inactive. Furthermore, the distribution of the ionomer will affect the ohmic resistance and the mass transport of the reactants and/or products in the CL. With respect to this, some effort has been made in optimizing the distributions of ionomer content and Pt loading across the entire thickness of cathode CL of fuel cell by maximizing the cell current density at a given potential.[116] It was found that the optimal distribution was a linearly increasing function from the gas diffusion layer side to the membrane side in the case of a single-variable optimization. When optimization was treated as a two-variable problem, the optimal distribution of ionomer content was still a linearly increasing function but the optimal distribution of Pt loading was a convex increasing function. This study pointed out that only the optimization of ionomer content distribution needed to be considered in the design of the CL. Passalacqua et al.[117] suggested that the optimum ionomer content in catalyst ink was located at about 30–35% NFP (Nafion® percentage which was defined as the percentage of the Nafion® per total catalyst and Nafion® weight), and was independent of the Pt loading on the carbon in the catalyst, the Pt content in the electrode, and the kind of perfluorosulfonic ionomer used. However, Sasikumar et al.[118] showed that the optimum Nafion® ionomer content was dependent on Pt loading. It seems that the optimal Nafion® ionomer content in catalyst ink is related to the eventual thickness of the equivalent Nafion® film in the CL, as shown in Figure 3.18. The Nafion® ionomer content in catalyst ink should be controlled appropriately. If the ionomer content is too low, the ionic conductivity of the CL is insufficient, leading to a low Pt utilization and high ohmic resistance; if the ionomer content is too high, the gas accessibility and electrical conductivity of the CL will be insufficient.

During the fuel cell operation, liquid water flooding can take place at the cathode CL of the fuel cell when the removal rate of water is lower than the rate of water produced by ORR and transported from anode to cathode by electro-osmotic drag. In this case, the pores of the CL are occupied by the liquid water, which limits the oxygen transport to the active sites of the CL and results in a lower fuel cell performance. Thus, introducing a hydrophobic component to the CL is also very important to create more channels or pores for the gas and liquid water transport. To optimize the composition of the CL, an optimal ratio of Catalyst:Ionomer:Teflon was found to be 1:0.875:0.875 for high fuel-cell performance.[119]

In addition, the cathode CL contamination should also be mentioned here as this contamination would cause a significant irreversible degradation in fuel cell performance. When the air stream of the fuel cell contains the pollutants such as SO_x, the pollutants can absorb on the surface of catalyst active sites, competing with oxygen adsorption, and leading performance degradation. Further, the presence of SO_x in the catalysts layer will reduce the pH, resulting in free acids in the electrode and causing potential shifts. For example, after the PEM fuel cell was exposed to 1 ppm SO_2/air for 100 h at 70 °C at a current density of 0.5 A cm^{-2}, 35% degradation for a fuel cell performance was observed, with a cell voltage decrease from 0.68 to 0.44 V.[120] A 78% cell performance decrease was also observed for air containing 5 ppm SO_2, while a 53% decrease was observed for air containing 2.5 ppm SO_2 at the same applied dosage,[121] but there

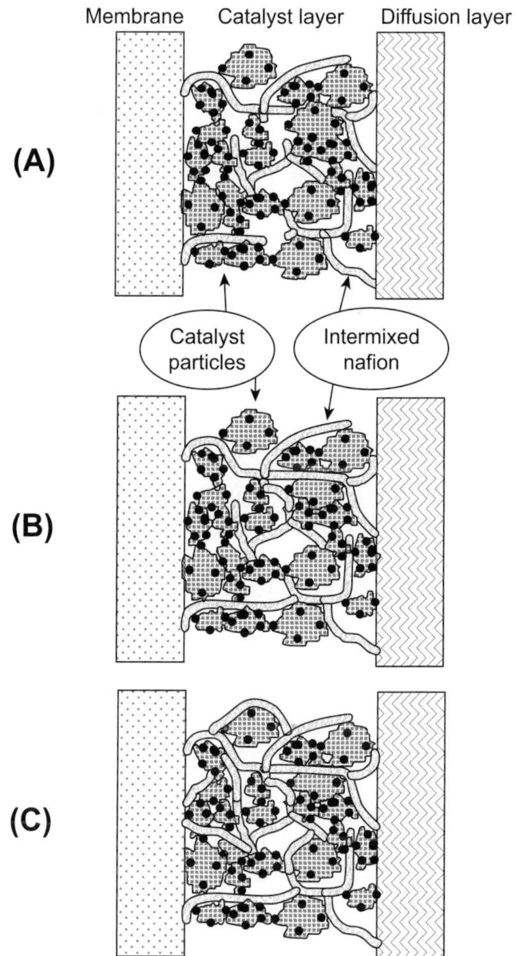

Figure 3.18 Schematic planar representation of the catalytic layer. (A) At low Nafion® ionomer content, not all the catalyst particles are connected to the membrane by a Nafion® ionomer bridge. (B) At the optimal Nafion® ionomer content, there are good connections of both ionic and electrical conductions. (C) When there is too much Nafion® ionomer, not all the catalyst particles are electronically connected to the diffusion layer.[117]

was no detrimental effect on cell performance with 500 ppb SO_2 in the air stream.[122] Small amounts of NO_x in the air can also cause the degradation of fuel cell performance. Knights et al.[123] found that the concentration of NO_x with even 115 ppb could cause a cell performance drop of more than 25 mV at 0.175 A cm^{-2}, and they also observed that the voltage dropped linearly as the NO_x concentration was increased from 0.016 to

0.115 ppm. Mohtadi et al.[121] and Yang et al.[124] also found that the performance decay was increased as NO_x concentration was increased in different NO_x content ranges, but they didn't get the linearly relationship between the performance loss and the NO_x content. However, it was observed that the NO_x contamination was recoverable by refresh the pure air for a long time.[121]

3.5.2. Catalyst Layer Fabrication and Characterization

Normally, CL preparation includes two steps: one is to prepare the catalyst ink, and the other is to prepare CL using this ink.

3.5.2.1. Catalyst Ink and Its Preparation

As discussed above, the hydrophobic properties of the cathode CL could avoid the water flooding. This requires the catalyst ink to contain some hydrophobic agent (such as Polytetrafluoro-ethylene (PTFE), or Teflon®) together with both catalyst and ionomer. However, if the PTFE content is too high, the catalyst particles could become wrapped, resulting in decreased electron conductivity and catalyst utilization. Normally, a content of 10−40% is the optimal content. On the other hand, the catalyst ink must contain ionomer for CL's ionic conductivity.[125]

Regarding the solvent used to prepare the catalyst ink, its properties in catalyst ink should be mentioned as it also plays an important role in determining the microstructure and catalytic activity of the CL. When ionomer such as Nafion® solution is mixed with solvent, the mixture may become a solution, a colloid, or a precipitate[126] due to the different dielectric constants of the solvent. When the dielectric constant is more than 10, a solution is formed; between three and 10, a colloidal solution is formed; and less than 3, precipitation occurs.[127] If the mixture is a solution (i.e., the "solution method"), excessive ionomer may cover the carbon surface, resulting in decreased Pt utilization. However, when the mixture is a colloid (the "colloidal method"), ionomer colloids adsorb on the catalyst powder and the size of the catalyst powder agglomerates increases, leading to an increased porosity of the CL. In this case, the mass transfer resistance could be diminished because of the continuous network of ionomers throughout the CL, which then improves the proton transport from the catalyst to the membrane.[126]

Furthermore, the catalyst ink deposition process requires a compromise between the catalyst ink's dielectric constant and other properties such as the viscosity, boiling point, and carbon wet capacity of the solvent.[128] For higher catalyst metal loading, higher viscosity of the ink is required as it is necessary to maintain a stable catalyst ink suspension. For example, glycerol has high viscosity and is generally added to catalyst ink to increase the paintability and stability. According to Chisaka et al.,[129] when the mass ratio of glycerol to carbon (rgc) was more than 5, the remaining glycerol was proportional to ratio of glycerol to carbon. They also found that the cell performance was decreased with increasing the rgc because of the high boiling point of the glycerol. The extra solvent could block the pores and reduce the number of reaction sites and also decrease gas transport.[130] Other factors such as the dispersion method and mixing procedure could also affect the ionomer distribution and catalyst utilization, leading to low performance of the CL. Lim et al.[131] reported that the catalyst ink made by a ball-milling process could give a higher current density in the kinetic-controlled region than that ink made by ultrasonication. Song et al.[132] showed that heat treatment of catalyst ink in an autoclave at 200 °C under pressurized conditions could enhance the penetration of the ionomer into the primary pores of the catalyst agglomerates, thus increasing catalyst the utilization and improving cell performance. Uchida et al.[133] showed that the dropping process was better than the paste process in decreasing catalyst loading. As shown in Figure 3.19 for a dropping process in which the PFSI colloid is adsorbed immediately after formation on the Pt/C catalyst before coagulation takes place, so the PFSI is finely dispersed on the Pt/C at a reduced thickness. Mass transport of gas to the reaction sites could also be accelerated through the dropping process. Consequently, reaction area is increased, and internal resistance as well as the supply of the reaction gas could be improved.

3.5.2.2. Catalyst Layer Fabrication and Characterization

As we discussed above, there are two major types of CL fabrication techniques. One is to apply the catalyst ink onto the gas diffusion layer to form a catalyzed diffusion medium (CDM), and the other is to apply the catalyst ink onto the PEM to form a CCM. Normally, applying the ink to the gas diffusion medium has the advantage of preserving the membrane from chemical attacks by the solvents in the catalyst ink. However, it seems that the CL does not come into close contact with the membrane and therefore the electrode is prone to delamination.[134] Regarding CCM, there are two ways of applying the catalyst ink to the membrane, namely the decal transferring process and the direct coating process. In the former, the CL is cast onto a PTFE blank

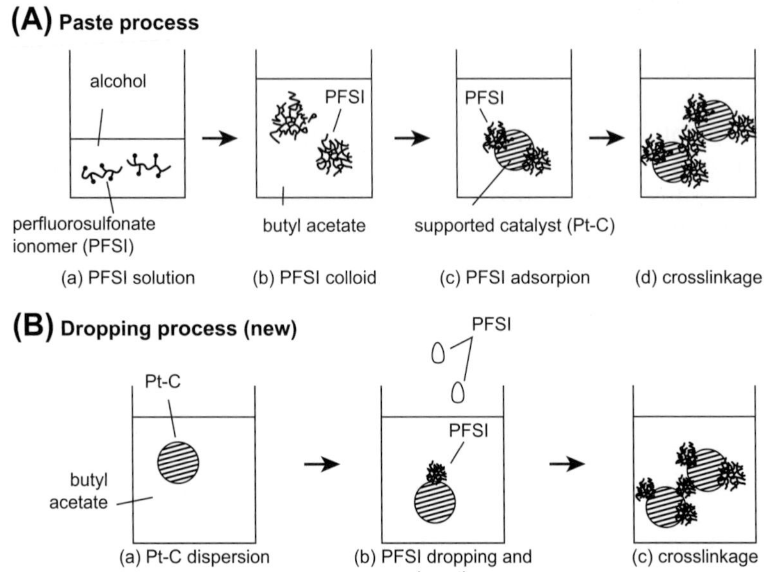

Figure 3.19 Schematic of the catalyst ink preparation process. (A) Paste process (B) Dropping process.[133]

first and then decaled to the membrane. In the latter, the membrane is coated directly with the CL. Frey et al.[135] suggested that the decal transfer process probably had a risk of uneven or incomplete catalyst transfer from the PTFE blank to the membrane. It was reported that the direct coating process could provide a better ionic connection between the membrane and the ionomer in the CL.[136] A typical process usually employed can be summarized as follows: (1) Stir a mixture of electrocatalyst (Pt/C catalyst) and solvent (e.g., ethanol) vigorously, and then disperse it in an ultrasonic machine; (2) Add a diluted PTFE emulsion and ionomer solution to the above mixture, followed by thorough ultrasonic stirring to form a catalyst ink; and (3) Transfer the catalyst ink onto a diffusion medium or membrane.

Transferring the catalyst ink onto a diffusion medium or membrane is usually realized by hand spreading or spraying. The spreading method spreads a PTFE-containing catalyst ink on a carbon diffusion layer and then heat treats it at 135 °C. In the spraying method, the catalyst ink is repeatedly sprayed on a sheet of carbon paper. Between each spraying, the carbon matrix is heated to evaporate the solvent to prevent the components from redissolving in the next spraying. Normally, the CL uniformity is not easily controlled by hand spreading and spraying.

In addition, the process is time-consuming and has poor reproducibility. Mechanical spraying is faster than both hand spreading and spraying, and allows a more uniform catalyst distribution. It has the potential for mass production with good reproducibility as well. However, a considerable amount of catalyst is often wasted in the feed lines, which could increase the cost of production. In order to increase the catalyst utilization and reduce the catalyst loading, several fabrication methods were further developed or modified. For example, Bevers et al.[137] described a catalyst powder deposition method, in which all the powder components were mixed in a fast running knife mill and then were applied onto the carbon matrix. Ionomer impregnation method was also developed in which the catalyst and PFSA suspension in alcohol and water are premixed before the CL was deposited. The gas diffusion layer/catalyst assembly could be formed by applying the catalyst ink onto the carbon matrix. During the impregnation reduction, the membrane was pre-exchanged to the Na^+ form.[138] One side of the membrane was faced to an aqueous Pt salt-containing solution for impregnation and the other side of the membrane was in contact with aqueous $NaBH_4$ reductant. The Pt precursor on the other side was reduced to metal Pt by diffused reductant through the membrane.

In the catalyst decaling method,[125,139] the catalyst ink was cast onto a PTFE blank for transferring to the membrane by hot pressing. A thin CL was left on the membrane when the PTFE blank was peeled away. Finally, the catalyzed membrane is rehydrated and ion-exchanged to the H^+ form by immersing it in a hot dilute H_2SO_4 solution. Moreover, low Pt loadings of catalysts layers could be prepared by evaporative deposition method, in which the Pt precursor was evaporatively deposited onto a membrane that was completely immersed in a reductive solution to get low Pt catalyst loading. A doctor-blade spreading technique was developed by Saab et al.[140,141] to prepare the CLs in a faster and highly reproducible fashion. The ink was coated onto laser jet transparency material with a doctor-blade device driven by an X—Y chart recorder time base. Machined aluminum slabs were used as the substrate support, and sit on the table of the X—Y chart recorder. They showed that with appropriate catalyst ink and transfer decal, doctor-blade spreading was not only significantly more precise and faster than hand painting, but also produced cells with nearly 25% higher current density in the critical voltage operating region. Specifically, some new technologies such as electrophoretic deposition (EPD),[142] electrospraying,[143,144] inkjet printing,[145] and rolling,[137,146,147] electrodeposition[148—150] and sputtering, have been developed to apply catalyst ink onto the gas diffusion layer or the membrane.

A brief presentation for several methods is as follows:

3.5.2.2.1. EPD Technique

Morikawa et al.[142] proposed an electrophoretic deposition (EPD) technique for preparing the CL, which didn't need any further mechanical pressing, and was suitable for a very thin or a very rigid membrane. In this method, the Nafion® membrane was set in the middle of the cell, and the left cell compartment was filled with $HClO_4$ aqueous solution and the right was filled with the catalyst ink. Two Pt electrodes were used as cathode and anode, respectively, to apply a high voltage to the cell. Figure 3.20 shows the schematic of the EPD cell. Using this method, they successfully prepared a CL with high Pt utilization that consisted of only fine carbon particles, and was smaller and more uniform than that prepared by the conventional painting method. The thickness of the CL was controlled by EPD duration or catalyst ink concentration.

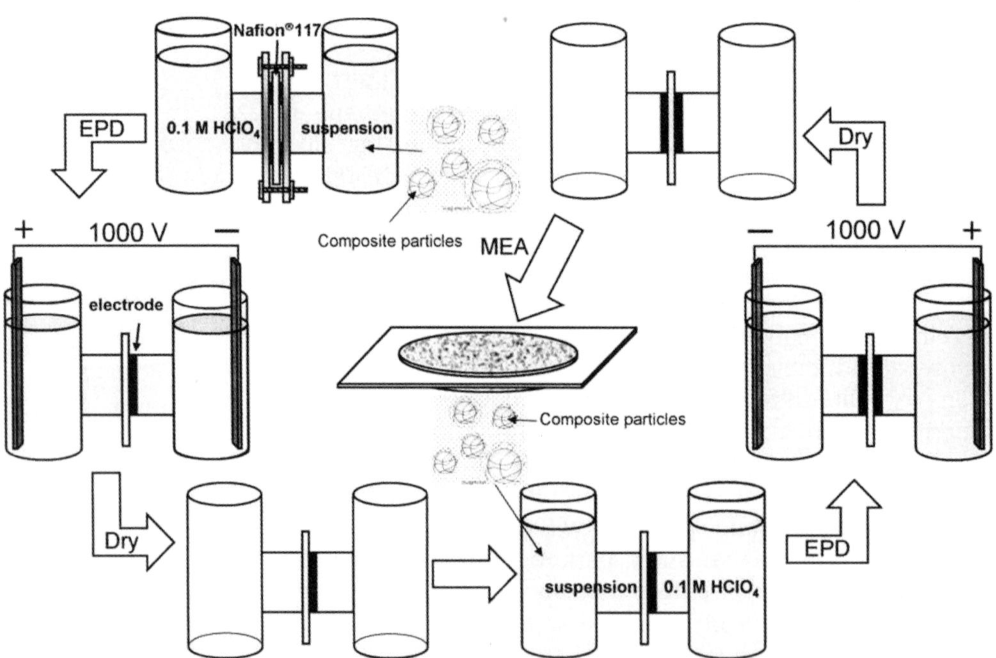

Figure 3.20 Schematic of EPD process for preparation of catalyst layer and MEA.[142]

3.5.2.2.2. Electrospraying Technique

An electrospraying technique was also developed for preparing the CL.[143,144] Figure 3.21 shows a schematic of the setup for electrospray and a detail of the electrospray process. This technique consists of applying a voltage (3300–4000 V) between a capillary tube, in which the ink is forced to flow using a high-pressure nitrogen stream over the carbon cloth substrate. The high electric field can make the solution generating a mist of highly charged droplets. During the spraying process, the droplets are reduced in size by evaporation of the solvent and/or by "Coulomb explosion". Further, a nitrogen-pressurized tank is used to force the catalyst ink, which is put in an ultrasonic bath, to pass through the capillary tube. The carbon support is then moved by means of an $X-Y$ axes-coordinated system. Compared to the CL prepared by a conventional technique, the electrospray-prepared CL shows an enhanced cell performance.

(A)

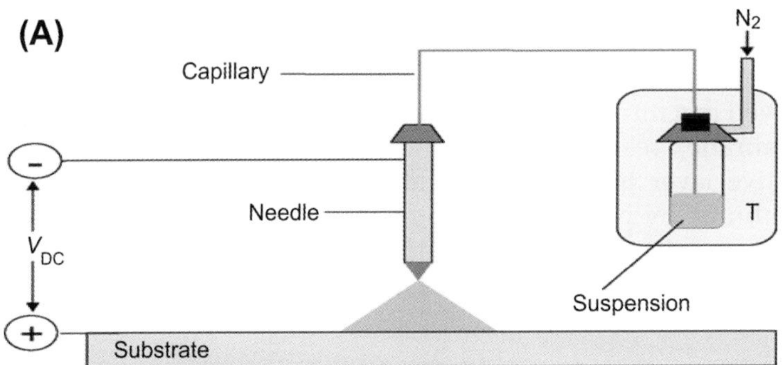

(B)

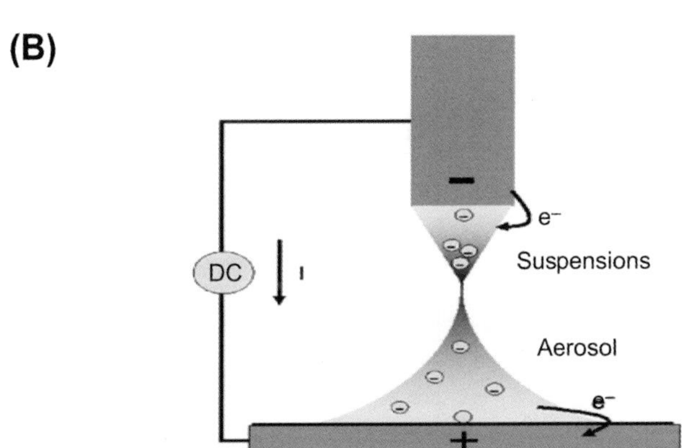

Figure 3.21 (A) Scheme of the setup for electrospray deposition of Pt/C suspensions. (B) Detail of the electrospray process in the needle to substrate space.[144] (For color version of this figure, the reader is referred to the online version of this book.)

3.5.2.2.3. Inkjet Printing Technique

Taylor et al.[145] employed an inkjet printing technique to deposit catalyst ink onto gas diffusion layers (GDLs). According to their description, a commercially available thermal inkjet printer was used. Standard black ink cartridges were used to print the catalyst inks. Carbon cloth was cut into squares and the squares were attached to standard white paper using double-sided tape. A black square of catalyst ink was printed onto the GDL using the "best print quality" feature in the printer settings to achieve the printer's best resolution. The paper containing the GDL was then reloaded into the paper tray and the printing was repeated until the desired loading was achieved. With this method, CLs with ultralow Pt loading can be easily fabricated and it also show higher Pt utilization compared with conventional method.

3.5.2.2.4. Dry Production Technique

A dry production technique based on rolling was also developed[137,147] and the schematic of the dry production technique was shown in Figure 3.22. In this technique, a carbon-supported catalyst with the desired amount of binder powder is mixed in a mill, and the mixed powder is then atomized and sprayed in a nitrogen stream through a slit nozzle directly onto a membrane, resulting in a uniformly distributed layer. The reactive layer is then fixed and thoroughly connected to the membrane by passing both through a calender. During all the steps, no solvents are used which avoids drying steps. In addition, the catalyst loading can be controlled to a low level.

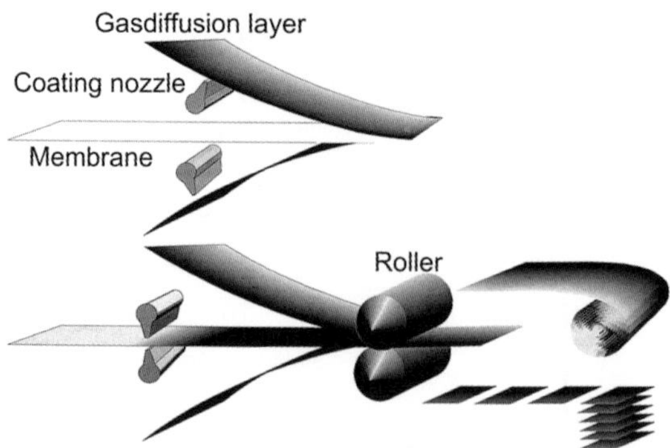

Figure 3.22 Scheme of the dry production technique for MEAs of PEM fuel cell or direct methanol fuel cell.[147]

Therefore, this technique can be scaled up and has the potential to meet industrial requirements.

3.5.2.2.5. Electrodeposition Method

Taylor et al.[148] reported the electrodeposition method to improve the utilization of Pt catalyst. According to their description, Pt ions diffused through a thin layer of Nafion® formed on the surface of uncatalyzed carbon electrode and were electrodeposited only in regions with ionic and electronic conductivity. Antonine et al.[151] impregnated carbon with H_2PtCl_6 and then applied electrochemical pulses to deposit Pt in a Nafion® active layer. However, Cl^- ions produced from the electrodeposition of Pt from H_2PtCl_6 remained in the active layer, which could poison the Pt and reduce its catalytic activity, according to Kim et al.[149,150] Therefore, they directly electrodeposited Pt on the surface of carbon blank electrode using a pulse electrodeposition technique. Their results indicated that pulse electrodeposition may be an attractive technique to achieve industry goals of reducing catalyst cost and increasing efficiency in PEM fuel cells.

3.5.2.2.6. Sputtering Technique

Sputtering technique is another fabrication method widely used for its potential to provide lower catalyst loading as well as thinner layers. The sputtering of CLs consists of a vacuum evaporation process that removes portions of a coating material (the target) and deposits a thin and resilient film of the target material onto an adjacent substrate, which can be either the GDL or the membrane.[152] O'Hayre et al.[153] directly sputtered Pt on the surface of a Nafion® membrane and showed that the performance of the sputtered Pt fuel cells was strongly dependent on the thickness of the sputtered CL. Cha et al.[154] obtained an ultrahigh utilization of Pt catalyst by alternating layers of sputtered Pt (5 nm thick) and a painted mixed electron- and proton-conducting layer of carbon black particles in an ink of the Nafion® ionomer. Repeating successive application of these layers up to five times could improve Pt utilization and enhance cell performance. Figure 3.23 shows a schematic of the sputtering process.

3.5.3. Catalyst Layer Characterization

In order to know the effect of the CL on fuel cell performance, the CL characterization is necessary. Usually, the CL is characterized by some ex situ physical measurements such as SEM,

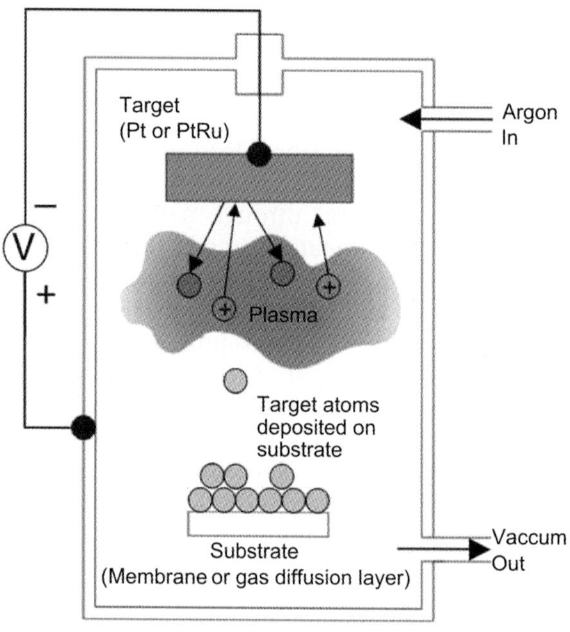

Figure 3.23 Schematic of sputtering in argon plasma for fabricating PEM fuel cell electrodes.[152]

TEM, XRD, XPS, Contact angle measurement, and so on. Several typical examples are briefly presented in this subsection:

3.5.3.1. Characterization by Electron Microscopy

Electron microscopy, such as scanning electron microscopy (SEM) and TEM, is a kind of powerful tool for catalyst structure and phase analysis. TEM with energy-dispersive X-ray (EDX) spectroscope has high resolution up to the atomic level and the capability for both imaging and element analysis. It has been widely used to measure the catalyst particle size and surface area, Pt dissolution and migration into the membrane, CL structure and phase changes, and MEA failures.[155] For example, Blom et al.[155] employed the diamond knife ultramicrotomy to create thin MEA sections suitable for TEM measurements both before and after fuel cell test. Some TEM imagines were used to compare the CL before the fuel cell test to that after the test. From the TEM imagine, the particle size, the interface between the CL and proton exchange membrane can be clearly observed, as shown in Figure 3.24.

SEM is another useful tool to analysis the catalyst structure and morphology.[156,157] Although SEM's resolution is lower than

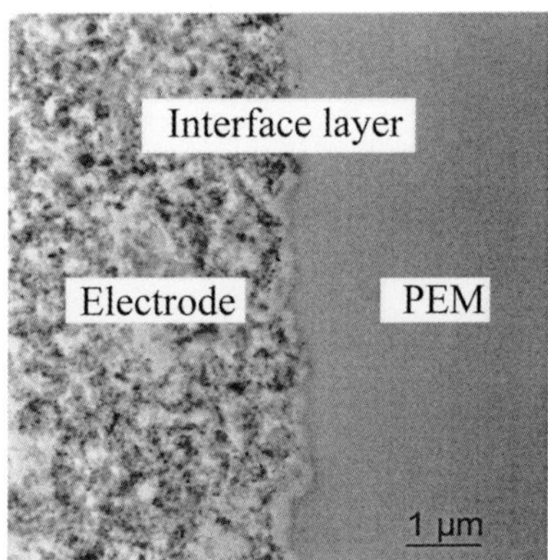

Figure 3.24 Micrograph of PEM fuel cell MEA after use in a fuel cell test stand for 325 h. Note the appearance of a new image feature at the interface between the membrane and the electrode.[155]

TEM's, sample preparation for SEM is much simpler than for TEM. Thus, SEM has also been widely used to characterize catalyst structure and analyze fuel cell failure. For example, Zhang et al.[158] using a field emission-scanning electron microscope (FESEM) to examine the catalysts layer morphology. Li et al.[159] employed an ex situ characterization tool to determine the ice/water distribution in the CL and sublayer at different stages of sub-zero Isothermal Constant Voltage (ICV) operation by FESEM to take the images of the MEA cross sections after the fuel cell test. The analysis method was designed to ensure that the entire experiment, from sample preparation to FESEM characterization, was carried out under subzero (°C) conditions so that the water was always kept in a frozen state without thawing; some examples of FESEM images are shown in Figure 3.25.

3.5.3.2. Contact Angle Measurement

Contact angle measurement has been broadly accepted for material surface analysis related to wetting, adhesion, and absorption. Yu and Ziegler[160] characterized the CLs using a CCM. They found that the first attachment of a water droplet at the CL surface showed a relatively high contact angle, and then the contact angle decreased with time, as shown in Figure 3.26. This could be due to water absorption into the polymer materials in the porous CL. They also found that CCM with a higher contact angle showed better fuel cell performance at 0.8 A cm^{-2} due to a lower mass transport resistance. Furthermore, they found that the contact angles differed slightly with water droplet location due to the nonuniform distribution of pore size in the CL. They

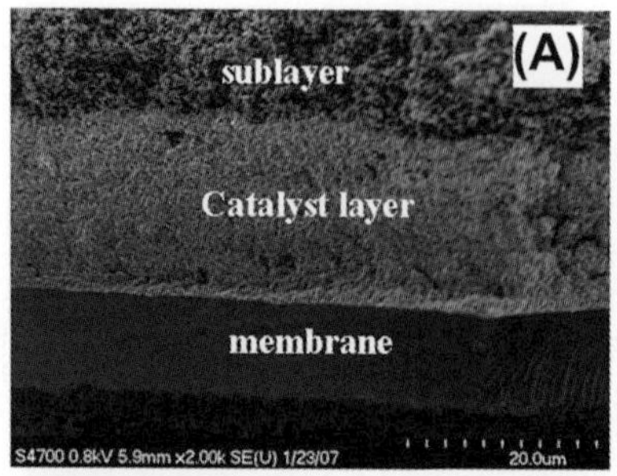

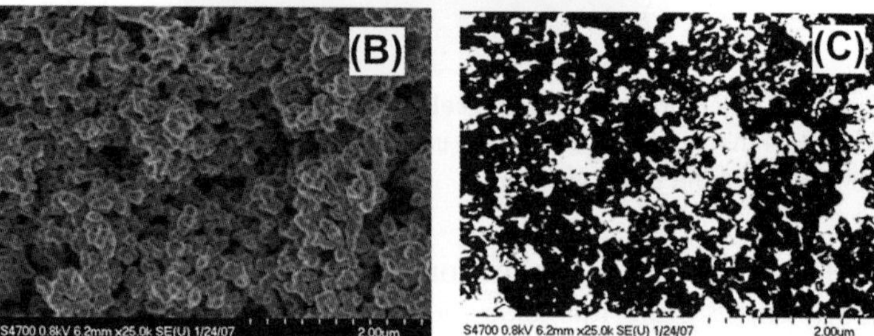

Figure 3.25 SEM image analysis. (A) SEM image of a cross section of the MEA2 under frozen state, (B) a close-up of sublyer in (A), and (C) same image of (A) was converted into binary image with histogram analysis of Image-Pro Plus software.[159]

also employed an environmental scanning electron microscope (ESEM) as a contact-angle analysis tool to investigate hydrophobicity and hydrophilicity on inhomogeneous materials. From Figure 3.27 it can be seen that while the water droplets increase in size, the cross section of the sample is always covered by water droplets. Therefore, to some extent, the water droplet at the surface could show the hydrophobicity and/or hydrophilicity of the bulk of the material. They also measured the contact angles of the CL before and after MEA evaluation, and found that the CL became more hydrophilic after MEA evaluation; thus, this

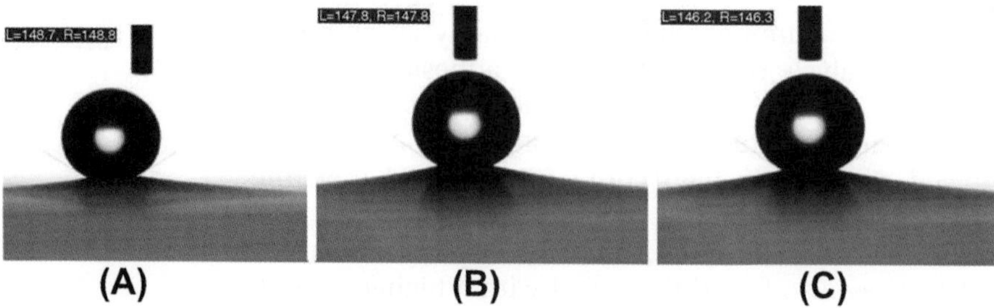

Figure 3.26 Time-dependent contact angle of a catalyst layer with wetting process (A) 0 s, i.e., the first attachment of the water droplet at the catalyst layer surface with a contact angle of 148.7°; (B) 60 s, the measured contact angle is 147.8°; (C) 100 s, the measured contact angle is 146.2°.[160] (For color version of this figure, the reader is referred to the online version of this book.)

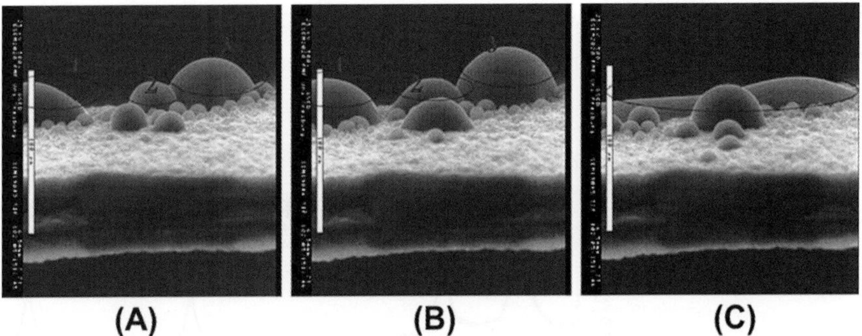

(A)   (B)   (C)

Figure 3.27 Water droplet growing and merging process under ESEM. (A) At the beginning, some small water droplets appear, i.e., water droplets one, two, and three. (B) Water droplets one, two, and three increase in size. (C) Water droplets one, two, and three merge together to form one big droplet. The contact angle of this water droplet is not be used.[160] (For color version of this figure, the reader is referred to the online version of this book.)

hydrophilic increase could be the cause of lower fuel cell performance.

3.5.3.3. Characterization by XRD Technique

XRD is a very useful tool for analysis of crystal structure and crystalline size. For example, Wilson et al.[161] performed the detailed X-ray diffraction (XRD) studies on PEM fuel cell electrodes using a Warren–Averbach Fourier transformation method to determine the weighted crystallite sizes after the test. Figure 3.28 shows a typical example of the X-ray powder diffraction pattern of a pristine Pt/Vulcan powder sample collected at normal incidence, in comparison with patterns of the anode and cathode in a cycled Pt/Vulcan MEA sample collected at 1° incidence. The average platinum crystal sizes (volume averaged) of pristine and cycled Pt/Vulcan samples were estimated using the Scherrer Eqn (3.8), by measuring the full width half maximum of the Pt X-ray powder diffraction peaks. Upon cycling between 0.6 and 1.0 V, the volume averaged platinum crystal size in the cycled MEA cathode increased significantly to 10.5 nm from 2.3 nm in the pristine Pt/Vulcan sample.

3.5.3.4. Characterization by XPS

X-ray photoelectron spectroscopy (XPS) can be used to identify the elements present on the surface of the sample. Thus, it can be used to identify Pt and carbon-surface chemical species that may present histories of chemical reactions or contamination in the CL. Wang et al.[163] studied XPS spectra of carbon and Pt before and after the fuel cell operation. They observed a

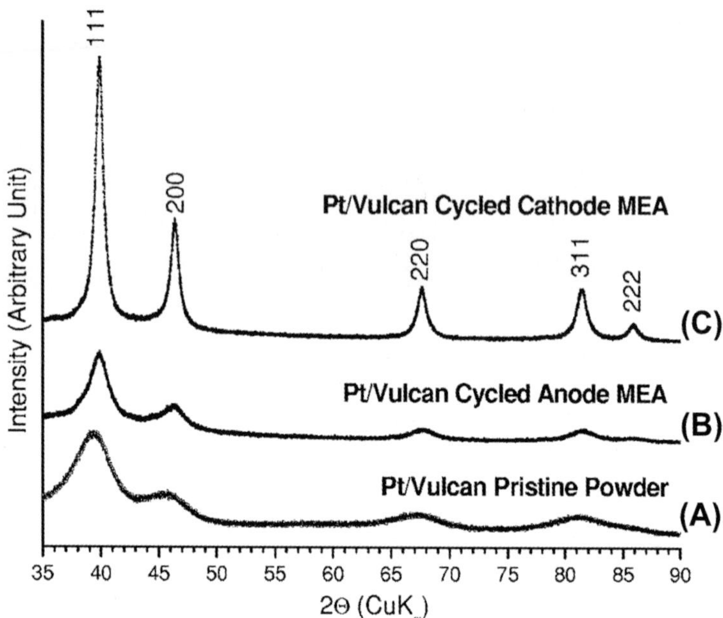

Figure 3.28 (A) X-ray powder diffraction pattern of the pristine Pt/Vulcan powder sample collected at normal incidence was compared to those of anode (B), and cathode (C) in the cycled Pt/Vulcan MEA sample collected at 1° incidence.[162]

significant increase in O 1 s peak value for each oxidized carbon support, a result of higher surface oxide content in the support surface due to electrochemical oxidation. However, sample preparation in XPS analysis is critical because these methods are very sensitive to a trace amount of contamination. Parry et al.[164] employed XPS technology to quantitatively characterize the cathode CLs aged in a PEM fuel cell with an anode operated under pure hydrogen and air and with 5 ppm CO-contaminated hydrogen. The detailed chemical XPS analysis of the aged samples demonstrated that the CL aging was mainly attributed to the oxidation of the carbon catalyst-support and a loss of the Nafion® ionomer in the cathode. It stated that XPS is a promising characterization tool to get the information for a chemical description of the MEA catalysts layer.

3.6. Chapter Summary

In order to give some basic knowledge and concepts, this chapter presents some fundamental concepts about the catalyst activity and stability of ORR electrocatalysts, which are the targeted research systems by rotating electrode technique. A detailed description about the electrocatalysts and CLs and their applications for ORR in terms of their types, structures, properties, catalytic activity/stability, as well as their research progress in the past several decades are also given. Furthermore, both the synthesis and characterization methods for ORR electrocatalysts, and the fabrication procedures for CLs, are also reviewed. It is believed that the technical breakthroughs in terms of both ORR activity and stability of the electrocatalysts and CLs should be heavily relied on the new and innovative material synthesis methods or procedures. It is our attention to present these useful information and knowledge in this chapter, which may benefit the readers for their continuing study throughout this entire book.

References

1. http://en.wikipedia.org/wiki/File:CatalysisScheme.png.
2. Lefèvre M, Proietti E, Jaouen F, Dodelet JP. Iron-based catalysts with improved oxygen reduction activity in polymer electrolyte fuel cells. *Science* 2009;**324**(5923):71–4.
3. US DOE Technical Plan (Fuel Cells), http://www1.eere.energy.gov/hydrogenandfuelcells/mypp/pdfs/fuel_cells.pdf.
4. Loukrakpam R, Luo J, He T, Chen Y, Xu Z, Njoki, et al. Nanoengineered PtCo and PtNi catalysts for oxygen reduction reaction: an assessment of the structural and electrocatalytic properties. *J Phys Chem C* 2011;**115**(5):1682–94.
5. Ignaszak A, Song C, Zhu W, Wang YJ, Zhang J, Bauer A, et al. Carbon–$Nb_{0.07}Ti_{0.93}O_2$ composite supported Pt–Pd electrocatalysts for PEM fuel cell oxygen reduction reaction. *Electrochim Acta* 2012;**75**(0):220–8.
6. Wang C, Markovic NM, Stamenkovic VR. Advanced platinum alloy electrocatalysts for the oxygen reduction reaction. *ACS Catal* 2012;**2**(5):891–8.
7. Zhang L, Lee K, Zhang J. The effect of heat treatment on nanoparticle size and ORR activity for carbon-supported Pd–Co alloy electrocatalysts. *Electrochim Acta* 2007;**52**(1):3088–94.
8. Vante NA, Jaegermann W, Tributsch H, Hoenle W, Yvon K. Electrocatalysis of oxygen reduction by chalcogenides containing mixed transition metal clusters. *J Am Chem Soc* 1987;**109**(11):3251–7.
9. Lee K, Zhang L, Zhang J. Ternary non-noble metal chalcogenide (W-Co-Se) as electrocatalyst for oxygen reduction reaction. *Electrochem Commun* 2007;**9**(7):1704–8.
10. Liu H, Song C, Tang Y, Zhang J. High-surface-area CoTMPP/C synthesized by ultrasonic spray pyrolysis for PEM fuel cell electrocatalysts. *Electrochim Acta* 2007;**52**(13):4532–8.
11. Wu G, More KL, Johnston CM, Zelenay P. High-performance electrocatalysts for oxygen reduction derived from polyaniline, iron, and cobalt. *Science* 2011;**332**(6028):443–7.
12. Zhang L, Kim J, Chen HM, Nan F, Dudeck K, Liu RS, et al. A novel CO-tolerant PtRu core–shell structured electrocatalyst with Ru rich in core and Pt rich in shell for hydrogen oxidation reaction and its implication in proton exchange membrane fuel cell. *J Power Sources* 2011;**196**(22):9117–23.

13. Zhang L, Kim J, Zhang J, Nan F, Gauquelin N, Botton GA, et al. Ti4O7 supported Ru@Pt core−shell catalyst for CO-tolerance in PEM fuel cell hydrogen oxidation reaction. *Appl Energy* 2013;**103**(0):507−13.
14. Wu J, Zhang J, Peng Z, Yang S, Wagner FT, Yang H. Truncated octahedral Pt_3Ni oxygen reduction reaction electrocatalysts. *J Am Chem Soc* 2010;**132**(14):4984−5.
15. Adzic R, Atanassova P, Atanassov P, More K, Myers D, Wieckowski A, Yan Y, et al. *U.S. Department of Energy Hydrogen Program 2008 Annual Merit Review Proceedings*. http://www.hydrogen.energy.gov/pdfs/review08/fc_3_zelenay.pdf.
16. Zelenay P, Chung H, Johnston C, Mack N, Nelson M, Turner P, et al. *FY 2011 progress report for the DOE hydrogen program*. p. 816. U.S. Department of Energy; February 2011. DOE/GO-102011-3178.
17. Wang Y-J, Wilkinson DP, Zhang J. Synthesis of conductive rutile-phased $Nb_{0.06}Ti_{0.94}O_2$ and its supported Pt electrocatalysts ($Pt/Nb_{0.06}Ti_{0.94}O_2$) for the oxygen reduction reaction. *Dalton Trans* 2012;**41**(4):1187−94.
18. Zhang J, Song C, Zhang J. Accelerated lifetime testing for proton exchange membrane fuel cells using extremely high temperature and unusually high load. *J Fuel Cell Sci Tech* 2011;**8**(5):051006.
19. Ignaszak A, Song C, Zhu W, Zhang J, Bauer A, Baker R, et al. Titanium carbide and its core-shelled derivative title=$TiC@TiO_2$ as catalyst supports for proton exchange membrane fuel cells. *Electrochim Acta* 2012;**69**(0):397−405.
20. Shim JH, Kim J, Lee C, Lee Y. Porous Pd layer-coated Au nanoparticles supported on carbon: synthesis and electrocatalytic activity for oxygen reduction in acid media. *Chem Mater* 2011;**23**(21):4694−700.
21. Kinoshita K. Particle size effects for oxygen reduction on highly dispersed platinum in acid electrolytes. *J Electrochem Soc* 1990;**137**(3):845−8.
22. Zhang S, Shao Y, Yin G, Lin Y. Stabilization of platinum nanoparticle electrocatalysts for oxygen reduction using poly(diallyldimethylammonium chloride). *J Mater Chem* 2009;**19**(42):7995.
23. Wang S, Yu D, Dai L. Polyelectrolyte functionalized carbon nanotubes as efficient metal-free electrocatalysts for oxygen reduction. *J Am Chem Soc* 2011;**133**(14):5182−5.
24. Zhang J, Sasaki K, Sutter E, Adzic RR. Stabilization of platinum oxygen-reduction electrocatalysts using gold clusters. *Science* 2007;**315**(5809):220−2.
25. Chen H, Wang Y, Dong S. An effective hydrothermal route for the synthesis of multiple PDDA-protected noble-metal nanostructures. *Inorganic Chem* 2007;**46**(25):10587−93.
26. Gasparotto LH, Ciapina EG, Ticianelli EA, Tremiliosi-Filho G. Electrodeposition of PVA-protected PtCo electrocatalysts for the oxygen reduction reaction in H_2SO_4. *J Power Sources* 2012;**197**:97−101.
27. Shim J, Lee CR, Lee HK, Lee JS, Cairns EJ. Electrochemical characteristics of Pt−WO_3/C and Pt−TiO_2/C electrocatalysts in a polymer electrolyte fuel cell. *J Power Sources* 2001;**102**(1−2):172−7.
28. Sasaki K, Zhang L, Adzic RR. Niobium oxide-supported platinum ultra-low amount electrocatalysts for oxygen reduction. *Phys Chem Chem Physics* 2008;**10**(1):159−67.
29. Chen L, Guo M, Zhang HF, Wang XD. Characterization and electrocatalytic properties of PtRu/C catalysts prepared by impregnation-reduction method using Nd_2O_3 as dispersing reagent. *Electrochim Acta* 2006;**52**(3):1191−8.
30. Zhou HP, Wu HS, Shen J, Yin AX, Sun LD, Yan CH. Thermally stable Pt/CeO_2 hetero-nanocomposites with high catalytic activity. *J Am Chem Soc* 2010;**132**(14):4998−9.

31. Yu HB, Kim JH, Lee HI, Scibioh MA, Lee J, Han J, et al. Development of nanophase CeO_2-Pt/C cathode catalyst for direct methanol fuel cell. *J Power Sources* 2005;**140**(1):59−65.
32. Wang X, Li X, Liu D, Song S, Zhang H. Green synthesis of Pt/CeO_2/graphene hybrid nanomaterials with remarkably enhanced electrocatalytic properties. *Chem Commun* 2012;**48**(23).
33. Ou DR, Mori T, Fugane K, Togasaki H, Ye F, Drennan J. Stability of ceria supports in $Pt-CeO_x$/C catalysts. *J Phys Chem C* 2011;**115**(39):19239−45.
34. Luches P, Pagliuca F, Valeri S. Morphology, stoichiometry, and interface structure of CeO_2 ultrathin films on Pt(111). *J Phys Chem C* 2011;**115**(21):10718−26.
35. Lee KH, Kwon K, Roev V, Yoo DY, Chang H, Seung D. Synthesis and characterization of nanostructured $PtCo$-CeO_x/C for oxygen reduction reaction. *J Power Sources* 2008;**185**(2):871−5.
36. Fugane K, Mori T, Ou DR, Suzuki A, Yoshikawa H, Masuda T, et al. Activity of oxygen reduction reaction on small amount of amorphous CeOx promoted Pt cathode for fuel cell application. *Electrochim Acta* 2011;**56**(11):3874−83.
37. Chen S, Wei Z, Guo L, Ding W, Dong L, Shen P, et al. Enhanced dispersion and durability of Pt nanoparticles on a thiolated CNT support. *Chem Commun* 2011;**47**(39):10984−6.
38. Zhou ZY, Kang X, Song Y, Chen S. Butylphenyl-functionalized palladium nanoparticles as effective catalysts for the electrooxidation of formic acid. *Chemical Commun* 2011;**47**(21):6075−7.
39. Bing Y, Neburchilov V, Song C, Baker R, Guest A, Ghosh D, et al. Effects of synthesis condition on formation of desired crystal structures of doped-TiO_2/carbon composite supports for ORR electrocatalysts. *Electrochim Acta* 2012;**77**(0):225−31.
40. Yuan Q, Zhou Z, Zhuang J, Wang X. Pd-Pt random alloy nanocubes with tunable compositions and their enhanced electrocatalytic activities. *Chem Commun* 2010;**46**(9):1491−3.
41. Jung DH, Bae SJ, Kim SJ, Nahm KS, Kim P. Effect of the Pt precursor on the morphology and catalytic performance of Pt-impregnated on Pd/C for the oxygen reduction reaction in polymer electrolyte fuel cells. *Int J Hydrogen Energy* 2011;**36**(15):9115−22.
42. Li Y, Zheng L, Liao S, Zeng J. Pt^Ru/C catalysts synthesized by a two-stage polyol reduction process for methanol oxidation reaction. *J Power Sources* 2011;**196**(24):10570−5.
43. Limpattayanate S, Hunsom M. Electrocatalytic activity of Pt–Pd electrocatalysts for the oxygen reduction reaction in proton exchange membrane fuel cells: Effect of supports. *Renewable Energy* 2014;**63**:205−11.
44. Qu J, Liu H, Ye F, Hu W, Yang J. Cage-bell structured Au-Pt nanomaterials with enhanced electrocatalytic activity toward oxygen reduction. *Int J Hydrogen Energy* 2012;**37**(17):13191−9.
45. Liu CW, Wei YC, Wang KW. Promotion of ceria-modified Pt-Au/C cathode catalysts for oxygen reduction reaction by H_2-induced surface segregation. *Chem Commun (Camb)* 2010;**46**(14):2483−5.
46. Yue Q, Zhang K, Chen X, Wang L, Zhao J, Liu J, et al. Generation of OH radicals in oxygen reduction reaction at Pt-Co nanoparticles supported on graphene in alkaline solutions. *Chem Commun (Camb)* 2010;**46**(19):3369−71.
47. Toda T, Igarashi H, Uchida H, Watanabe M. Enhancement of the electroreduction of oxygen on Pt alloys with Fe, Ni, and Co. *J Electrochem Soc* 1999;**146**(10):3750−6.

48. Mukerjee S, Srinivasan S, Soriaga MP, McBreen J. Role of structural and electronic properties of Pt and Pt alloys on electrocatalysis of oxygen reduction. *J Electrochem Soc* 1995;**142**(5):1409–22.

49. Salgado JRC, Antolini E, Gonzalez ER. Structure and activity of carbon-supported Pt–Co electrocatalysts for oxygen reduction. *J Phys Chem B* 2004;**108**(46):17767–74.

50. Zhang J, Yang H, Fang J, Zou S. Synthesis and oxygen reduction activity of shape-controlled Pt3Ni nanopolyhedra. *Nano Lett* 2010;**10**(2):638–44.

51. Beard BC, Ross JPN. Characterization of a titanium-promoted supported platinum electrocatalyst. *J Electrochem Soc* 1986;**133**(9):1839–45.

52. Zhou Z-Y, Kang X, Song Y, Chen S. Enhancement of the electrocatalytic activity of Pt nanoparticles in oxygen reduction by chlorophenyl functionalization. *Chem Commun* 2012;**48**(28):3391–3.

53. Liu Y, Li D, Stamenkovic VR, Soled S, Henao JD, Sun S. Synthesis of Pt_3Sn alloy nanoparticles and their catalysis for electro-oxidation of CO and methanol. *ACS Catal* 2011:1719–23.

54. Stamenkovic V, Schmidt TJ, Ross PN, Markovic NM. Surface composition effects in electrocatalysis: kinetics of oxygen reduction on well-defined Pt_3Ni and Pt_3Co alloy surfaces. *J Phys Chem B* 2002;**106**(46):11970–9.

55. Paulus UA, Wokaun A, Scherer GG, Schmidt TJ, Stamenkovic V, Radmilovic V, et al. Oxygen reduction on carbon-supported Pt–Ni and Pt–Co alloy catalysts. *J Phys Chem B* 2002;**106**(16):4181–91.

56. Stamenkovic VR, Fowler B, Mun BS, Wang G, Ross PN, Lucas CA, et al. Improved oxygen reduction activity on $Pt_3Ni(111)$ via increased surface site availability. *Science* 2007;**315**(5811):493–7.

57. Mani P, Srivastava R, Strasser P. Dealloyed binary PtM_3 (M = Cu, Co, Ni) and ternary $PtNi_3M$ (M = Cu, Co, Fe, Cr) electrocatalysts for the oxygen reduction reaction: performance in polymer electrolyte membrane fuel cells. *J Power Sources* 2011;**196**(2):666–73.

58. Min MK, Cho J, Cho K, Kim H. Particle size and alloying effects of Pt-based alloy catalysts for fuel cell applications. *Electrochim Acta* 2000;**45**(25–26):4211–7.

59. Jalan V, Taylor EJ. Importance of interatomic spacing in catalytic reduction of oxygen in phosphoric acid. *J Electrochem Soc* 1983;**130**(11):2299–302.

60. Paffett MT, Beery JG, Gottesfeld S. Oxygen reduction at $Pt_{0.65}Cr_{0.35}$, $Pt_{0.2}Cr_{0.8}$ and roughened platinum. *J Electrochem Soc* 1988;**135**(6):1431–6.

61. Toda T, Igarashi H, Watanabe M. Enhancement of the electrocatalytic O_2 reduction on Pt–Fe alloys. *J Electroanal Chem* 1999;**460**(1–2):258–62.

62. Marković NM, Adžić RR, Cahan BD, Yeager EB. Structural effects in electrocatalysis: oxygen reduction on platinum low index single-crystal surfaces in perchloric acid solutions. *J Electroanal Chem* 1994;**377**(1–2):249–59.

63. Antolini E, Salgado JRC, Gonzalez ER. Oxygen reduction on a $Pt_{70}Ni_{30}/C$ electrocatalyst prepared by the borohydride method in H_2SO_4/CH_3OH solutions. *J Power Sources* 2006;**155**(2):161–6.

64. Gasteiger HA, Kocha SS, Sompalli B, Wagner FT. Activity benchmarks and requirements for Pt, Pt-alloy, and non-Pt oxygen reduction catalysts for PEMFCs. *Appl Catal B: Environ* 2005;**56**(1–2):9–35.

65. Lee K, Savadogo O, Ishihara A, Mitsushima S, Kamiya N, Ota KI. Methanol-tolerant oxygen reduction electrocatalysts based on Pd-3D transition metal alloys for direct methanol fuel cells. *J Electrochem Soc* 2006;**153**(1):A20–4.

66. Bagotzky VS, Tarasevich MR, Radyushkina KA, Levina OA, Andrusyova SI. Electrocatalysis of oxygen reduction process on metal-chelates in acid electrolyte. *J Power Sources* 1978;**2**(3):233–40.

67. Ladouceur M, Lalande G, Guay D, Dodelet JP, Dignard-Bailey L, Trudeau ML, et al. Pyrolyzed cobalt phthalocyanine as electrocatalyst for oxygen reduction. *J Electrochem Soc* 1993;**140**(7):1974–81.

68. Wiesener K, Ohms D, Neumann V, Franke R. N-4 macrocycles as electrocatalysts for the cathodic reduction of oxygen. *Mater Chem Phys* 1989;**22**(3–4):457–75.

69. Sun GQ, Wang JT, Savinell RF. Iron(III) tetramethoxyphenylporphyrin (FeTMPP) as methanol tolerant electrocatalyst for oxygen reduction in direct methanol fuel cells. *J Appl Electrochem* 1998;**28**(10):1087–93.

70. Bouwkamp-Wijnoltz AL, Visscher W, van Veen JAR. The selectivity of oxygen reduction by pyrolysed iron porphyrin supported on carbon. *Electrochim Acta* 1998;**43**(21–22):3141–52.

71. Lalande G, Cote R, Tamizhmani G, Guay D, Dodelet JP, Dignard-Bailey L, et al. Physical, chemical and electrochemical characterization of heat-treated tetracarboxylic cobalt phthalocyanine adsorbed on carbon-black as electrocatalyst for oxygen reduction in polymer electrolyte fuel-cells. *Electrochim Acta* 1995;**40**(16):2635–46.

72. Gouerec P, Savy M. Oxygen reduction electrocatalysis: ageing of pyrolyzed cobalt macrocycles dispersed on an active carbon. *Electrochim Acta* 1999;**44**(15):2653–61.

73. Gojkovic SL, Gupta S, Savinell RF. Heat-treated iron(III) tetramethoxyphenyl porphyrin chloride supported on high-area carbon as an electrocatalyst for oxygen reduction—part II. Kinetics of oxygen reduction. *J Electroanal Chem* 1999;**462**(1):63–72.

74. Chu D, Jiang R. Novel electrocatalysts for direct methanol fuel cells. *Solid State Ionics* 2002;**148**(3–4):591–9.

75. Biloul A, Gouerec P, Savy M, Scarbeck G, Besse S, Riga J. Oxygen electrocatalysis under fuel cell conditions: behaviour of cobalt porphyrins and tetraazaannulene analogues. *J Appl Electrochem* 1996;**26**(11):1139–46.

76. Zhong W, Dong X, Chen S, Lin W. A new non-noble metal electro-catalyst for the reduction of oxygen. *Battery* 2003;**33**(4):245–8.

77. van Veen JAR, Colijn HA, van Baar JF. On the effect of a heat treatment on the structure of carbon-supported metalloporphyrins and phthalocyanines. *Electrochim Acta* 1988;**33**(6):801–4.

78. Bouwkamp-Wijnoltz AL, Visscher W, Van Veen JAR, Boellaard E, Van der Kraan AM, Tang SC. On active-site heterogeneity in pyrolyzed carbon-supported iron porphyrin catalysts for the electrochemical reduction of oxygen: an in situ Mossbauer study. *J Phys Chem B* 2002;**106**(50):12993–3001.

79. Lefèvre M, Dodelet JP, Bertrand P. Molecular oxygen reduction in PEM fuel cells: evidence for the simultaneous presence of two active sites in Fe-based catalysts. *J Phys Chem B* 2002;**106**(34):8705–13.

80. Zhang Jiujun, editor. *PEM fuel cell catalysts and catalyst layers—fundamentals and applications*. London UK: Springer; 2008.

81. Kuttiyiel KA, Sasaki K, Choi Y, Su D, Liu P, Adzic RR. Bimetallic IrNi core platinum monolayer shell electrocatalysts for the oxygen reduction reaction. *Energy & Environ Sci* 2012;**5**(1):5297.

82. Sarkar A, Murugan AV, Manthiram A. Synthesis and characterization of nanostructured Pd–Mo electrocatalysts for oxygen reduction reaction in fuel cells. *J Phys Chem C* 2008;**112**(31):12037–43.

83. Garsuch A, Michaud X, Wagner G, Klepel O, Dahn, JR. Templated Ru/Se/C electrocatalysts for oxygen reduction. *Electrochim Acta* 2009;**54**(4):1350–4.

84. Guo S, Dong S, Wang E. Raspberry-like hierarchical Au/Pt nanoparticle assembling hollow spheres with nanochannels: an advanced nanoelectrocatalyst for the oxygen reduction reaction. *J Phys Chem C* 2009;**113**(14):5485−92.

85. Sarkar A, Murugan AV, Manthiram A. Pt-encapsulated Pd−Co nanoalloy electrocatalysts for oxygen reduction reaction in fuel cells. *Langmuir* 2009;**26**(4):2894−903.

86. Jang JH, Kim J, Lee YH, Kim IY, Park MH, Yang CW, et al. One-pot synthesis of core−shell-like Pt$_3$Co nanoparticle electrocatalyst with Pt-enriched surface for oxygen reduction reaction in fuel cells. *Energy & Environ Sci* 2011;**4**(12):4947.

87. Zhang L, Lee K, Zhang JJ. The effect of heat treatment on nanoparticle size and ORR activity for carbon-supported Pd-Co alloy electrocatalysts. *Electrochim Acta* 2007;**52**(9):3088−94.

88. Bezerra CWB, et al. A review of heat-treatment effects on activity and stability of PEM fuel cell catalysts for oxygen reduction reaction. *J Power Sources* 2007;**173**(2):891−908.

89. Bezerra CW, Zhang L, Liu H, Lee K, Marques AL, Marques EP, et al. Electrochemical reduction of oxygen: an alternative method to prepare active CoN$_4$ catalysts. *Electrochim Acta* 1999;**45**(3):379−86.

90. Faubert G, Côté R, Guay D, Dodelet JP, Denes G, Poleunis C, et al. Activation and characterization of Fe-based catalysts for the reduction of oxygen in polymer electrolyte fuel cells. *Electrochim Acta* 1998;**43**(14−15):1969−84.

91. Gupta S, Tryk D, Bae I, Aldred W, Yeager E. Heat-treated polyacrylonitrile-based catalysts for oxygen electroreduction. *J Appl Electrochem* 1989;**19**(1):19−27.

92. Fournier J, Lalande G, Guay D, Dodelet JP. Activation of various Fe-based precursors on carbon black and graphite supports to obtain catalysts for the reduction of oxygen in fuel cells. *J Electrochem Soc* 1997;**144**(1):218−26.

93. Ohms D, Herzog S, Franke R, Neumann V, Wiesener K, Gamburcev S, et al. Influence of metal-ions on the electrocatalytic oxygen reduction of carbon materials prepared from pyrolyzed polyacrylonitrile. *J Power Sources* 1992;**38**(3):327−34.

94. Lalande G, Cote R, Guay D, Dodelet JP, Weng LT, Bertrand P. Is nitrogen important in the formulation of Fe-based catalysts for oxygen reduction in solid polymer fuel cells? *Electrochim Acta* 1997;**42**(9):1379−88.

95. Wei G, Wainright JS, Savinell RF. Catalytic activity for oxygen reduction reaction of catalysts consisting of carbon, nitrogen and cobalt. *J New Mater Electrochem Syst* 2000;**3**(2):121−9.

96. Tamizhmani G, Dodelet JP, Guay D, Lalande G, Capuano GA. Electrocatalytic activity of Nafion-impregnated pyrolyzed cobalt phthalocyanine—a correlative study between rotating-disk and solid polymer electrolyte fuel-cell electrodes. *J Electrochem Soc* 1994;**141**(1):41−5.

97. Bogdanoff P, Herrmann I, Hilgendorff M, Dorbandt I, Fiechter S, Tributsch H. Probing structural effects of pyrolysed CoTMPP-based electrocatalysts for oxygen reduction via new preparation strategies. *J New Mater Electrochem Syst* 2004;**7**(2):85−92.

98. Ye SY, Vijh AK. Non-noble metal-carbonized aerogel composites as electrocatalysts for the oxygen reduction reaction. *Electrochem Commun* 2003;**5**(3):272−5.

99. Bashyam R, Zelenay P. A class of non-precious metal composite catalysts for fuel cells. *Nature* 2006;**443**(7107):63−6.

100. Weeber AW, Bakker H. Amorphization by ball milling—a review. *Phys B-Condens Matter* 1988;**153**(1−3):93−135.

101. Jeon I-Y, Choi H-J, Jung S-M, Seo J-M, Kim M-J, Dai L, et al. Large-scale production of edge-selectively functionalized graphene nanoplatelets via ball milling and their use as metal-free electrocatalysts for oxygen reduction reaction. *J Am Chem Soc* 2013;**135**(4):1386—93.
102. Liang H-P, Guo Y-G, Zhang H-M, Hu J-S, Wan L-J, Bai C-L. Controllable AuPt bimetallic hollow nanostructures. *Chem Commun* 2004;(13):1496.
103. Marković NM, Schmidt TJ, Stamenković V, Ross PN. Oxygen reduction reaction on Pt and Pt bimetallic surfaces: a selective review. *Fuel Cells* 2001;**1**(2):105—16.
104. Zhang Z, Wang X, Cui Z, Liu C, Lu T, Xing W. Pd nanoparticles supported on WO_3/C hybrid material as catalyst for oxygen reduction reaction. *J Power Sources* 2008;**185**(2):941—5.
105. Ignaszak A, Teo C, Ye S, Gyenge ED. $Pt-SnO_2-Pd/C$ electrocatalyst with enhanced activity and durability for the oxygen reduction reaction at low Pt loading: the effect of carbon support type and activation. *J Phys Chem C* 2010;**114**(39):16488—504.
106. Zhang J, Mo Y, Vukmirovic MB, Klie R, Sasaki K, Adzic RR. Platinum monolayer electrocatalysts for O_2 reduction: Pt monolayer on Pd(111) and on carbon-supported Pd nanoparticles. *J Phys Chem B* 2004;**108**(30):10955—64.
107. Abdelkareem MA, Morohashi N, Nakagawa N. Factors affecting methanol transport in a passive DMFC employing a porous carbon plate. *J Power Sources* 2007;**172**(2):659—65.
108. Abdelkareem MA, Nakagawa N. DMFC employing a porous plate for an efficient operation at high methanol concentrations. *J Power Sources* 2006;**162**(1):114—23.
109. Achmad F, Kamarudin SK, Daud WRW, Majlan EH. Passive direct methanol fuel cells for portable electronic devices. *Appl Energy* 2011;**88**(5):1681—9.
110. Bae B, Kho BK, Lim T-H, Oh I-H, Hong S-A, Ha HY. Performance evaluation of passive DMFC single cells. *J Power Sources* 2006;**158**(2):1256—61.
111. Baglio V, Stassi A, Matera FV, Di Blasi A, Antonucci V, Aricò AS. Optimization of properties and operating parameters of a passive DMFC mini-stack at ambient temperature. *J Power Sources* 2008;**180**(2):797—802.
112. Chen R, Zhao TS. Performance characterization of passive direct methanol fuel cells. *J Power Sources* 2007;**167**(2):455—60.
113. Chen R, Zhao TS. A novel electrode architecture for passive direct methanol fuel cells. *Electrochem Commun* 2007;**9**(4):718—24.
114. Chu D, Jiang R. Effect of operating conditions on energy efficiency for a small passive direct methanol fuel cell. *Electrochim Acta* 2006;**51**(26):5829—35.
115. Lee JK, Lee J, Han J, Lim TH, Sung YE, Tak Y. Influence of Au contents of AuPt anode catalyst on the performance of direct formic acid fuel cell. *Electrochim Acta* 2008;**53**(9):3474—8.
116. Song DT, Wang QP, Liu ZS, Eikerling M, Xie Z, Navessin T, et al. A method for optimizing distributions of Nafion and Pt in cathode catalyst layers of PEM fuel cells. *Electrochim Acta* 2005;**50**(16—17):3347—58.
117. Passalacqua E, Lufrano F, Squadrito G, Patti A, Giorgi L. Nafion content in the catalyst layer of polymer electrolyte fuel cells: effects on structure and performance. *Electrochim Acta* 2001;**46**(6):799—805.
118. Sasikumar G, Ihm JW, Ryu H. Dependence of optimum Nafion content in catalyst layer on platinum loading. *J Power Sources* 2004;**132**(1—2):11—7.
119. Friedmann R, Nguyena TV. Optimization of the cathode catalyst layer composition using a novel 2-step preparation method. *ECS Trans* 2008;**16**(2):2021—9.

120. Jing F, Hou M, Shi W, Fu J, Yu H, Ming P, et al. The effect of ambient contamination on PEMFC performance. *J Power Sources* 2007;**166**(1): 172−6.

121. Mohtadi R, Lee Wk, Van Zee JW. Assessing durability of cathodes exposed to common air impurities. *J Power Sources* 2004;**138**(1−2):216−25.

122. Moore JM, Adcock PL, Lakeman JB, Mepsted GO. The effects of battlefield contaminants on PEMFC performance. *J Power Sources* 2000;**85**(2): 254−60.

123. Knights S, Jia N, Chuy C, Zhang J. In: Fuel cell reactant supply: effects of reactant contaminants. Fuel cell seminar 2005: fuel cell progress, challenges and markets. Palm Springs, (CA); Burnaby (BC): Ballard Power Systems; November 14−18, 2005. p. 121−25.

124. Yang D, Ma J, Xu L, Wu M, Wang H. The effect of nitrogen oxides in air on the performance of proton exchange membrane fuel cell. *Electrochim Acta* 2006;**51**(19):4039−44.

125. Wilson MS, Gottesfeld S. Thin-film catalyst layers for polymer electrolyte fuel-cell electrodes. *J Appl Electrochem* 1992;**22**(1):1−7.

126. Shin SJ, Lee JK, Ha HY, Hong SA, Chun HS, Oh IH. Effect of the catalytic ink preparation method on the performance of polymer electrolyte membrane fuel cells. *J Power Sources* 2002;**106**(1−2):146−52.

127. Uchida M, Aoyama Y, Eda N, Ohta A. New preparation method for polymer-electrolyte fuel-cells. *J Electrochem Soc* 1995;**142**(2):463−8.

128. Fernandez R, Ferreira-Aparicio P, Daza L. PEMFC electrode preparation: influence of the solvent composition and evaporation rate on the catalytic layer microstructure. *J Power Sources* 2005;**151**:18−24.

129. Chisaka M, Daiguji H. Effect of glycerol on micro/nano structures of catalyst layers in polymer electrolyte membrane fuel cells. *Electrochim Acta* 2006;**51**(23):4828−33.

130. Yang TH, Yoon YG, Park GG, Lee WY, Kim CS. Fabrication of a thin catalyst layer using organic solvents. *J Power Sources* 2004;**127**(1−2):230−3.

131. Lim C, Allen RG, Scott K. Effect of dispersion methods of an unsupported Pt-Ru black anode catalyst on the power performance of a direct methanol fuel cell. *J Power Sources* 2006;**161**(1):11−8.

132. Song JM, Suzuki S, Uchida H, Watanabe M. Preparation of high catalyst utilization electrodes for polymer electrolyte fuel cells. *Langmuir* 2006;**22**(14):6422−8.

133. Uchida M, Fukuoka Y, Sugawara Y, Ohara H, Ohta A. Improved preparation process of very-low-platinum-loading electrodes for polymer electrolyte fuel cells. *J Electrochem Soc* 1998;**145**(11):3708−13.

134. Xiong L, Manthiram A. High performance membrane-electrode assemblies with ultra-low Pt loading for proton exchange membrane fuel cells. *Electrochim Acta* 2005;**50**(16−17):3200−4.

135. Frey T, Linardi M. Effects of membrane electrode assembly preparation on the polymer electrolyte membrane fuel cell performance. *Electrochim Acta* 2004;**50**(1):99−105.

136. Chun YG, Kim CS, Peck DH, Shin DR. Performance of a polymer electrolyte membrane fuel cell with thin film catalyst electrodes. *J Power Sources* 1998;**71**(1−2):174−8.

137. Bevers D, Wagner N, von Bradke M. Innovative production procedure for low cost PEFC electrodes and electrode/membrane structures. *Int J Hydrogen Energy* 1998;**23**(1):57−63.

138. Fedkiw PS, Her WH. An impregnation-reduction method to prepare electrodes on Nafion Spe. *J Electrochem Soc* 1989;**136**(3):899−900.

139. Wilson MS, Gottesfeld S. High-performance catalyzed membranes of ultra-low Pt loadings for polymer electrolyte fuel-cells. *J Electrochem Soc* 1992;**139**(2):L28−30.
140. Saab AP, Garzon FH, Zawodzinski TA. Determination of ionic and electronic resistivities in carbon/polyelectrolyte fuel-cell composite electrodes. *J Electrochem Soc* 2002;**149**(12):A1541−6.
141. Bender G, Zawodzinski TA, Saab AP. Fabrication of high precision PEFC membrane electrode assemblies. *J Power Sources* 2003;**124**(1):114−7.
142. Morikawa H, Tsuihiji N, Mitsui T, Kanamura K. Preparation of membrane electrode assembly for fuel cell by using electrophoretic deposition process. *J Electrochem Soc* 2004;**151**(10):A1733−7.
143. Benitez R, Soler J, Daza L. Novel method for preparation of PEMFC electrodes by the electrospray technique. *J Power Sources* 2005;**151**:108−13.
144. Chaparro AM, Benitez R, Gubler L, Scherer GG, Daza L. Study of membrane electrode assemblies for PEMFC, with cathodes prepared by the electrospray method. *J Power Sources* 2007;**169**(1):77−84.
145. Taylor AD, Kim EY, Humes VP, Kizuka J, Thompson LT. Inkjet printing of carbon supported platinum 3-D catalyst layers for use in fuel cells. *J Power Sources* 2007;**171**(1):101−6.
146. Gulzow E, Kaz T, Reissner R, Sander H, Schilling L, von Bradke M. Study of membrane electrode assemblies for direct methanol fuel cells. *J Power Sources* 2002;**105**(2):261−6.
147. Gulzow E, Schulze M, Wagner N, Kaz T, Reissner R, Steinhilber G, et al. Dry layer preparation and characterisation of polymer electrolyte fuel cell components. *J Power Sources* 2000;**86**(1−2):352−62.
148. Taylor EJ, Anderson EB, Vilambi NRK. Preparation of high-platinum-utilization gas-diffusion electrodes for proton-exchange-membrane fuel-cells. *J Electrochem Soc* 1992;**139**(5):L45−6.
149. Kim H, Popov BN. Development of novel method for preparation of PEMFC electrodes. *Electrochem Solid State Lett* 2004;**7**(4):A71−4.
150. Kim HS, Subramanian NP, Popov BN. Preparation of PEM fuel cell electrodes using pulse electrodeposition. *J Power Sources* 2004;**138**(1−2):14−24.
151. Antoine O, Durand R. In situ electrochemical deposition of Pt nanoparticles on carbon and inside Nafion. *Electrochem Solid State Lett* 2001;**4**(5):A55−8.
152. Litster S, McLean G. PEM fuel cell electrodes. *J Power Sources* 2004;**130**(1−2):61−76.
153. O'Hayre R, Lee SJ, Cha SW, Prinz FB. A sharp peak in the performance of sputtered platinum fuel cells at ultra-low platinum loading. *J Power Sources* 2002;**109**(2):483−93.
154. Cha SY, Lee WM. Performance of proton exchange membrane fuel cell electrodes prepared by direct deposition of ultrathin platinum on the membrane surface. *J Electrochem Soc* 1999;**146**(11):4055−60.
155. Blom DA, Dunlap JR, Nolan TA, Allarda LF. Preparation of cross-sectional samples of proton exchange membrane fuel cells by ultramicrotomy for TEM. *J Electrochem Soc* 2003;**150**(4):A414−8.
156. Baturina OA, Wnek GE. Characterization of proton exchange membrane fuel cells with catalyst layers obtained by electrospraying. *Electrochem Solid State Lett* 2005;**8**(6):A267−9.
157. Li J, He P, Wang K, Davis M, Ye S. Characterization of catalyst layer structural changes in PEMFC as a function of durability testing. *ECS Trans* 2006;**3**(1):743−51.

158. Zhang F-Y, Spernjak D, Prasad AK, Advani SG. In situ characterization of the catalyst layer in a polymer electrolyte membrane fuel cell. *J Electrochem Soc* 2007;**154**(11):B1152−7.

159. Li J, Lee S, Roberts J. Ice formation and distribution in the catalyst layer during freeze-start process—CRYO- SEM investigation. *ECS Trans* 2007;**11**(1):595−605.

160. Yu HM, Ziegler C, Oszcipok M, Zobel M, Hebling C. Hydrophilicity and hydrophobicity study of catalyst layers in proton exchange membrane fuel cells. *Electrochim Acta* 2006;**51**(7):1199−207.

161. Wilson MS, Garzon FH, Sickafus KE, Gottesfeld S. Surface-area loss of supported platinum in polymer electrolyte fuel-cells. *J Electrochem Soc* 1993;**140**(10):2872−7.

162. Ferreira PJ, la O GJ, Shao-Horn Y, Morgan D, Makharia R, Kocha S, et al. Instability of Pt/C electrocatalysts in proton exchange membrane fuel cells—a mechanistic investigation. *J Electrochem Soc* 2005;**152**(11):A2256−71.

163. Wang JJ, Yin GP, Shao YY, Zhang S, Wang ZB, Gao YZ. Effect of carbon black support corrosion on the durability of Pt/C catalyst. *J Power Sources* 2007;**171**(2):331−9.

164. Parry V, Berthomé G, Joud J-C, Lemaire O, Franco AA. XPS investigations of the proton exchange membrane fuel cell active layers aging: characterization of the mitigating role of an anodic CO contamination on cathode degradation. *J Power Sources* 2011;**196**(5):2530−8.

4

ELECTROCHEMICAL OXYGEN REDUCTION REACTION

Fengzhan Si[a,b], Yuwei Zhang[a,b], Liang Yan[a,b], Jianbing Zhu[a,b], Meiling Xiao[a,b], Changpeng Liu[b], Wei Xing[a] and Jiujun Zhang[c]

[a]*State Key Laboratory of Electroanalytical Chemistry, Changchun Institute of Applied Chemistry, Changchun, PR China,* [b]*Laboratory of Advanced Power Sources, Changchun Institute of Applied Chemistry, Changchun, PR China,* [c]*Energy, Mining and Environment, National Research Council of Canada, Vancouver, BC, Canada*

CHAPTER OUTLINE

Rotating Electrode Methods and Oxygen Reduction Electrocatalysts. http://dx.doi.org/10.1016/B978-0-444-63278-4.00004-5

4.1. Introduction

The oxygen reduction reaction (ORR) is probably the most important reaction in life processes such as biological respiration, and in energy-converting systems such as fuel cells and metal–air batteries. ORR in aqueous solutions occurs mainly by two pathways: the direct four-electron transfer pathway from O_2 to H_2O, and the two-electron transfer pathway from O_2 to hydrogen peroxide (H_2O_2). In nonaqueous aprotic solvents and/or in alkaline solutions, the one-electron transfer pathway from O_2 to superoxide (O^{2-}) could also occur. In proton exchange membrane (PEM) fuel cells, including both hydrogen/oxygen and methanol/oxygen (direct methanol) fuel cells, ORR is the reaction occurring at the cathode. Unfortunately, the ORR kinetics is normally very slow. In order to speed up the ORR kinetics to reach a practical usable level in a fuel cell, a cathode ORR catalyst is needed. At the current stage of technology, platinum (Pt)-based materials are the most practical catalysts. Because these Pt-based catalysts are too expensive for making commercially viable fuel cells, extensive research over the past several decades has focused on developing alternative catalysts, including nonnoble metal catalysts.[1] The alternative electrocatalysts explored include noble metals and alloys, carbon materials, quinone and derivatives, transition metal macrocyclic compounds, transition metal chalcogenides, and transition metal carbides.

To further improve the ORR performance of Pt-based catalysts and explore new alternative non-Pt ORR catalysts, besides innovative catalyst synthesis, characterizing catalyst activity for new catalyst design, performance optimization and down-selection using rotating electrode technology including rotating disk electrode (RDE) and rotating ring-disk electrode (RRDE) techniques is necessary. In order to have a meaningful and accurate characterization, some understanding on ORR catalysis and its fundamentals including reaction thermodynamics, kinetics, and mechanisms is important. In this chapter, we will focus on the ORR kinetics and mechanisms catalyzed by electrocatalysts.

4.2. Electrochemical Thermodynamics and Electrode Potential of ORR

Table 4.1 presents the selected ORR processes and their corresponding thermodynamic electrode potentials at standard conditions. All the potentials are against the standard hydrogen

Table 4.1. Thermodynamic Electrode Potentials of Oxygen Reduction Reaction in Two Aqueous Electrolyte Solutions at 25 °C and 1.0 atm[1,2]

Electrochemical Oxygen Reduction Reactions	Thermodynamic Electrode Potential (V vs SHE)	Electrolyte Condition
$O_2 + 4H^+ + 4e^- \leftrightarrow 2H_2O$	1.229	Acidic aqueous electrolyte solution
$O_2 + 2H^+ + 2e^- \leftrightarrow H_2O_2$	0.70	
$H_2O_2 + 2H^+ + 2e^- \leftrightarrow 2H_2O$	1.76	
$O_2 + H_2O + 4e^- \leftrightarrow 4OH$	0.401	Alkaline aqueous electrolyte solution
$O_2 + H_2O + 2e^- \leftrightarrow HO_2^- + OH^-$	−0.065	
$HO_2^- + H_2O + 2e^- \leftrightarrow 3OH^-$	0.867	

potential, which is defined as zero in 1.0 M proton aqueous solution at any temperature and 1.0 atm hydrogen gas pressure. In this table, two cases of electrolyte solution are selected: one is the acidic aqueous solution, and the other is the alkaline aqueous solution.

The thermodynamic electrode potentials listed in Table 4.1 are those at standard conditions (25 °C, 1.0 atm). However, if these reactions are not at standard conditions, the electrode potentials can be expressed as the Nernst forms. For example, for the reaction

$$O_2 + 4H^+ + 4e^- \leftrightarrow 2H_2O \quad E^o_{O_2/H_2O} = 1.229 \text{ V vs SHE} \quad (4\text{-I})$$

the thermodynamic reversible electrode potential $(E^r_{O_2/H_2O})$ at nonstandard conditions can be expressed as Eqn (4.1) if assuming O_2 in Reaction (4-I) is a gas, and H_2O is a liquid:

$$E^r_{O_2/H_2O} = E^o_{O_2/H_2O} + \frac{RT}{n_{O_2}F} \ln\left(P_{O_2} C_{H^+}^4\right) \quad (4.1)$$

where $E^o_{O_2/H_2O}$ is the thermodynamic electrode potential as listed in Table 4.1, P_{O_2} is the partial pressure of O_2 gas (atmospheres), C_{H^+} is the concentration of proton (moles per cubic decimeter), n_{O_2} (=4) is the electron transfer number of Reaction (4-I), F is the Faraday constant (96,487 C mol^{-1}), R is the universal gas constant (8.314 J K^{-1} mol^{-1}), and T is the temperature (Kelvin). Using Eqn (4.1), the reversible electrode potentials can be calculated at other different O_2 partial pressures, different temperatures, and in different solutions containing different proton concentrations. In a similar way, the reversible electrode potentials for other reactions listed in Table 4.1 can also be expressed as the Nernst form as Reaction (4-I).

All the reactions listed in Table 4.1 are called half-electrochemical reactions. In reality, if there are no count half-electrochemical reactions to supply electrons, these reactions cannot occur. In an electrochemical cell, there are two half-electrochemical reactions, one is called the anode reaction, and the other is called the cathode reaction. For example, in a H_2/O_2 fuel cell, there are two half-electrochemical reactions:

$$H_2 \leftrightarrow 2H^+ + 2e^- \text{ (Anode reaction)} \quad E^o_{H_2/H^+}$$

$$= 0.00 \text{ V vs SHE (Standard Hydrogen Electrode)} \tag{4-II}$$

$$\frac{1}{2}O_2 + 2H^+ + 2e^- \leftrightarrow H_2O \text{ (Cathode reaction)} \tag{4-III}$$

$$E^o_{O_2/H_2O} = 1.229 \text{ V vs SHE}$$

This cathode reaction is actually the same as Reaction (4-I). Both Reactions (4-II) and (4-III) have their own thermodynamic electrode potentials, $E^o_{H_2/H^+}$ and $E^o_{O_2/H_2O}$, at standard conditions. As indicated by Reaction (4-II), the anode electrode potential is 0.000 V vs SHE, and by Reaction (4-III), the cathode electrode potential is 1.229 V vs SHE. If these two half-electrochemical reactions are combined together to form a fuel cell, the overall cell reaction would be

$$\frac{1}{2}O_2 + H_2 \leftrightarrow H_2O \quad E^o_{cell} = 1.229 \text{ V} \tag{4-IV}$$

As shown, this overall electrochemical reaction has a thermodynamic cell voltage (E^o_{cell}) of 1.229 V at reversible conditions. Accordingly, the standard free energy change of the overall reaction $(\Delta G^o_{cell} = -nFE^o_{cell}$, where n is the electron transfer number of the overall reaction (here $n = 2$), and F is the Faraday's constant, 96,745 C mol^{-1}). If the ΔG^o_{cell} is negative, the corresponding cell reaction should have spontaneity in standard conditions.

In a similar way as Eqn (4.1), the thermodynamic reversible electrode potential for anode $(E^r_{H_2/H^+})$ at nonstandard conditions can be expressed as Eqn (4.2) if assuming O_2 in Reaction (4-I) is a gas, and H_2O is a liquid:

$$E^r_{H_2/H^+} = E^o_{H_2/H^+} + \frac{RT}{n_{H_2}F}\ln\left(\frac{C^2_{H^+}}{P_{H_2}}\right) \tag{4.2}$$

where $E^{\text{o}}_{\text{H}_2/\text{H}^+}$ is the thermodynamic electrode potential of Reaction (4-II), P_{H_2} is the partial pressure of H_2 gas (atmosphere), n_{H_2} (=2) is the electron transfer number of Reaction (4-II), C_{H^+}, R, and T have the same meanings as that in Reaction (4-I). If combining Eqns (4.1) and (4.2), the thermodynamic or fuel cell voltage or theoretical open circuit voltage ($E^{\text{o}}_{\text{cell}}$) can be obtained:

$$E^{\text{o}}_{\text{cell}} = E^{\text{r}}_{\text{O}_2/\text{H}_2\text{O}} - E^{\text{r}}_{\text{H}_2/\text{H}^+} = E^{\text{o}}_{\text{O}_2/\text{H}_2\text{O}} - E^{\text{o}}_{\text{H}_2/\text{H}^+} + \frac{RT}{nF}\ln\left(P^{1/2}_{\text{O}_2}P_{\text{H}_2}\right)$$

$$(4.3)$$

where n (=2) is the electron transfer number of the fuel cell reaction (Reaction (4-IV), and other symbols have their meanings defined in Eqns (4.1) and (4.2).

As discussed, ORR equations listed in Table 4.1 and the description above are all the cases for thermodynamically reversible processes. In reality, all reactions have limited reaction rates, which are either slower or faster, depending on the natures of the reactions. Furthermore, the ORR processes are actually not as simple as expressed by those reactions listed in Table 4.1. For each reaction, there should be a reaction mechanisms associated with it. This reaction mechanism may include several elementary reaction steps. Particularly, for ORR electrocatalyzed by a catalyst, the reaction mechanism may be even more complicated. Therefore, when ORR is discussed, both its reaction mechanism and kinetics must be explored in order to obtain fundamental understanding and full pictures.

4.3. Electrochemical Kinetics and Mechanism of ORR

In Chapter 2, the fundamentals of both chemical and electrochemical kinetics have been presented, which will be used as a basis for our discussion about the electrochemical kinetics and mechanism of ORR in this chapter.

4.3.1. ORR on Platinum Catalyst

ORR on platinum (Pt) catalyst is the most important system that has been given the most extensive study in the literature. Here we take this system as an example to show how to obtain electrode kinetic information of ORR.

Normally, ORR catalyzed by Pt catalyst occurs predominately through a four-electron transfer pathway to water (or to

OH^- in alkaline solutions). However, there is still a small amount of intermediates, such as peroxide, which can be detected using RRDE technique. Therefore, the ORR catalyzed by a Pt catalyst has a complex reaction mechanism. In general, the first step of the mechanism is the adsorption of O_2 on the Pt surface. The adsorbed oxygen can be electrochemically reduced through several elementary steps, as shown in Reactions (4-V)−(4-XI)[3]:

$$Pt + O_2 \leftrightarrow Pt - O_{2(ads)} \tag{4-V}$$

$$Pt - O_{2(ads)} + H^+ + e^- \rightarrow Pt - O_2H_{(ads)}$$
$$(Rate - determining\ step) \tag{4-VI}$$

$$x\left(Pt - O_2H_{(ads)} + e^- + H^+ \rightarrow Pt - O_2H_{2(ads)}\right) \tag{4-VII}$$

$$x\left(Pt - O_2H_{2(ads)} \rightarrow Pt + H_2O_2\right) \tag{4-VIII}$$

$$(1 - x)\left(Pt + Pt - O_2H_{(ads)} + H^+ + e^- \rightarrow 2Pt - OH_{(ads)}\right) \tag{4-IX}$$

$$2(1 - x)\left(Pt - OH_{(ads)} + H^+ + e^- \rightarrow Pt + H_2O\right) \tag{4-X}$$

Note this proposed mechanism is for that in acidic solution, and the subscript "ads" indicates the corresponding species are adsorbed on the electrode surface or catalyst metal active site. In this mechanism, Reaction (4-V) represents the adsorption/desorption of O_2 at the Pt surface, which is the fast reaction, and we assume that it is always at an equilibrium state. Reaction (4-VI) represents the first electron transfer of the ORR, which is considered the slowest reaction in the whole mechanism (i.e., the rate-determining step), producing the intermediate $Pt-O_2H_{(ads)}$. A portion of such an intermediate can be further reduced to produce H_2O_2 through Reactions (4-VII) and (4-VIII). The other portion will be further reduced to produce H_2O through Reactions (4-IX) and (4-X). It can be seen that the first portion of the reaction forms a two-electron transfer ORR pathway to produce H_2O_2, and the other portion forms a four-electron transfer ORR pathway to produce H_2O. In fact, the ORR normally has a mixed two- and four-electron transfer pathway, giving an overall electron transfer number of less than four. The relative proportions of two- and four-electron transfer pathways are strongly dependent on Reactions (4-VIII) and (4-X). The proportion of Reaction (4-VIII) can be expressed as x in the mechanism above, and the proportion of Reaction (4-X) can be expressed as $(1 - x)$. If $x = 1$, the mechanism will be a complete two-electron transfer pathway, while if $x = 0$, the mechanism will become a totally four-electron pathway. In practical situations on a Pt catalyst, the x value is much smaller than 1, giving a four-electron dominated transfer pathway to produce H_2O, with less than 2% H_2O_2.

The ORR mechanism expressed by Reactions (4-V)−(4-X) may be too complicated to be quantitatively treated to obtain the reaction rate expression. In order to obtain the relationship between the current and the electrode potential, some reasonable assumptions may be made to simplify this mechanism. We may make two assumptions, one is to assume $x = 0$ so that the whole mechanism only goes through a four-electron pathway to produce water (this may be true for ORR catalyzed by Pt), and the other is to combine the fast Reactions (4-IX) and (4-X) together into one reaction. In this way, the ORR mechanism may be simplified as the following three reactions:

$$Pt + O_2 \underset{\overleftarrow{k}_{10}}{\overset{\overrightarrow{k}_{10}}{\leftrightarrow}} Pt - O_{2(ads)} \tag{4-XI}$$

$$Pt - O_{2(ads)} + H^+$$

$$+ e^- \underset{\overleftarrow{k}_{20}}{\overset{\overrightarrow{k}_{20}}{\leftrightarrow}} Pt{-}O_2H_{(ads)} \quad (Rate-determining\ step) \tag{4-XII}$$

$$Pt - O_2H_{(ads)} + 3H^+ + 3e^- \underset{\overleftarrow{k}_{30}}{\overset{\overrightarrow{k}_{30}}{\leftrightarrow}} Pt + 2H_2O \tag{4-XIII}$$

In all three reaction equations above, $\overrightarrow{k}_{10}$ and $\overleftarrow{k}_{10}$ represent the chemical reaction rate constants in the forward and backward directions, respectively, which are independent of the electrode potential. For electrochemical reactions (4-XII) and (4-XIII), the rate constants $\overrightarrow{k}_{20}, \overleftarrow{k}_{20}, \overrightarrow{k}_{30}$, and $\overleftarrow{k}_{30}$ are all dependent on the electrode potential. According to Eqns (2.17) and (2.18), these potential-dependent rate constants can be expressed as follows:

$$\overrightarrow{k}_{20} = \overrightarrow{k}_{20}^{\,o} \exp\left(-\frac{(1 - \alpha_{20})n_{\alpha20}FE}{RT}\right) \tag{4.4}$$

$$\overleftarrow{k}_{20} = \overleftarrow{k}_{20}^{\,o} \exp\left(\frac{\alpha_{20}n_{\alpha20}FE}{RT}\right) \tag{4.5}$$

$$\overrightarrow{k}_{30} = \overrightarrow{k}_{30}^{\,o} \exp\left(-\frac{(1 - \alpha_{30})n_{\alpha30}FE}{RT}\right) \tag{4.6}$$

$$\overleftarrow{k}_{3O} = \overleftarrow{k}_{3O}^{o} \exp\left(\frac{\alpha_{3O} n_{\alpha 3O} FE}{RT}\right) \tag{4.7}$$

where $\overrightarrow{k}_{2O}^{o}$, $\overleftarrow{k}_{2O}^{o}$, $\overrightarrow{k}_{3O}^{o}$, and $\overleftarrow{k}_{3O}^{o}$ are the potential-independent forward and backward reaction rate constants for Reactions (4-XII) and (4-XIII), α_{2O} and α_{3O} are the electron transfer coefficients, $n_{\alpha 2O}$ and $n_{\alpha 3O}$ are the electron transfer numbers for Reactions (4-XII) and (4-XIII), respectively, and E is the electrode potential.

It can be seen that Reactions (4-XII) and (4-XIII) are all the reactions occurred on Pt surface. The concentrations of Pt-related surface species or surface reaction sites such as Pt–O$_2$, Pt–O$_2$H, and Pt in Reactions (4-XII) and (4-XIII) may be expressed as the corresponding surface coverage, such as θ_{Pt-O_2}, θ_{Pt-O_2H}, and θ_{Pt}. Figure 4.1 shows the schematic diagram of a platinum surface with adsorption of O$_2$, and O$_2$H.

Three surface coverages shown in Figure 4.1 should have a relationship:

$$\theta_{Pt-O_2} + \theta_{Pt-O_2H} + \theta_{Pt} = 1 \tag{4.8}$$

By assuming Reaction (4-XI) is a fast reaction and always in an equilibrium state, both θ_{Pt-O_2} and θ_{Pt-O_2H} can be derived into Eqns (4.9) and (4.10), respectively[3]:

$$\theta_{Pt-O_2} = \frac{\overrightarrow{k}_{1O}}{\overleftarrow{k}_{1O}} C_{O_2} \frac{\overleftarrow{k}_{2O}^{o} \exp\left(\frac{\alpha_{2O} n_{\alpha 2O} FE}{RT}\right) + \overrightarrow{k}_{3O}^{o} C_{H^+}^{3} \exp\left(-\frac{(1-\alpha_{3O}) n_{\alpha 3O} FE}{RT}\right)}{\overrightarrow{k}_{2O}^{o}\dfrac{\overrightarrow{k}_{1O}}{\underset{k_{1O}}{\overleftarrow{}}} C_{O_2} C_{H^+} \exp\left(-\frac{(1-\alpha_{2O}) n_{\alpha 2O} FE}{RT}\right) + \overrightarrow{k}_{2O}^{o}\left(1 + \dfrac{\overrightarrow{k}_{1O}}{\underset{k_{1O}}{\overleftarrow{}}} C_{O_2}\right)\exp\left(\frac{\alpha_{2O} n_{\alpha 2O} FE}{RT}\right)}$$

$$+ \overleftarrow{k}_{3O}^{o} C_{H_2O}^{2} \exp\left(\frac{\alpha_{3O} n_{\alpha 3O} FE}{RT}\right) + \overrightarrow{k}_{3O}^{o} C_{H^+}^{3}\left(1 + \dfrac{\overrightarrow{k}_{1O}}{\underset{k_{1O}}{\overleftarrow{}}} C_{O_2}\right)\exp\left(-\frac{(1 - \alpha_{3O}) n_{\alpha 3O} FE}{RT}\right) \tag{4.9}$$

$$\theta_{Pt-O_2} = 1 - \left(1 + \dfrac{\overrightarrow{k}_{1O}}{\underset{k_{1O}}{\overleftarrow{}}} C_{O_2}\right) \frac{\overleftarrow{k}_{2O}^{o} \exp\left(\frac{\alpha_{2O} n_{\alpha 2O} FE}{RT}\right) + \overrightarrow{k}_{3O}^{o} C_{H^+}^{3} \exp\left(-\frac{(1-\alpha_{3O}) n_{\alpha 3O} FE}{RT}\right)}{\overrightarrow{k}_{2O}^{o}\dfrac{\overrightarrow{k}_{1O}}{\underset{k_{1O}}{\overleftarrow{}}} C_{O_2} C_{H^+} \exp\left(-\frac{(1-\alpha_{2O}) n_{\alpha 2O} FE}{RT}\right) + \overleftarrow{k}_{2O}^{o}\left(1 + \dfrac{\overrightarrow{k}_{1O}}{\underset{k_{1O}}{\overleftarrow{}}} C_{O_2}\right)\exp\left(\frac{\alpha_{2O} n_{\alpha 2O} FE}{RT}\right)}$$

$$+ \overleftarrow{k}_{3O}^{o} C_{H_2O}^{2} \exp\left(\frac{\alpha_{3O} n_{\alpha 3O} FE}{RT}\right) + \overrightarrow{k}_{3O}^{o} C_{H^+}^{3}\left(1 + \dfrac{\overrightarrow{k}_{1O}}{\underset{k_{1O}}{\overleftarrow{}}} C_{O_2}\right)\exp\left(-\frac{(1 - \alpha_{3O}) n_{\alpha 3O} FE}{RT}\right) \tag{4.10}$$

where C_{O_2} is the concentration or partial pressure of oxygen, C_{H^+} is the concentration of protons in the aqueous phase, other parameters have the same meanings as defined above.

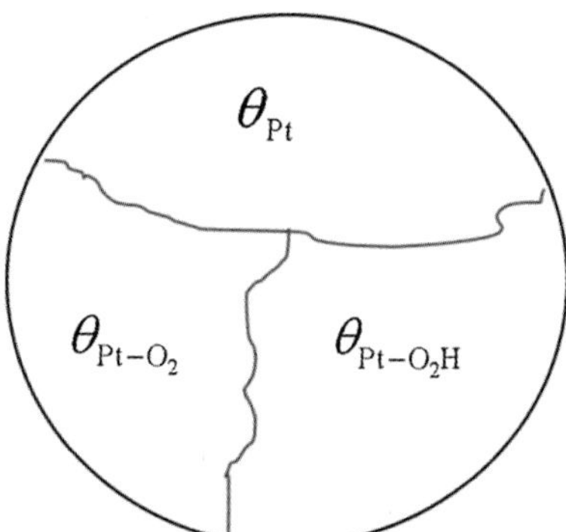

Figure 4.1 Schematic diagram of a platinum surface with adsorption of O$_2$, and O$_2$H. (For color version of this figure, the reader is referred to the online version of this book.)

If the electrode potential E in Eqns (4.9) and (4.10) is replaced by the equilibrium electrode potential $E^r_{O_2/H_2O}$, θ_{Pt} and θ_{Pt-O_2H} will become θ^o_{Pt} and $\theta^o_{Pt-O_2H}$, which can be expressed as Eqns (4.11) and (4.12), respectively:

$$\theta^o_{Pt-O_2} = \frac{\overrightarrow{k}_{10}}{\overleftarrow{k}_{10}} C_{O_2} \frac{\overleftarrow{k}^o_{20}\exp\left(\frac{\alpha_{20}n_{\alpha20}FE^r_{O_2/H_2O}}{RT}\right) + \overrightarrow{k}^o_{30}C^3_{H^+}\exp\left(-\frac{1-\alpha_{30}n_{\alpha30}FE^r_{O_2/H_2O}}{RT}\right)}{\overrightarrow{k}^o_{20}\frac{\overrightarrow{k}_{10}}{\overleftarrow{k}_{10}}C_{O_2}C_{H^+}\exp\left(-\frac{(1-\alpha_{20})n_{\alpha20}FE^r_{O_2/H_2O}}{RT}\right) + \overleftarrow{k}^o_{20}\left(1+\frac{\overrightarrow{k}_{10}}{\overleftarrow{k}_{10}}C_{O_2}\right)\exp\left(\frac{\alpha_{20}n_{\alpha20}FE^r_{O_2/H_2O}}{RT}\right) + \overleftarrow{k}^o_{30}C^2_{H_2O}\exp\left(\frac{\alpha_{30}n_{\alpha30}FE^r_{O_2/H_2O}}{RT}\right)}$$

$$+\overrightarrow{k}^o_{30}C^3_{H^+}\left(1+\frac{\overrightarrow{k}_{10}}{\overleftarrow{k}_{10}}C_{O_2}\right)\exp\left(-\frac{(1-\alpha_{30})n_{\alpha30}FE^r_{O_2/H_2O}}{RT}\right) \quad (4.11)$$

$$\theta^o_{Pt-O_2H} = 1-\left(1+\frac{\overrightarrow{k}_{10}}{\overleftarrow{k}_{10}}C_{O_2}\right)\frac{\overleftarrow{k}^o_{20}\exp\left(\frac{\alpha_{20}n_{\alpha20}FE^r_{O_2/H_2O}}{RT}\right) + \overrightarrow{k}^o_{30}C^3_{H^+}\exp\left(-\frac{1-\alpha_{30}n_{\alpha30}FE^r_{O_2/H_2O}}{RT}\right)}{\overrightarrow{k}^o_{20}\frac{\overrightarrow{k}_{10}}{\overleftarrow{k}_{10}}C_{O_2}C_{H^+}\exp\left(-\frac{(1-\alpha_{20})n_{\alpha20}FE^r_{O_2/H_2O}}{RT}\right) + \overleftarrow{k}^o_{20}\left(1+\frac{\overrightarrow{k}_{10}}{\overleftarrow{k}_{10}}C_{O_2}\right)\exp\left(\frac{\alpha_{20}n_{\alpha20}FE^r_{O_2/H_2O}}{RT}\right) + \overleftarrow{k}^o_{30}C^2_{H_2O}\exp\left(\frac{\alpha_{30}n_{\alpha30}FE^r_{O_2/H_2O}}{RT}\right)}$$

$$+\overrightarrow{k}^o_{30}C^3_{H^+}\left(1+\frac{\overrightarrow{k}_{10}}{\overleftarrow{k}_{10}}C_{O_2}\right)\exp\left(-\frac{(1-\alpha_{30})n_{\alpha30}FE^r_{O_2/H_2O}}{RT}\right) \quad (4.12)$$

According to Butler–Volmer equation presented in Chapter 2, the net ORR current density (i_{O_2/H_2O}) can be expressed as Eqn (4.13):

$$i_{O_2/H_2O} = n_O F\left[\overrightarrow{k}^o_{20}\theta_{Pt-O_2}C_{H^+}\exp\left(-\frac{(1-\alpha_{20})n_{\alpha20}FE}{RT}\right)\right.$$
$$\left.- \overleftarrow{k}^o_{20}\theta_{Pt-O_2H}\exp\left(\frac{\alpha_{20}n_{\alpha20}FE}{RT}\right)\right] \quad (4.13)$$

The exchange current density ($i^o_{O_2/H_2O}$) for the ORR can also be derived as Eqn (4.14):

$$i^o_{O_2/H_2O} = n_O F \overrightarrow{k}^o_{2O} \theta^o_{Pt-O_2} C_{H^+} \exp\left(-\frac{(1-\alpha_{2O})n_{\alpha 2O}FE^r_{O_2/H_2O}}{RT}\right)$$

$$= n_O F \overleftarrow{k}^o_{2O} \theta^o_{Pt-O_2H} \exp\left(\frac{\alpha_{2O}n_{\alpha 2O}FE^r_{O_2/H_2O}}{RT}\right) \quad (4.14)$$

where $\theta^o_{Pt-O_2H}$ and $\theta^o_{Pt-O_2}$ are the surface coverage of the Pt$-$O$_2$H and Pt$-$O$_2$ sites at the equilibrium cathode potential ($E^r_{O_2/H_2O}$). This exchange current density can also be expressed in another form. For example, from Eqn (4.13), the equilibrium cathode potential can be expressed as Eqn (4.15):

$$E^r_{O_2/H_2O} = \frac{2.303RT}{n_{\alpha 2O}F}\log\left(\frac{\overrightarrow{k}^o_{2O}}{\overleftarrow{k}^o_{2O}}\right) + \frac{2.303RT}{n_{\alpha 2O}F}\log\left(\frac{\theta^o_{Pt-O_2} C_{H^+}}{\theta^o_{Pt-O_2H}}\right) \quad (4.15)$$

Combining Eqns (4.14) and (4.15), an alternative expression of the exchange current density can be obtained:

$$i^o_{O_2/H_2O} = n_O F \left(\overrightarrow{k}^o_{2O}\right)\left(\overleftarrow{k}^o_{3O}\right)^{1-\alpha_{2O}}\left(\theta^o_{Pt-O_2}\right)^{\alpha_{2O}}(C_{H+})^{\alpha_{2O}}\left(\theta^o_{Pt-O_2H}\right)^{1-\alpha_{2O}}$$

$$(4.16)$$

As discussed in Chapter 2, the exchange current density is one of the most import parameters defining how fast the rate of

Table 4.2. ORR Exchange Current Densities on Various Electrode Materials or Catalysts

Electrode Material/ Catalyst	ORR Exchange Current Density, (A cm^{-2})	Electron Transfer Coefficiency	Electron Transfer Number in the Rate-Determining Step	Measurement Conditions	Reference
Pt	2.8×10^{-7}	0.48	—	Pt/Nafion interface at 30 °C	4
PtO/Pt	1.7×10^{-10}	0.46	—	Pt/Nafion interface at 30 °C	4
FePc	1.3×10^{-7}	—	—	pH 1.3 solution	5
PtFe/C	2.15×10^{-7}	0.55	1	0.5 M H$_2$SO$_4$ at 60 °C	6
PtW$_2$C/C	4.7×10^{-7}	0.45	2	0.5 M H$_2$SO$_4$ at 25 °C	7
	5.0×10^{-5}	0.47	1		
Ru$_x$Se$_y$	2.22×10^{-8}	0.52	1	0.5 M H$_2$SO$_4$ at 25 °C	8
Ru$_x$Fe$_y$Se$_z$	4.47×10^{-8}	0.51	1	0.5 M H$_2$SO$_4$ at 25 °C	9

electrochemical reaction would be. Table 4.2 shows the typical ORR exchange current densities catalyzed by Pt-based catalysts together with those catalyzed by other catalysts. It can be seen that Pt-based catalysts can give the highest values of the exchange current density among others.

To obtain the ORR exchange current density and the electron transfer coefficiency, Tafel equation is normally used for calculation based on the experiment data. If the second term at the right hand of Eqn (4.13) is much smaller than the first term, the Tafel equation for ORR can be obtained by combining Eqns (4.14) and (4.15):

$$
E = E^{r}_{O_2/H_2O} + \frac{2.303RT}{(1-\alpha_{2O})n_{\alpha2O}F}\log(i^{o}_{O_2/H_2O})
$$

$$
- \frac{2.303RT}{(1-\alpha_{2O})n_{\alpha2O}F}\log\left(i_{O_2/H_2O}\right) \tag{4.17}
$$

In experiments, the electrode potential (E) at different current densities (i_{O_2/H_2O}) can be measured. Then the plot of E vs $\log(i_{O_2/H_2O})$ can be obtained. From the Tafel slope $\left(= \frac{2.303RT}{(1-\alpha_{2O})n_{\alpha2O}F}\right)$, the kinetic parameter $(1-\alpha_{2O})n_{\alpha2O}$ can be calculated according to Eqn (4.17), and from the intercept $\left(= E^{r}_{O_2/H_2O} + \frac{2.303RT}{(1-\alpha_{2O})n_{\alpha2O}F}\log(i^{o}_{O_2/H_2O})\right)$, the exchange current density ($i^{o}_{O_2/H_2O}$) can be calculated if $E^{r}_{O_2/H_2O}$ is known. Therefore, two important electrode kinetic parameters, the exchange current density ($i^{o}_{O_2/H_2O}$), and the product of electron transfer and electron transfer coefficient ($(1-\alpha_{2O})n_{\alpha2O}$) (where $n_{\alpha2O}=1$ for ORR), can be experimentally measured based on Tafel Eqn (4.17).

Regarding the electron transfer coefficient (α_{2O}) of ORR, it can be expressed to be temperature dependent:

$$
\alpha_{2O} = 0.001678T \tag{4.18}
$$

According to the mass transfer theory presented in Chapter 2, the mass transfer effect on ORR kinetics can be reflected in the following equation:

$$
i_{O_2/H_2O} = i^{o}_{O_2/H_2O}\left[\left(1 - \frac{i_{O_2/H_2O}}{\overrightarrow{I}_{O_2/H_2O}}\right)\exp\left(-\frac{(1-\alpha_{2O})n_{\alpha2O}FE - E^{r}_{O_2/H_2O}}{RT}\right)\right.
$$

$$
\left. - \left(1 - \frac{i_{O_2/H_2O}}{\overleftarrow{I}_{O_2/H_2O}}\right)\exp\left(\frac{\alpha_{2O}n_{\alpha2O}FE - E^{r}_{O_2/H_2O}}{RT}\right)\right]
$$

$$
\tag{4.19}
$$

where $\overrightarrow{I}^{\,\mathrm{l}}_{\mathrm{O_2/H_2O}}$ and $\overleftarrow{I}^{\,\mathrm{l}}_{\mathrm{O_2/H_2O}}$ are the mass transfer limiting current densities for the ORR (forward) and the water oxidation reaction (backward), respectively. If the overpotential is large enough and the mass transfer is taken into consideration, Eqn (4.19) can be simplified as a Tafel form:

$$E = E^{\mathrm{r}}_{\mathrm{O_2/H_2O}} + \frac{2.303RT}{(1-\alpha_{2\mathrm{O}})n_{\alpha 2\mathrm{O}}F}\log(i^{\mathrm{o}}_{\mathrm{O_2/H_2O}})$$

$$- \frac{2.303RT}{(1-\alpha_{2\mathrm{O}})n_{\alpha 2\mathrm{O}}F}\log\left(\frac{i_{\mathrm{O_2/H_2O}}\,\overrightarrow{I}^{\,\mathrm{l}}_{\mathrm{O_2/H_2O}}}{\overrightarrow{I}^{\,\mathrm{l}}_{\mathrm{O_2/H_2O}} - i_{\mathrm{O_2/H_2O}}}\right) \tag{4.20}$$

It is worth pointing out that Eqns (4.13)−(4.20) are for the ORR on a smooth planar electrode or catalyst surface rather than in a porous matrix catalyst layer. It is expected that the situation in the catalyst layer may be more complicated than on the planar surface. However, it is believed that with modification using the apparent parameters as well as the real electrochemical active surface, the equations are still valid for quantitative treatment of experimental data.

4.3.2. ORR on Pt alloys

According to theoretical study (see Section 4.5), the catalytic activity of Pt toward ORR strongly depends on its O_2 adsorption energy, the dissociation energy of the O−O bond, and the binding energy of OH on the Pt surface. The electronic structure of the Pt catalyst (Pt d-band vacancy) and the Pt−Pt interatomic distance (geometric effect) can strongly affect these energies.[6] Theoretical calculations on O_2 and OH-binding energy on several metals have predicted that Pt should have the highest catalytic activity among other metals with the ORR activity of Pt > Pd > Ir > Rh, which is in agreement with the experimental results. Regarding the Pt alloy catalysts, calculations have also predicted that PtM (M = Fe, Co, Ni, etc.) alloys should have higher catalytic activity than pure Pt, which has again been proven by experiments.[10] The activity enhancement that occurs when Pt is alloyed with other metals can be explained by the change in electronic structure (the increased Pt d-band vacancy) and in geometric effect (Pt−Pt interatomic distance). Normally, alloying can cause a lattice contraction, leading to a more favorable Pt−Pt distance for the dissociative adsorption of O_2. The d-band vacancy can be increased after alloying, producing a strong metal−O_2 interaction then weakening the O−O bonds. For example, Figure 4.2 shows

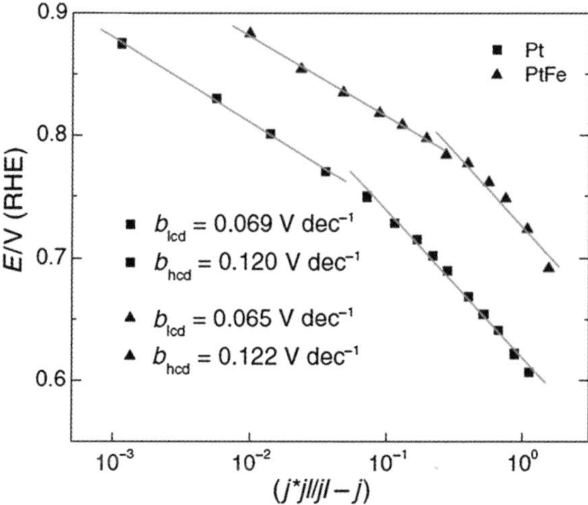

Figure 4.2 Tafel plots for ORR in 0.5 M H$_2$SO$_4$ on different catalysts. b_{lcd}: Tafel slope at low current density; b_{hcd}: Tafel slope at high current density.[6] (For color version of this figure, the reader is referred to the online version of this book.)

the Tafel plots of Pt and PtFe alloys for the ORR. The Tafel slopes for the two catalysts are the same. However, the exchange current density of oxygen reduction on PtFe is higher than on Pt. In the Tafel region of 120 mV dec^{-1}, the exchange current density for a Pt catalyst is 1.63×10^{-8} A cm^{-2}, while that for PtFe catalysts is 2.15×10^{-7} A cm^{-2}, indicating a ninefold enhancement.[6]

4.3.3. Difference in ORR Activity Catalyzed by Pt, Pt-Skeleton, and Pt-Skin Surfaces

It has been recognized that different Pt surfaces such as Pt-skeleton, and Pt-skin surfaces could give different ORR catalytic activities.[11–14] Figure 4.3 shows an example for Pt$_3$Co catalyst, from which two important observations can be made. First, the activity increases in the order Pt < Pt-skeleton < Pt-skin. Second, the kinetic parameters (e.g., the electron transfer number Figure 4.3(D)) and the Tafel slope (90–110 mV dec^{-1}), and the activation energy (20–25 KJ mol^{-1}) are almost identical for all three surfaces.[14] From these results and the corresponding literature,[14,15] it can be concluded that (1) the Pt-skin surface is the most active for the ORR; (2) the reaction mechanism on the Pt-skin and Pt-skeleton surfaces is the same as that proposed for polycrystalline Pt, i.e., a "series" 4e$^-$ reduction; (3) the differences in the catalytic activities strongly correlates with the availability of surface atoms to adsorb O$_2$ (Pt–O$_2$); and (4) the differences in the

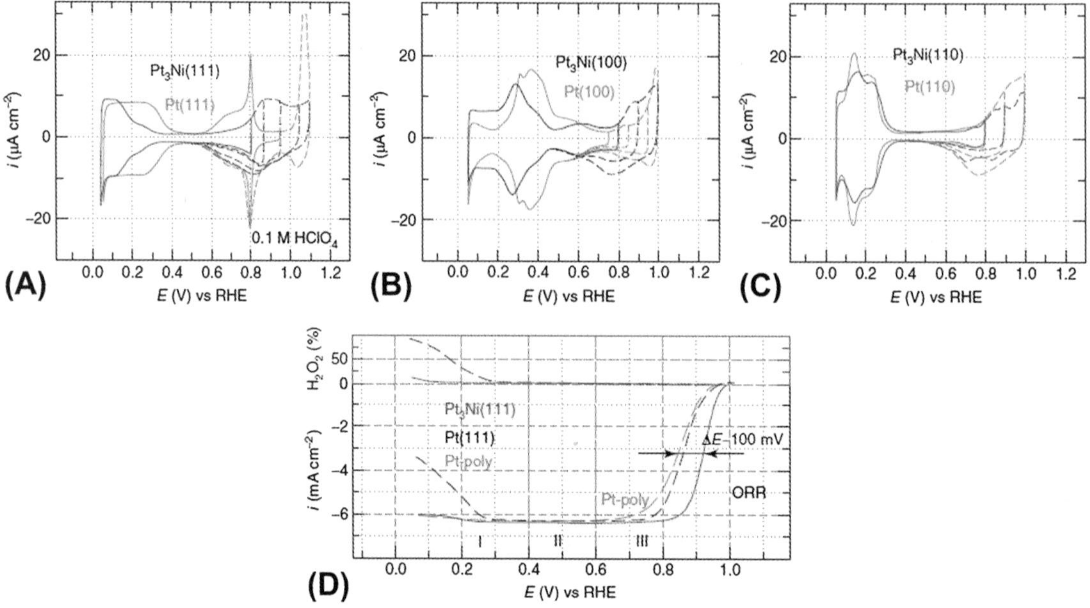

Figure 4.3 Cyclic voltammograms of Pt₃Ni(hkl) vs Pt(hkl) surfaces in 0.1 M HClO₄ at 298 K. (A) (111); (B) (100); and (C) (110) orientations. (D) Polarization curves obtained from rotational disk electrode (RDE) measurements in 0.1 M HClO₄ at 333 K at 1600 rpm with 50 mV s⁻¹: Pt₃Ni(111) (bold), Pt(111) (long dashed), and Pt-poly (dashed).[14] Reprinted from Science, 315, 493 (2007), Vojislav R. Stamenkovic., Improved oxygen reduction activity on Pt₃Ni(111) via increased surface site availability, 493–497, © 2007, with permission from Sciencemag.

binding energies of O_2 and the reaction intermediates[16] are rather small. In tandem with the analysis of other Pt₃M (M = Ni, Co, Fe, V, and Ti) surfaces, the results can provide an excellent basis for both understanding the factors that control the kinetics of the ORR on Pt₃M electrodes as well for rationalizing the volcano relationships between kinetic rates and the position of the d-band center, as shown in Figure 4.4.

The volcano curves shown in Figure 4.4 give the fundamental relationship between the experimentally determined surface electronic structure (the d-band center) and activity, establishing the electrocatalytic trends for ORR on Pt₃M bimetallic surfaces. This "volcano-type" behavior indicates that the maximum catalytic activity is governed by a balance between strong adsorption energies of reaction intermediates and low surface coverage by reaction products and spectator species. This, in turn, depends on the position of the metal d state relative to the Fermi level. Using these activity trends, a fundamental approach for the screening of new catalysts for the ORR, that is, by looking for surfaces that bind oxygen a little weaker than Pt or for surfaces with a downshift of the Pt d state relative to the Fermi level, can be established.

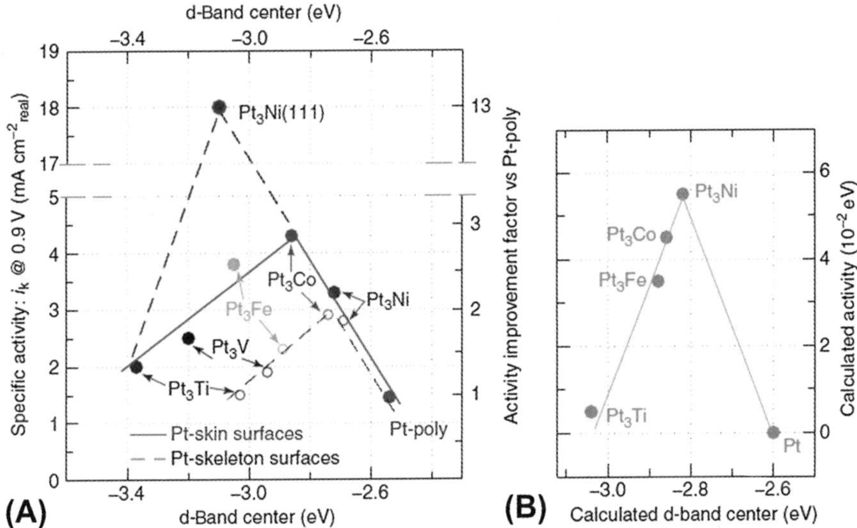

Figure 4.4 Trends in electrocatalysis: relationships between experimentally measured specific activities for the ORR on Pt$_3$M surfaces in 0.1 M HClO$_4$ at 0.9 V and 333 K vs the d-band center position for (A) the Pt-skin and Pt-skeleton surfaces. Results for each alloy were collected from five independent RDE measurements and error range is expressed by the size of circles. (B) Activities and values for d-band center position predicted from (Density functional theory) calculations for (111) oriented skin surfaces.[14] Reprinted from Science, 315, 493 (2007), Vojislav R. Stamenkovic., Improved oxygen reduction activity on Pt3Ni(111) via increased surface site availability, 493–497, © 2007, with permission from Sciencemag.

4.4. ORR on Carbon Materials

ORR on carbon is normally catalyzed by the carbon surface through strong interactions.[17] It was reported that the attachment (by preadsorption or chemical bonding) of various quinones to basal plane graphite could result in a large enhancement of the oxygen reduction current. In the literature, different mechanisms were proposed by different work, with one key difference between using graphite and glassy carbon (GC). Normally, the ORR is the first order in oxygen. Tafel slopes are typically around −120 mV dec^{-1} for graphite and −60 mV dec^{-1} for GC, although variations around these values were reported for different carbons, and even values as low as −30 mV dec^{-1} were also reported in strong alkali (i.e., 6 M KOH).[18]

Generally, the ORR mechanism on carbon materials is complicated, involving several elementary steps. The first step was thought to involve adsorption and electron transfer to form superoxide (O$_2^-$). Following this, either further reduction or disproportionation of superoxide would happen.[19] A step involving a transformation between two different forms of adsorbed superoxide was also proposed.[17] In alkaline solutions (pH > 10), two waves were seen in oxygen reduction polarization curves, both corresponding to oxygen reduction to peroxide, but most probably at different sites on the carbon surface.[20] Different electrolyte cations, D_2O vs H_2O, and selectively modification of different carbon surface groups, were all reported to shift the oxygen reduction voltammetric wave.[19]

It is commonly recognized that the carbon surface chemistry plays a complex role in the ORR, which should be taken into account when considering the ORR mechanism. Some of the important surface features of carbon are shown in Figure 4.5.

4.4.1. ORR on Graphite and Glassy Carbon

Two mechanisms have been proposed for carbon-catalyzed ORR in alkaline solution. On a GC electrode, the following reaction mechanism may be proposed based on the literature[17]:

$$GC + O_2 \leftrightarrow GC - O_{2(ads)} \tag{4-XIV}$$

$$GC - O_{2(ads)} + e^- \rightarrow GC - O_{2(ads)^*}^- \tag{4-XV}$$

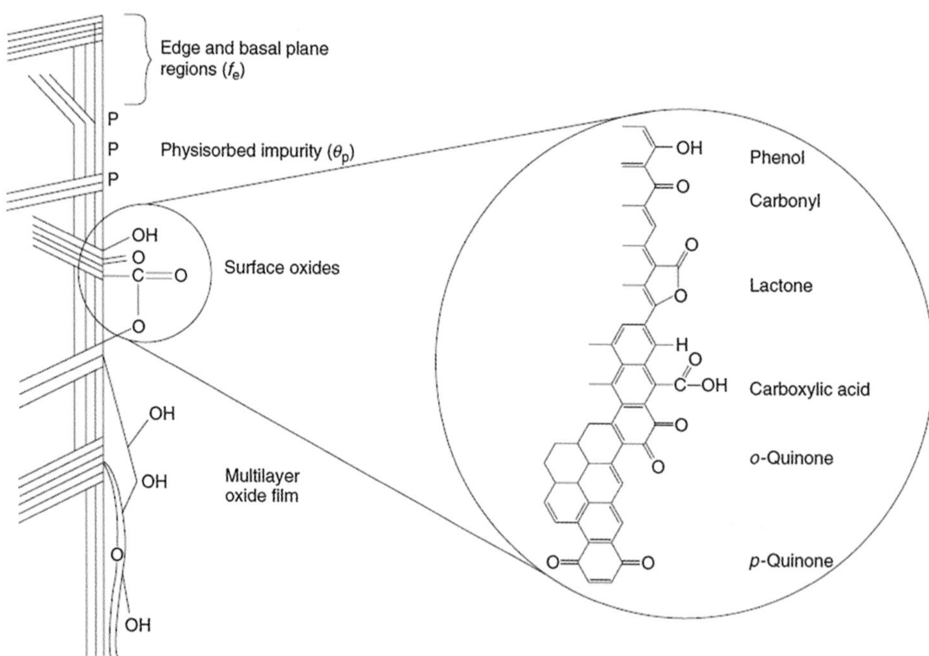

Figure 4.5 Schematic showing some of the carbon surface features of relevance to electrochemistry, and details of some different possible oxygen-containing surface functional groups. Reproduced from Ref. 21 by permission of Marcel Dekker, Inc.

$$GC - O_{2(ads)^*} \rightarrow GC - O_{2(ads)}^- \qquad (4\text{-}XVI)$$

$$GC - O_{2(ads)}^- + H_2O \rightarrow GC - O_2H_{ads} + OH^- \qquad (4\text{-}XVII)$$

$$GC - O_2H_{ads} + e^- \rightarrow GC - (O_2H)_{ads}^- \qquad (4\text{-}XVIII)$$

$$GC - (O_2H)_{ads}^- \rightarrow GC + HO_2^- \qquad (4\text{-}XIX)$$

The reactant and product in Reaction (4-XVI) are two different forms of the superoxide ion on the carbon surface. The left one is a relatively inert form adsorbed on an inert graphite site, and the right one is the same species, but migrating to an active site according to Reaction (4-XVI). It was confirmed that this reaction was the rate-determining step. However, Taylor et al.[16,22] found that the rate-determining step was dependent on pH. At pH > 10, Reaction (4-XVI) was the rate-determining step, and at pH < 10, Reaction (4-XV) was the rate-determining step. On pyrolytic graphite (PG) electrodes, the mechanism could also be proposed based on literature[17]:

$$PG + O_2 \leftrightarrow PG - O_{2(ads)} \qquad (4\text{-}XX)$$

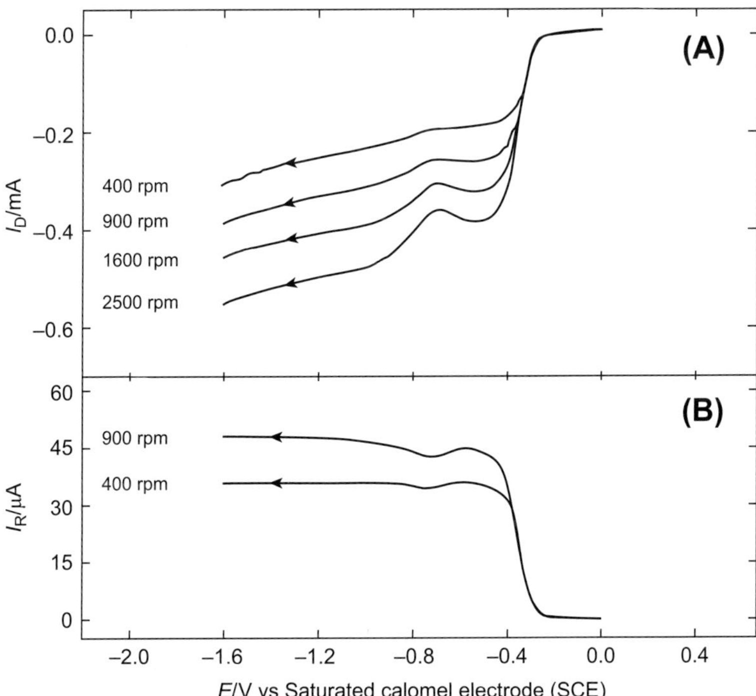

Figure 4.6 (A) Current–potential curves for O_2-saturated 1.0 M NaOH solution at a polished glassy carbon disk electrode, $T = 298$ K; potential scan rate, 5 mV s^{-1}; (B) response at a Pt ring held at 0.0 V vs SCE during the potential scan at the disk. Rotation rates are given on the figure.[23]

$$PG - O_{2(ads)} + e^- \rightarrow PG - O_{2(ads)}^- \qquad (4\text{-}XXI)$$

$$2PG - O_{2(ads)}^- + H_2O \rightarrow PG + O_2 + HO_2^- + OH^- \qquad (4\text{-}XXII)$$

In this mechanism, the rate-determining step was believed to be Reaction (4-XXI).

Normally, for ORR current–potential curves recorded on either a graphite or a GC electrode surface, two waves can be observed, both of which are attributed to the 2-electron transfer reduction of O_2, producing H_2O_2. Figure 4.6 shows the RRDE results for O_2 reduction on a GC electrode. On the disk electrode, the first steep wave appears at the formal potential $(E_{1/2})$ of $-0.34\,V$ vs SCE and the second one at $E_{1/2} = -0.78\,V$. The oxidation current at the ring shows that both waves correspond to H_2O_2 production. This two-electron reduction occurs at two different potential ranges, suggesting that there are two kinds of ORR active sites on GC. In a further experiment with 1 M NaOH solution containing 5 mM H_2O_2, no reduction wave was observed. This result confirmed that H_2O_2 could not be further reduced on a GC electrode.[23]

For ORR activity at a graphite electrode, it was found to be dependent on graphite planes. The activity on an edge plane was higher than that on a basal plane, as shown in Figure 4.7.[24] This reflects the difference in electrocatalytic activity of these two orientations. This catalytic activity difference is mainly due to the presence of more functional groups on an edge plane than on a basal plane.

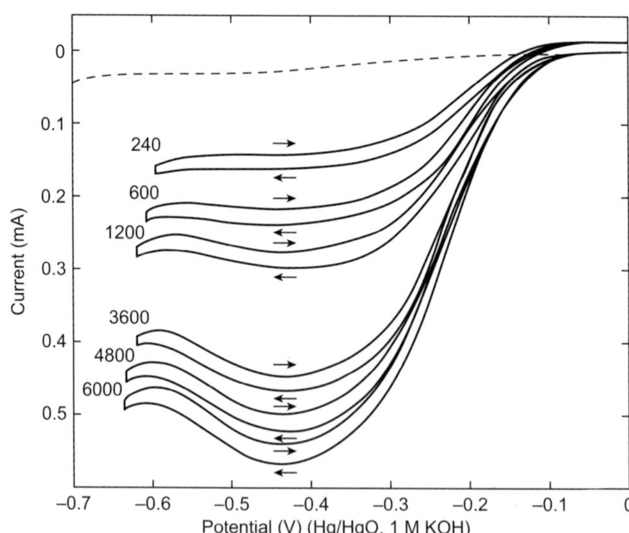

Figure 4.7 Current–potential curves for O_2-saturated 1 M KOH solution on high-pressure-annealed pyrolytic graphite: — edge plane, — basal plane. Reprinted from Ref. 24, with permission from Elsevier.

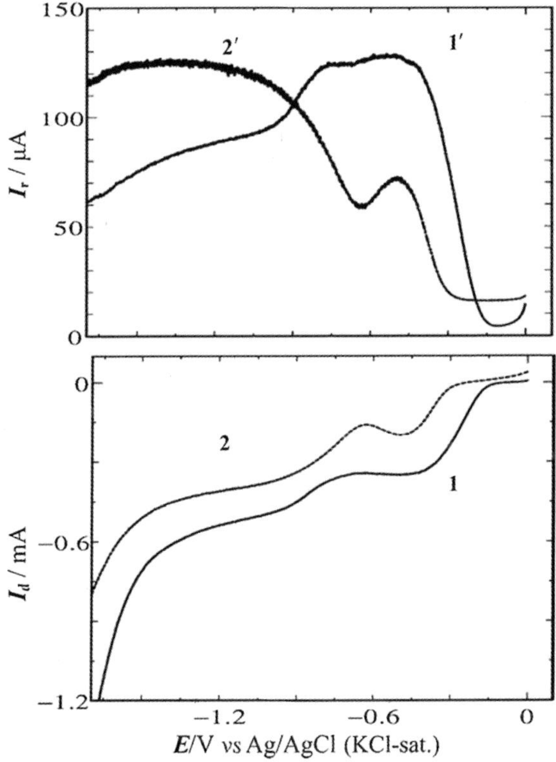

Figure 4.8 RRDE current–voltage curves for O$_2$ reduction at the PDDA (Phthalic acid Diethylene glycol Diacrylate)/MWCNTs/GC (curve 1) and bare GC (curve 2) disk electrodes in O$_2$-saturated 0.10 M KOH solution. Curves 1′ and 2′ represent the current for the oxidation of HO$_2^-$ produced at the corresponding disk electrodes. Potential scan rate, 10 mV s^{-1}; electrode rotating rate, 400 rpm. The Pt ring electrode was polarized at +0.50 V for the oxidation of HO$_2^-$.[26]

4.4.2. ORR on Carbon Nanotubes

Carbon nanotubes are a new kind of material that is expected to have extensive applications. Recently, ORR catalyzed by carbon nanotubes has been investigated by several groups,[25,26] which found that the catalyzed ORR was dependent on the preparation method for the carbon nanotube films. Zhang et al.[25] prepared multiwalled carbon nanotube (MWCNT) films by a dihexadecyl hydrogen phosphate method on a GC electrode and found that the films could catalyze O$_2$ reduction by two two-electron processes, producing OH$^-$. As shown in Figure 4.8, at a potential range of −0.4 to −0.8 V vs Ag/AgCl, the first reduction current plateau on the RDE curve can be observed, and attributed to a two-electron process producing H$_2$O$_2$; at potentials more negative than -0.9 V, a second plateau appeared, which could be attributed to the further reduction of H_2O_2, producing OH^-. Jurmann et al.[26] prepared some similar films using the same method on a highly oriented PG electrode, and reported that the first two-electron process could be observed at the potential range of -0.4 to -0.6 V vs SCE, while the second plateau was not clearly discerned. A mixture process of two- and four-electron reduction could be observed at more negative potentials. They also found that on MWCNT films prepared by the poly(diallyldimethylammonium chloride) method, the electron transfer number of ORR was in the range of 3–3.5 at the potential range of -0.4 to -1.2 V Figure 4.8 shows the results, together with those obtained on a bare GC electrode for comparison. It can be seen that on a GC electrode, the ring current increases with increasing disk current in the potential range of -0.4 to -1.2 V. However, on an MWCNT-modified electrode, the ring current decreases with increasing disk current in the similar potential range, indicating less H_2O_2 is produced.

4.4.3. ORR on Heteroatom-Doped Carbons

Doping carbon with heteroatoms such as nitrogen can change the carbon's properties. For example, nitrogen (N)-doped carbon shows a high oxidation resistance capability and higher catalytic activity toward ORR. Maldonado and Stevenson[27] found that N-doped carbon fiber showed an improved catalytic activity by shifting the ORR potential up by 70 mV (Figure 4.9), and the electron transfer number of ORR catalyzed by N-doped carbon fiber was close to four. They discussed that N-doped carbon can catalyze not only O_2 reduction reaction but also H_2O_2 decomposition.[27]

The active sites in N-doped carbon are carbon atoms adjacent to the N atom. Using a cluster model, Sidik et al.[28] did theoretical calculations and found that the carbon radical sites adjacent to the substitutional N showed strong bonding ability to adsorbed OOH, favoring the production of H_2O_2,[28] which could improve its catalytic activity.

4.4.4. ORR on Pretreated Carbon Surface

Pretreatment of carbon materials can significantly improve their catalytic activity toward ORR and change their electrochemical behavior. A variety of treatment methods have been used, including polishing the electrode, radio frequency plasma treatment, heating at low pressures, in-situ laser irradiation,

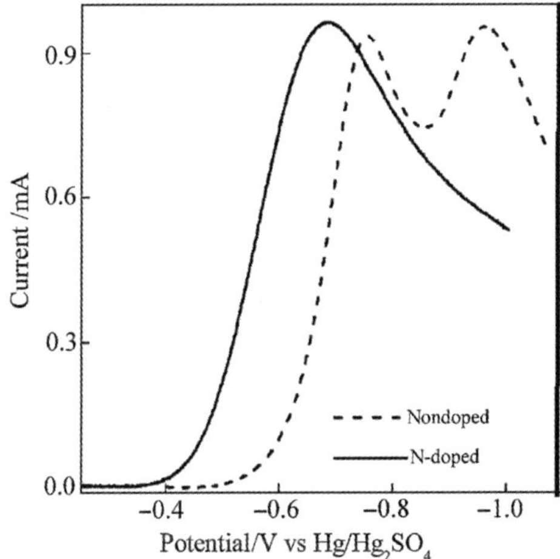

Figure 4.9 Background subtracted voltammetric responses of a nondoped CNF (Carbon Nanofiber) electrode (dashed line) and an N-doped CNF electrode (solid line) immersed in an O₂-saturated 1 M KNO₃ solution. Scan rate = 0.1 V s^{-1}.[27]

vacuum heat treatment, chemical oxidation, as well as electrochemical oxidation.[29–32]

It is not fully understood why pretreatment of carbon surfaces can improve their catalytic activity. Pretreatment does create more surface functional groups, and exposes fresh carbon edges, microparticles, and defects.[30,31,33] Jia et al.[32] found that pretreatment of carbon supports and Pt-loaded carbon supports could result in a decrease in ionic resistance, indicating that surface functional groups such as carboxylic acid were generated. Pretreatment could also increase the surface area of carbon by generating porous surface films.[34] Sullivan et al.[34] found that surface oxidation of GC at 1.95 V vs SCE led to the formation of porous carbon film, and an increase in film thickness with linearly increasing oxidation time. Overall, pretreatment of carbon can change its surface state. The surface groups or changed surface state facilitate ORR by acting as an electron transfer mediator or by increasing adsorption sites for O_2^- species, which enhances the kinetics of ORR. Figure 4.10 shows the current–potential curves of an unoxidized GC electrode that had then been oxidized at 2 V vs RHE in 0.5 M H_2SO_4 at different times. It can be seen that increasing oxidization time can result in a positive shift of ORR onset potential and an increase in ORR reduction current as well.

Pretreated carbon can also affect the activity of other catalysts supported on it. For example, pretreatment of a carbon support

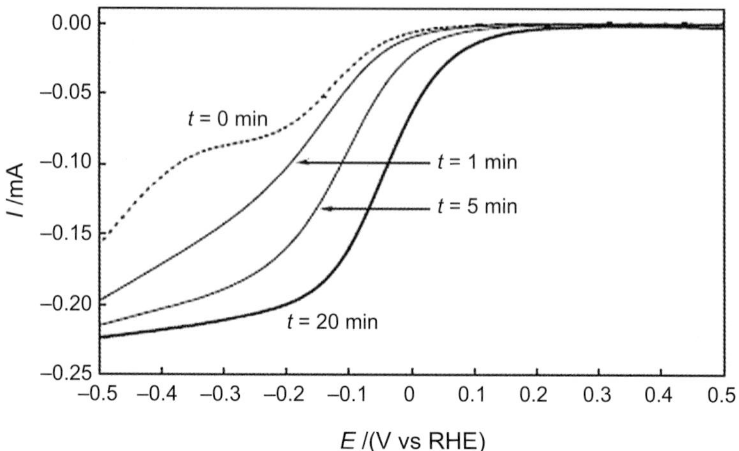

Figure 4.10 Current—potential curves for O$_2$ reduction reaction at a bare glassy carbon (GC) RDE in O$_2$-saturated 0.5 M H$_2$SO$_4$. GC surface oxidation time: 0 min (dashed line); 1 min (thin line); 5 min (thick line); 20 min (very thick line). Scan rate, 10 mV s^{-1}; electrode rotation speed, 2000 rpm. Reprinted from Ref. 31, with permission from Elsevier.

followed by deposition of Pt on it or treatment of carbon-supported Pt could result in significant increases in the ORR catalytic activity of Pt.[32] Jia et al.[32] reported that an untreated catalyst (Pt deposited on untreated carbon supports) gave a kinetic current of 0.3 ± 0.1 mA at 0.6 V vs SCE, while Pt deposited on pretreated carbon showed a kinetic current of 1.6 ± 0.7 mA, and posttreated carbon-supported Pt catalyst (refluxing carbon-supported Pt catalyst in nitric acid) could give a kinetic current of 7.7 ± 1.7 mA.

4.4.5. ORR on Graphene

Graphene, a monolayer of sp^2 bonded carbon atoms in a honeycomb lattice, has created a surge in research activities during the last 6—7 years owing to its ballistic transport, chemical inertness, high thermal conductivity, optical transmittance, as well as super hydrophobicity at nanometer scale. Studies showed that graphene containing nitrogen atom could improve ORR activity up to three times higher than platinum.[35] It was believed that the superior catalytic capabilities of these N-doped carbon nanomaterials should be directly related to their unique nano-structure. Among the nitrogen-doped structures, pyrrolic and pyridinic nitrogen were reported to play an important role in the enhanced ORR activity in both alkaline[35–38] and acidic[39,40] solutions. It was predicted that the Stone—Wales defects might be able to alter the electronic properties (band structure and density of states) of graphene,[41–43] modify its chemical reactivity toward adsorbates, and result in impact on its catalytic properties.

Although some theoretical work was done on ORR pathways on N-doped graphene,[44–46] the role of materials structures, including N distribution and defects played in ORR, still remained unclear.

4.5. ORR on Macrocyclic Transition Metal Complexes

Transition metal macrocyclic complexes can catalyze the O_2 reduction reaction through a two-electron or four-electron transfer pathway to produce either H_2O_2 or H_2O. Sometimes, they can also catalyze ORR through a mixed pathway of two- and four-electron transfer reduction. In rare cases, transition metal macrocyclic complexes can also catalyze a one-electron O_2 reduction, producing superoxide ions. Normally, the catalytic reaction mechanism proposed is a modified "redox catalysis" procedure.[47,48] In the first step, the adduct between oxygen and the metal ion center of the macrocyclic compound is formed, followed by an intraadduct electron transfer from the metal ion to the oxygen. The addition of protons from the electrolyte, together with the electron transfer, then produces H_2O_2. The H_2O_2 is either the final product or can be further reduced to produce water, depending on the individual transition metal macrocyclic compounds used. The mechanism can be summarized as follows[49]:

$$[LMe^{II}] + O_2 \leftrightarrow LMe^{\delta+}\ldots\ldots O_2^{\delta}{}_-] \tag{4-XXIII}$$

$$[LMe^{\delta+}\cdots\cdots O_2^{\delta}{}_-] + H^+ \rightarrow [LMe^{III}\cdots\cdots O_2H]^+ \tag{4-XXIV}$$

$$[LMe^{III}\cdots\cdots O_2H]^+ + H^+ + 2e^- \rightarrow [LMe^{II}] + H_2O_2 \tag{4-XXV}$$

$$H_2O_2 + 2e^- + 2H^+ \rightarrow 2H_2O \tag{4-XXVI}$$

$$H_2O_2 \rightarrow H_2O + 1/2O_2 \tag{4-XXVII}$$

where L represents the ligand and Me is the metal center.

The catalytic activity of transition metal macrocyclic compounds toward ORR strongly depends on the individual transition metal center and the macrocyclic ligand, as well as the size of the π-electron system. Compounds studied previously include a variety of metal centers and ligands. The transition metals used in macrocyclic catalysts include Fe, Co, Ni, and Cu, and the macrocyclic ligands include chelating atoms N_4, N_2O_2, N_2S_2, O_4, and S_4. The conjugate π-electron derivative compounds usually include metal phthalocyanine and porphyrin, as well as their derivatives.[47,48]

The activity of these compounds changes with respect to the central metal ions in the following order: Fe > Co > Ni > Cu. For a metal center, the chelating atoms of the macrocyclic ring can also change the ORR activity. For example, the active sequence of Fe complexes is as follows: $N_4 > N_2O_2 > N_2S_2 > O_4 \approx S_4$ (inactive); for Co centers: $N_2O_2 > N_4 > N_2S_2 > O_4 \approx S_4$; for Cu centers: $N_4 > O_4 > N_2O_2 > N_2S_1 > S_4$ (inactive); and for Ni centers: $O_4 > N_2O_2 > N_2S_2 > N_4$. With respect to these active sequences, recent research has mainly focused on Fe and Co centers, and for macrocyclic rings, on the N_4 system. In this section, the discussion will focus on the Fe-N_4 and Co-N_4 systems.

M-N_4 complexes can strongly and irreversibly adsorb on a graphite electrode surface to form a monolayer or multilayers of ORR catalysts. This adsorption can create a well-defined electrode surface, and then provide a theoretical treatable situation for fundamental understanding of the catalyst activity and mechanism.[50–52]

For Fe-based N_4 catalysts (Fe-N_4), the ORR normally takes a four-electron transfer pathway to produce H_2O, as reported by Zagal et al.[50] Shi et al.[51] found that oxygen reduction catalyzed by Fe-tetrakis(4-*N*-methylpyridyl) porphyrin could produce a mixture of H_2O and H_2O_2. With respect to the ligand effect, Shigehara et al.[52] found that ligands had a strong effect on the ORR activity catalyzed by Fe-porphyrins; both Fe-meso-tetraphenylporphine and Fe-protoporphyrin IX could catalyze a four-electron oxygen reduction, while Fe-meso-tetra(3-pyridyl)porphine could only catalyze a two-electron oxygen reduction.[51,52] Nonetheless, in most cases, Fe-N_4 can catalyze four-electron oxygen reduction and the product is water.

In general, mononuclear Co-N_4 complexes can only catalyze two-electron O_2 reduction reaction. Zagal et al.[50] studied Co- and Fe-tetrasulfonate phthalocyanines and found that the Co complex only catalyzed a two-electron oxygen reduction process. However, bi-nuclear Co-N_4 complexes or Co-N_4 dimers have some different behaviors from those of mononuclear Co-N_4 complexes. They can catalyze not only two-electron but also four-electron O_2 reduction reactions. A planar bi-cobalt complex such as $(Co_2TAPH)^{4+}$, as shown in Figure 4.11(A), can catalyze a four-electron reduction of O_2 in alkaline solutions. This catalytic activity was found to be due to the interaction between O_2 and the Co metal centers. This interaction could effectively weaken the O—O bond, rendering it easily being broken. Furthermore, face-to-face di-Co-N_4 complexes also favor a four-electron transfer pathway, depending on the Co—Co distance in the molecules

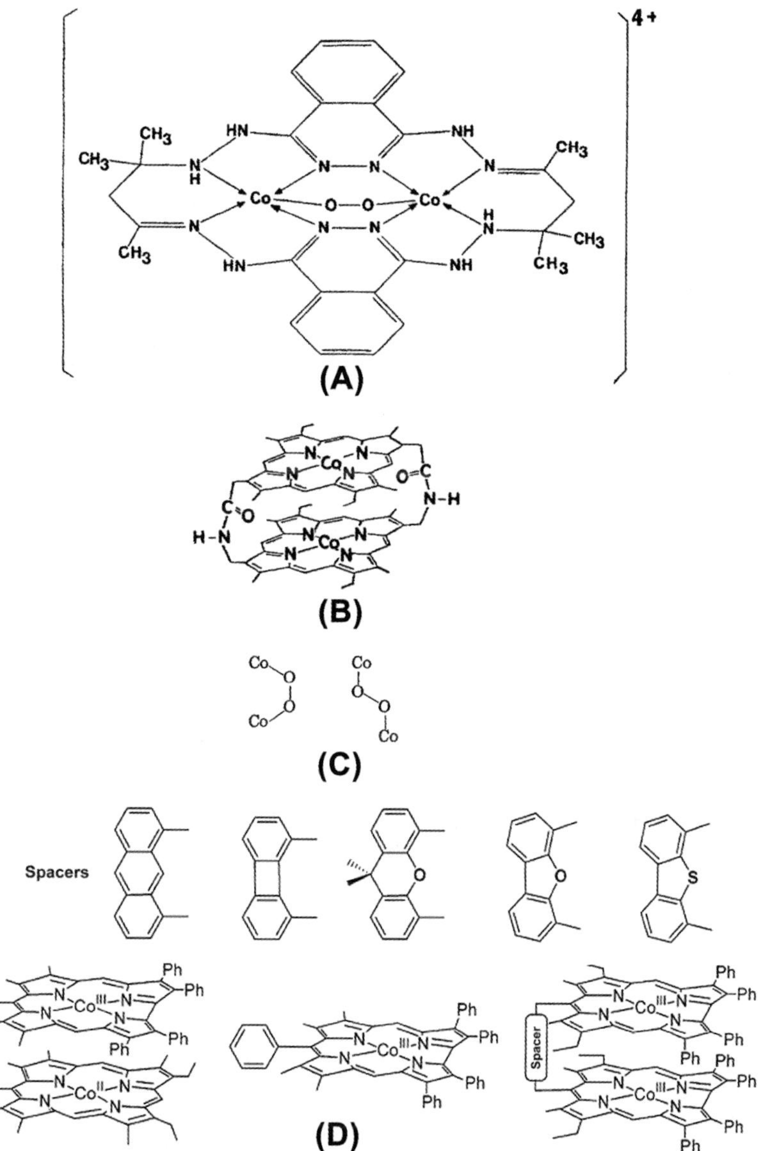

Figure 4.11 (A) Dimetal complex $(Co_2TAPH)_4^+(NO_3)_4$. (TAPH = 6,7,8,9,12,19, 20,21,22,25-decahydro-8,8,10,21,21,23,-hexamethyl-5,26:13,18-*bis*(azo)-dibez(1,2,6,7,12,13,17,18)oxaazacyclodocosine)[17]; (B) Face-to-face Co—Co 4-porphyrin. *(Reprinted from Ref. 17, with permission from Elsevier.)* (C) *Cis* and *trans* Co-O-O-Co facial configuration[17]; and (D) Co-corrole compounds for O_2 reduction reaction.[53]

(Figure 4.11(B)). When the Co—Co distance is around 4 Å, the O—O bridge between the Co—Co centers can be formed as shown in Figure 4.11(C), resulting in a four-electron transfer of ORR. If the Co—Co distance is larger or smaller, only the two-electron transfer pathway to peroxide is favored. This can be explained by the formation of a *cis* or *trans* configuration (Figure 4.11(C)). The *cis* configuration favors the four-electron ORR to H_2O, while the *trans* configuration does not.[54]

Recently, Kadish et al.[53] synthesized three series of Co corroles (shown in Figure 4.11(D)) and investigated their catalytic activity toward the O_2 reduction reaction.[53] The mixed valent Co(II)/Co(III) complexes, $(PCY)Co_2$, and the biscorrole complexes, $(BCY)Co_2$, both contain two Co(III) ions in their air-stable forms. It was found that all these complexes could catalyze the direct four-electron pathway for O_2 reduction to H_2O in aqueous acidic electrolyte. The most efficient catalysis process was observed when the complex had an anthracene spacer. The four-electron transfer pathway was further confirmed by RRDE measurement, in which only a relatively small amount of hydrogen peroxide was detected at the ring electrode in the vicinity of $E_{1/2}$: 0.47 V vs SCE for $(PCA)Co_2$ and 0.39 V for $(BCA)Co_2$. The cobalt(III) mono-corrole, $(Me_4Ph_5Cor)Co$, could also catalyze ORR at $E_{1/2} = 0.38$ V, with the final products being an approximate 50% mixture of H_2O_2 and H_2O.

It has been recognized that the high-temperature pyrolysis of M—Nx/C catalysts in an inert environment can result in heat annealment and active site formation, leading to catalytic materials with both higher ORR activity and stability when compared to their nonpyrolyzed counterparts. This occurrence was reviewed recently.[55,56] Figure 4.12 shows that different pyrolysis temperatures in inert environments have varying effects on both ORR activity and stability of the synthesized catalysts. Normally, M—Nx/C catalysts displaying the highest activity toward ORR can be synthesized at pyrolysis temperatures between 500 and 800 °C in inert environments (N_2 or Ar). In this temperature range, it has been proposed that a thermal annealment process occurs between the carbon, nitrogen, and metal groups, creating some catalytically active sites responsible for the ORR activity observed through electrochemical measurements. This observed ORR activity is commonly attributed to the presence of coordinated metal—nitrogen chelate complexes or their remaining fragments well dispersed and annealed on the surface of the carbon support material. As previously discussed, the exact nature of the catalytically active sites formed during pyrolysis is a subject of extensive debate. At temperatures approaching and exceeding

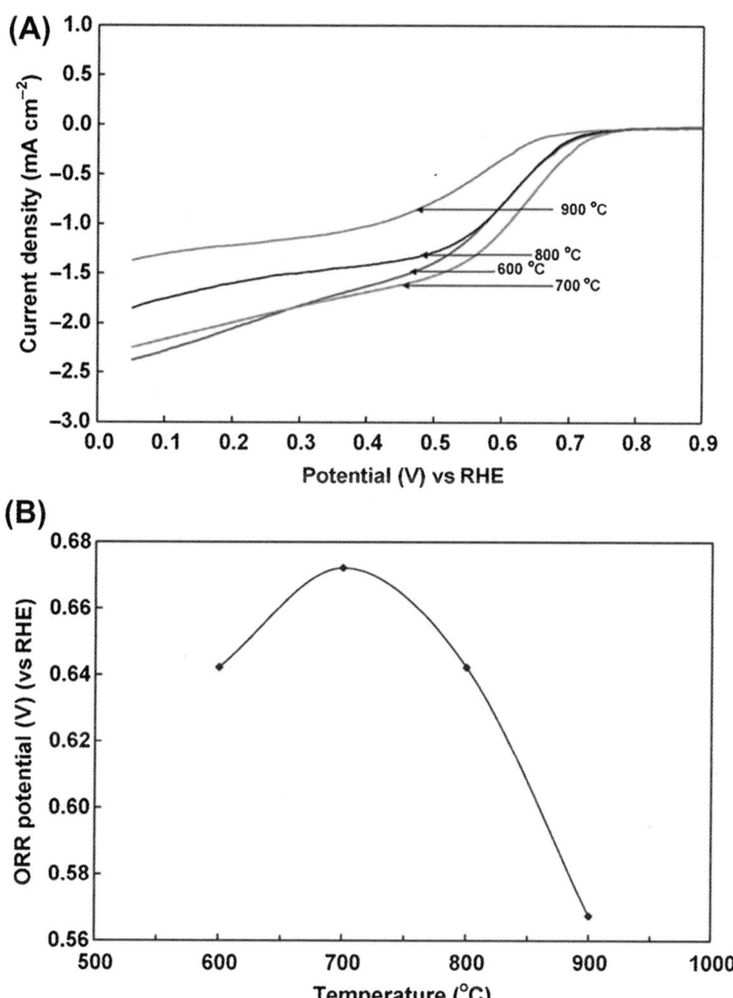

Figure 4.12 (A) ORR polarization curves obtained at an electrode rotation of 400 rpm in 0.5 M H_2SO_4 for a pyrolyzed CoTPTZ/C catalyst and (B) pyrolysis temperature effect on the ORR potential obtained at a current density of 0.5 mA cm^{-2}. (For color version of this figure, the reader is referred to the online version of this book.) Reprinted from Ref. 57, with permission from Elsevier.

800 °C, the metal−nitrogen chelate bonds are known to decompose and a reduction in ORR activity was commonly observed. Various reports have attributed this decrease in ORR activity to either (1) formation and growth of relatively inert metallic or metal oxide/carbide particles at these excessive temperatures, or (2) a decrease in the surface nitrogen content of the pyrolyzed M−Nx/C catalysts.

4.6. Fundamental Understanding of ORR Mechanisms

To help understand the ORR mechanism, some theoretical approaches are necessary. The molecular orbital diagram of oxygen molecule is shown in Figure 4.13.[58] According to Hund's rule, in the ground state, O_2 possesses two unpaired electrons located in a doubly degenerate π^* antibonding orbital. This corresponds to a triplet state. The bonding orbitals of oxygen molecule can be ascribed to the $3\sigma_g$ orbital with two electrons, the doubly degenerate $1\pi_u$ and $1\pi_g^*$ orbitals, where the $1\pi_u$ orbital has double occupancy while the $1\pi_g^*$ orbital has single occupancy. The resulting bond order is two. When O_2 undergoes reduction, the electrons added occupy antibonding orbitals, decreasing the bond order of O−O. This increases the O−O bond distance and at the same time the vibrational frequency decreases. The excess of bonding over nonbonding electrons in the diagram of Figure 4.13 is four. This explains the high stability of the O_2 molecule and its relatively low reactivity, in spite of its high oxidizing power.[59] The plausible 1:1 and 2:1 metal-dioxygen complex structures are given in Figure 4.14. Geometries 1 and 2 have been shown to give similar bonding patterns: each exhibits one σ and one π interaction. The Griffith's model[60] (1) involves a side-on interaction of O_2 with the metal. The bonding can be viewed as arising from two contributions: σ type of bond formed by the overlap between a mainly π orbital of oxygen and d_z^2 (and s)

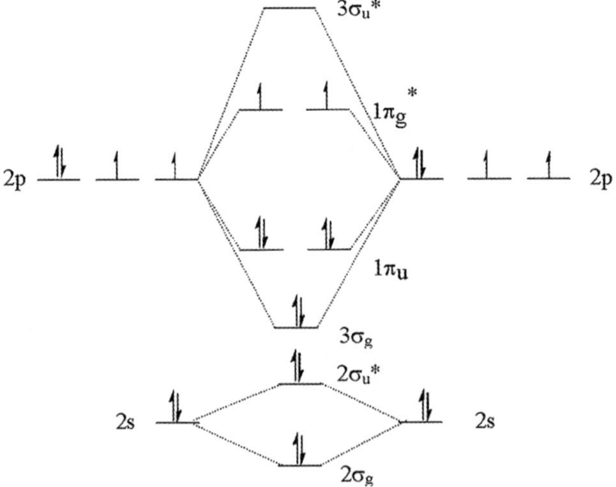

Figure 4.13 Molecular orbital diagram of oxygen molecule. Reproduced from Ref. 58.

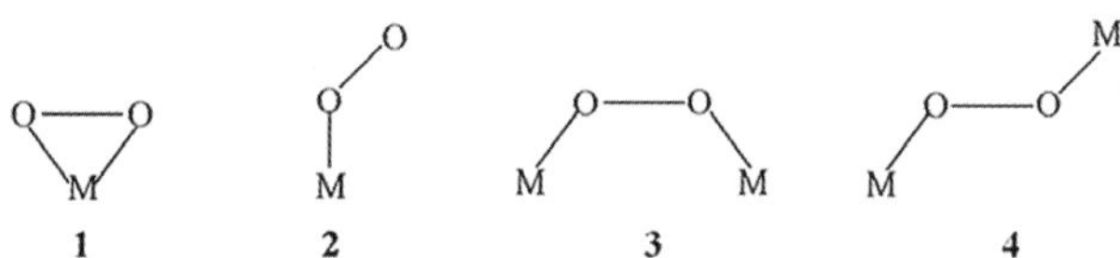

Figure 4.14 Possible configurations of dioxygen interaction with a metal in a complex.

orbital on the metal and a π backbond interaction between the metal dπ orbitals and the partially occupied π^* antibonding orbital on O_2; The Pauling model[61] (2) involves an end-on interaction of O_2 with the metal, a σ-bond is formed by the donation of electron density from the σ-rich orbital of dioxygen to an acceptor orbital d_z^2 on the metal, whereas the π interaction is produced between the metal d-orbitals (d_{xz}, d_{yz}) and π^* orbital of dioxygen, with charge transfer from the metal to the O_2 molecule. It indicates that the preference of geometries 1 and 2 is determined by the relative donating abilities of the filled π and σ orbitals of dioxygen molecule, respectively.

In 2:1 complexes of metal-dioxygen, the bonding arises from the interaction between the d-orbitals on the metal with π^* and π orbitals combination on O_2.[62] It means that the singlet or triplet nature of dioxygen orbitals determines the bridge- or trans mode of interaction of dioxygen with metal. The bridge interaction was proposed by Yeager[63] and it is likely to occur on noble metals such as Pt, face-to-face porphyrins[64] where O_2 is reduced to water with little or no peroxide formed. A trans configuration is likely to occur on metal porphyrins, metal phthalocyanines, and dimeric metal complexes.[65,66] Griffith type of interaction could also lead to the rupture of the O$-$O bond. Some organometallic complexes[67] are known to form adducts with O_2 with a side-on interaction. However, these compounds do not exhibit any catalytic activity for O_2 reduction, probably because the interaction leads to a stable adduct. Depending on the mode of adsorption of oxygen molecule on metal surface, different mechanistic steps will occur.[22,68–70] The ORR pathways in both acidic and alkaline medium are given in Table 4.3.

Based on all these points above, various authors proposed different schemes for the ORR pathway.[71–73] Among the schemes proposed, a modified scheme proposed by Wroblowa et al.[73] seems to be the interesting one to describe the complicated reaction pathway for the reduction of O_2 at the metal surface (refer to Figure 4.15). In this scheme, at high potentials, $k_1 > k_2$, the portion of direct reduction of oxygen to water will be more than the two-electron reduction to peroxide; at intermediate potentials, the decrease in k_1/k_2 ratio will indicate an increase in the

Table 4.3. Mechanistic ORR Pathways in Acidic and Basic Media

Model of Interaction	ORR Pathway in Acid Medium	ORR Pathway in Alkali Medium
Bridge (or) trans	$O_2 + 2e^- + 2H^+ \rightarrow 2OH_{ads}$ $2OH_{ads} + 2H^+ + 2e^- \rightarrow 2H_2O$ Overall direct reaction: $O_2 + 4H^+ + 4e^- \rightarrow 2H_2O$; $E^0_{O_2/H_2O} = 1.23$ V vs NHE	$O_2 + 2e^- + 2H_2O \rightarrow 2OH_{ads} + 2OH^-$ $2OH_{ads} + 2e^- \rightarrow 2OH^-$ Overall direct reaction: $O_2 + 2H_2O + 4e^- \rightarrow 4OH^-$ $E^0_{O_2/OH^-} = 0.401$ V vs NHE
End-on	$O_2 + e^- + H^+ \rightarrow HO_{2,ads}$ $HO_{2,ads} + e^- + H^+ \rightarrow H_2O$ Overall direct reaction: $O_2 + 2H^+ + 2e^- \rightarrow H_2O_2$; $E^0_{O_2/H_2O} = 0.682$ V vs NHE With $H_2O_2 + 2H^+ + 2e^- \rightarrow 2H_2O$; $E^0_{H_2O_2/H_2O} = 1.77$ V vs NHE	$O_2 + H_2O + e^- \rightarrow HO_{2,ads} + OH^-$ $HO_{2,ads} + e^- \rightarrow HO_2^-$ Overall direct reaction: $O_2 + H_2O + 2e^- \rightarrow HO_2^- + OH^-$; $E^0_{O_2/HO_2^-} = 0.076$ V vs NHE With $HO_2^- + H_2O + 2e^- \rightarrow 3OH^-$; $E^0_{HO_2^-/OH^-} = 0.88$ V vs NHE

portion of two-electron reduction to peroxide; and at lower potentials k_1/k_2 becomes lower than 1. However, if k_3 increases, a further reduction of peroxide to water will happen before it escapes into solution.

On most of the electrocatalysts, oxygen reduction takes place by the formation of high-energy intermediate, peroxide, which is then further reduced to H_2O. This is probably a consequence of the high stability of the O–O bond, which has a dissociation energy of 494 kJ mol^{-1}. In contrast, the dissociation energy of the O–O bond in H_2O_2 is only 146 kJ mol^{-1}.[74] In order to obtain maximum efficiency and to avoid corrosion of carbon supports and other materials by peroxide, it is desired to achieve a

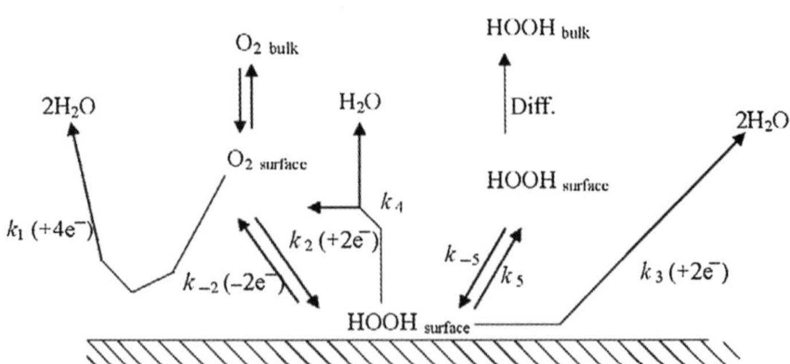

Figure 4.15 Schematic presentation of ORR pathway suggested by Wroblowa et al.[39]

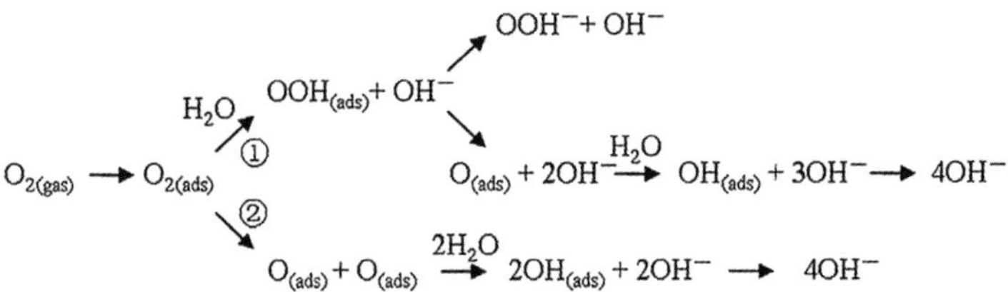

Figure 4.16 Reaction scheme of ORR on N-graphene in alkaline solution, where ① presents an associative mechanism and ② a dissociative mechanism.[46]

four-electron reduction. The two-electron reduction is of practical interest for the production of hydrogen peroxide. Therefore, exploring suitable electrocatalysts that can promote the direct four-electron reduction of oxygen molecule is an important task in both fuel cells and metal−air batteries.

Regarding the theoretical study of ORR on graphene electrode, some modeling and calculation have been carried out.[35,44,46] For example, Yu et al.[46] considered the possible reaction pathways on graphene as illustrated in Figure 4.16. The free energy profiles are shown in Figure 4.17(A). The first step is the adsorption of O_2 on Site three followed by two paths; $O_{2(ads)}$ can accept a proton from H_2O to form $OOH_{(ads)}$ or directly dissociate to yield two $O_{(ads)}$, i.e., the associative or dissociative mechanisms, respectively. The dissociation barrier of $O_{2(ads)}$ in the presence of water is 1.56 eV, which is remarkably higher than that of the associative reduction step to form $OOH_{(ads)}$ with an effective barrier of 0.51 eV, as can be seen from Figure 4.17(A). A similar result is obtained on S2 (Figure 4.17(B)). It is clear that at the working temperature of alkaline fuel cell (∼ 350 K), the barrier of 1.56 eV is too high for $O_{2(ads)}$ dissociation to occur with a reasonable rate on the N-graphene surface. Therefore, the more energetically favored associative mechanism should be dominating in the reduction of O_2. The barrier of the $O_{2(ads)}$ hydrogenation to form $OOH_{(ads)}$ is 0.42 eV from approach (2) and 0.43 eV from approach (3), which is slightly higher than that reported previously on Pt(111) (∼ 0.3 eV) in the presence of water,[75] but still moderate for low-temperature reactions. Following hydrogenation, $OOH_{(ads)}$ may desorb from the surface to give OOH^-, or alternatively the O−O bond of $OOH_{(ads)}$ may break to yield $O_{(ads)}$ and OH^-. Calculation shows that the $OOH_{(ads)}$ desorption is energetically unfavorable with an energy loss of 0.22 eV (endothermic), as illustrated in Figure 4.17(A). In contrast, the O−O bond scission is strongly exothermic with overall barriers of 0.56 and 0.46 eV. This is in agreement with experimental results that only traces of peroxide are produced.[35]

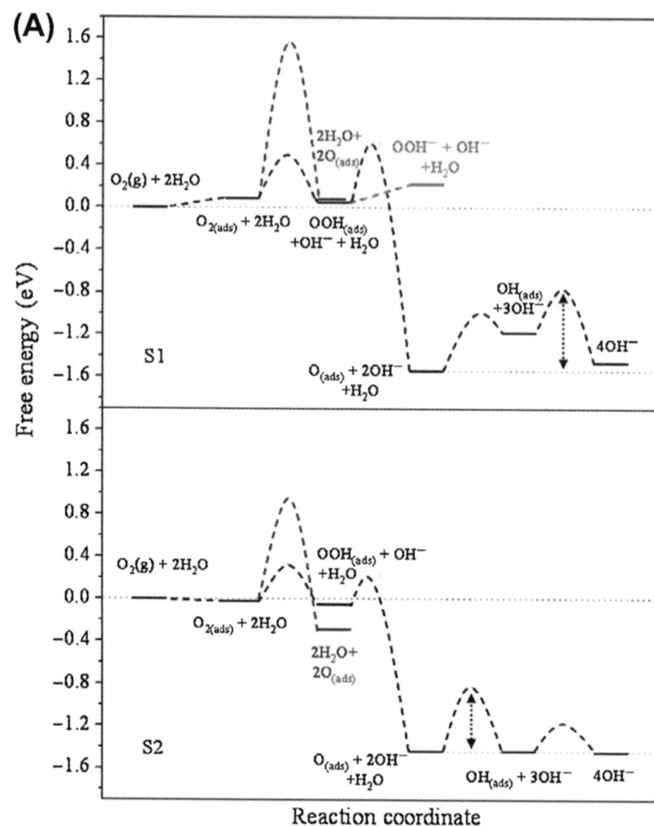

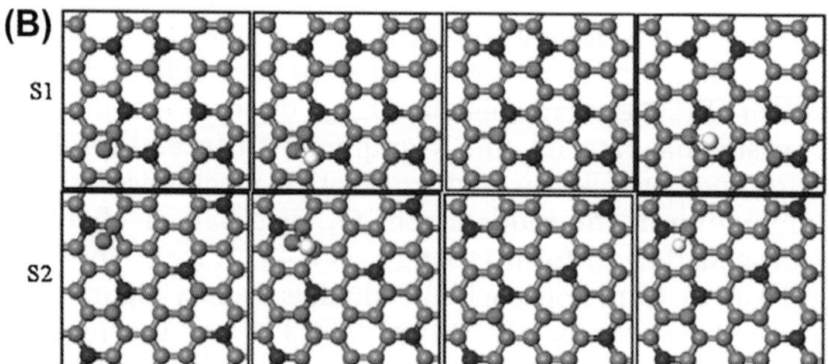

Figure 4.17 (A) Free energy diagram for O_2 reduction on S1 and S2 under the condition of 0.04 V and pH = 14. The black line indicates the intermediates and reaction barriers of associative mechanism. The blue line indicates the dissociative barrier of O_2 and the intermediate. The red line indicates the formation of OOH^-. (B) Intermediate structures of associative mechanism on S1 and S2. From the left to right: $O_{2(ads)}$, $OOH_{(ads)}$, $O_{(ads)}$, and $OH_{(ads)}$. The gray, blue, red, and white spheres represent C, N, O, and H atoms, respectively.[46] (For interpretation of the references to color in this figure legend, the reader is referred to the online version of this book.)

In the subsequent step, the barrier of hydrogenation of $O_{(ads)}$ was calculated to be almost the same (~ 0.55 eV) from all the approaches. In the last step, the removal of $OH_{(ads)}$ needs to overcome a barrier of 0.42 eV, giving rise to the overall barrier of 0.78 eV to remove $O_{(ads)}$. Using the kinetic analysis model proposed by Kozuch and Shaik,[76,77] the step of the removal of $O_{(ads)}$

is the rate-determining step. This is similar to previous theoretical results on Pt(111), where the rate of OH removal determined the overall ORR activity.[78,79]

Regarding the ORR on transition metal macrocyclic compounds, theoretical approaches have predicted that not only the metal center can affect the ORR catalytic activity but also the substituents on the macrocyclic rings can also play a role. For example, Co-phthalocyanine complexes with electron donating substituents should show improved ORR catalytic activity because the substituents can increase the binding energy between O_2 and the metal center(s).[80] Calculation indicated that the catalytic activity of the transition metal macrocyclic complexes is due to the partial electron transition between the filled d_{xz}, d_{yz}, and empty d_z^2 orbitals of the transition metals, and the antibonding π orbitals of O_2. Based on molecular orbital theory, Alt et al.[81] explained that in the interaction between O_2 and a transition metal, electron transition occurs first from oxygen into the empty d_z^2 orbital, forming a π-bond, lowering the antibonding π orbitals and raising the energy of the d_{xz} and d_{yz} orbitals of the transition metals, thus allowing the electron transition from these filled orbitals to the antibonding π-orbital, and resulting in enhanced interaction. The order of catalytic ORR activity for both the tetramethoxyphenylporphyrin and the dihydrodibenzotetraazaannulene systems is Co > Fe > Ni ≈ Cu. This order can be well explained in terms of molecular orbital theory.[48,81]

4.7. Importance of ORR in Fuel Cells

Fuel cells have received much attention as possible economically viable power sources for many applications. Fuel cells are electrochemical devices that convert chemical energy into electrical energy directly without going through the routes of thermal and mechanical energy and thus can be more efficient than other energy conversion devices. As for the hydrogen/oxygen polymer electrolyte membrane (PEM) fuel cell, hydrogen gas is oxidized at the anode, electrons are carried through an electrical circuit where useful work can be done, and proton (in acidic fuel cells) is transported through the membrane to the cathode, or hydroxide ion (in alkaline fuel cells) is transported through the membrane to the anode. At the cathode, oxygen is reduced to complete the overall fuel cell reaction. Therefore, in fuel cells, ORR is the reaction occurring at the cathode. Normally, the ORR kinetics is much slower than that of anode fuel oxidation. In order to speed up the ORR kinetics to reach a practical usable level in a fuel cell, high-performing cathode ORR catalyst is needed. At the current stage of technology, platinum (Pt)-based materials are the most practical catalysts. Because these Pt-based catalysts are too expensive for making commercially viable fuel cells, extensive research over the past several decades has focused on developing alternative catalysts, including nonnoble metal catalysts.[47] These electrocatalysts include noble metals and alloys, carbon materials, quinone and derivatives, transition metal macrocyclic compounds, transition metal chalcogenides, and transition metal carbides.

4.8. Chapter Summary

In the exploring catalysts for fuel cell cathode ORR, besides new catalyst synthesis, catalyst characterization using electrochemical measurements such as RDE and RRDE techniques seems necessary in order to validate the catalytic ORR activity of the synthesized catalysts, and the down-selection of new catalyst designs.

In this chapter, the fundamentals of ORR including thermodynamics and electrode kinetics are presented. The ORR kinetics including reaction mechanisms catalyzed by different electrode materials and catalysts including Pt, Pt alloys, carbon materials, and nonnoble metal catalysts are discussed based on literature that have both experimental and theoretical approaches. It is our belief that these fundamentals of ORR are also necessary in order to perform the meaningful characterization of catalytic ORR activity using both RDE and RRDE techniques. It is also our intention to give useful knowledge and information to the reader for their continuing journey to the rest of this book.

References

1. Zhang J. *PEM fuel cell electrocatalysts and catalyst layers*. London: Springer-Verlag; 2008.
2. Bard AJ, Faulkner LR. *Electrochemical methods: fundamentals and applications*. New York: Wiley; 1980.
3. Zhang J, Wu J, Zhang H, Zhang J. *Pem fuel cell testing and diagnosis, chapter 1*. Elsevier; 2013.
4. Parthasarathy A, Srinivasan S, Appleby AJ, Martin CR. Temperature dependence of the electrode kinetic of oxygen reduction at the platinum/Nafion interface-a microelectrode investigation. *J Electrochem Soc* 1992;**139**:2530−7.
5. Savy M, Andro P, Bernard C, Magner G. Studies of oxygen reduction on the monomeres and polymeres-i. Principles, fundamentals and choice of the central ion. *Electrochim Acta* 1973;**18**:191−7.

6. Stassi A, D'Urso C, Baglio V, Di Blasi A, Antonucci V, Arico AS, et al. Electrocatalytic behaviour for oxygen reduction reaction of small nanostructured crystalline bimetallic Pt-M supported catalysts. *J Appl Electrochem* 2006;**36**:1143−9.

7. Meng H, Shen P. Tungsten carbide nanocrystal promoted Pt/C electrocatalysts for oxygen reduction. *J Phys Chem B* 2005;**109**:22705−9.

8. González-Huerta RG, Chávez-Carvayar JA, Solorza-Feria O. Electrocatalysis of oxygen reduction on carbon supported Ru-based catalysts in a polymer electrolyte fuel cell. *J Power Sources* 2006;**153**:11−7.

9. Gochi-Ponce Y, Alonso-Nuñez G, Alonso-Vante N. Synthesis and electrochemical characterization of a novel platinum chalcogenide electrocatalyst with an enhanced tolerance to methanol in the oxygen reduction reaction. *Electrochem Commun* 2006;**8**:1487−91.

10. Norskov JK, Rossmeisl J, Logadotir A, Lindqvist L, Kitchin JR, Bligaard T, et al. Origin of the overpotential for oxygen reduction at a fuel-cell cathode. *J Phys Chem B* 2004;**108**:17886−92.

11. Appleby AJ. Electrocatalysis of aqueous dioxygen reduction. *J Electroanal Chem* 1993;**357**:117−79.

12. Markovic NM, Gasteiger HA, Ross PN. Oxygen reduction on platinum low-index single-crystal surfaces in sulfuric acid solution: rotating ring-pt (hkl) disk studies. *J Phys Chem* 1995;**99**:3411−5.

13. Stamenkovic VR, Markovic NM. *Oxygen reduction on platinum bimetallic alloy catalysts. Handbook of fuel cells*; 2010.

14. Stamenkovic VR, Schmidt TJ, Ross PN, Markovic NM. Surface composition effects in electrocatalysis: kinetics of oxygen reduction on well-defined Pt_3Ni and Pt_3Co alloy surfaces. *J Phys Chem B* 2002;**106**:11970−9.

15. Stamenkovic VR, Mun BS, Arenz M, Mayrhofer KJ, Lucas CA, Wang G, et al. Trends in electrocatalysis on extended and nanoscale Pt-bimetallic alloy surfaces. *Nat Mat* 2007;**6**:241−7.

16. Markovic NM, Ross PN. Surface science studies of model fuel cell electrocatalysts. *Surf Sci Rep* 2002;**45**:121−229.

17. Yeager E. Dioxygen electrocatalysis: mechanism in relation to catalyst structure. *J Mol Catal* 1986;**38**:5−25.

18. Appleby AJ, Marie J. Kinetics of oxygen reduction on carbon materials in alkaline-solution. *Electrochim Acta* 1979;**24**:195−202.

19. Xu J, Huang WH, McCreery RL. Isotope and surface preparation effects on alkaline dioxygen reduction at carbon electrodes. *J Electroanal Chem* 1996;**410**:235−42.

20. Taylor RJ, Humffray AA. Electrochemical studies on glassy carbon electrodes II. Oxygen reduction in solutions of high pH (pH > 10). *J Electroanal Chem* 1975;**64**:63−84.

21. McCreery RL. Carbon electrodes: structural effects on electron transfer kinetics. *Electroanal Chem* 1991;**17**:221−374.

22. Tarasevich MR, Sadkowski A, Yeager E, Coway BE, Bockris J, Khan S, et al. Comprehensive treatise of electrochemistry. *New York* 1983;**7**:310−98.

23. Baez VB, Pletcher D. Preparation and characterization of carbon/titanium dioxide surfaces-the reduction of oxygen. *J Electroanal Chem* 1995;**382**:59−64.

24. Morcos I, Yeager E. Kinetic studies of the oxygen−peroxide couple on pyrolytic graphite. *Electrochim Acta* 1970;**15**:953−75.

25. Zhang M, Yan Y, Gong K, Mao L, Guo Z, Chen Y. Electrostatic layer-by-layer assembled carbon nanotube multilayer film and its electrocatalytic activity for oxygen reduction. *Langmuir* 2004;**20**:8781−5.

26. Jürmann G, Tammeveski K. Electroreduction of oxygen on multi-walled carbon nanotubes modified highly oriented pyrolytic graphite electrodes in alkaline solution. *J Electroanal Chem* 2006;**597**:119−26.

27. Maldonado S, Stevenson KJ. Influence of nitrogen doping on oxygen reduction electrocatalysis at carbon nanofiber electrodes. *J Phys Chem B* 2005;**109**:4707−16.

28. Sidik RA, Anderson AB, Subramanian NP, Kumaraguru SP, Popov BN. O_2 reduction on graphite and nitrogen-doped graphite: experiment and theory. *J Phys Chem B* 2006;**110**:1787−93.

29. Hu I, Karweik DH, Kuwana T. Activation and deactivation of glassy carbon electrodes. *J Electroanal Chem* 1985;**188**:59−72.

30. Sljukic B, Banks CE, Compton RG. An overview of the electrochemical reduction of oxygen at carbon-based modified electrodes. *J Iran Chem Soc* 2005;**2**:1−25.

31. Maruyama J, Abe I. Cathodic oxygen reduction at the interface between Nafion (R) and electrochemically oxidized glassy carbon surfaces. *J Electroanal Chem* 2002;**527**:65−70.

32. Jia N, Martin RB, Qi Z, Lefebvre MC, Pickup PG. Modification of carbon supported catalysts to improve performance in gas diffusion electrodes. *Electrochim Acta* 2001;**46**:2863−9.

33. Hu I, Karweik DH, Kuwana T. Activation and deactivation of glassy carbon electrodes. *J Electroanal Chem Interfacial Electrochem* 1985;**188**:59−72.

34. Sullivan MG, Kotz R, Haas O. Thick active layers of electrochemically modified glassy carbon − electrochemical impedance studies. *J Electrochem Soc* 2000;**147**:308−17.

35. Qu L, Liu Y, Baek JB, Dai L. Nitrogen-doped graphene as efficient metal-free electrocatalyst for oxygen reduction in fuel cells. *ACS Nano* 2010;**4**:1321−6.

36. Gong K, Du F, Xia Z, Durstock M, Dai L. Nitrogen-doped carbon nanotube Arrays with high electrocatalytic activity for oxygen reduction. *Science* 2009;**323**:760−4.

37. Luo Z, Lim S, Tian Z, Shang J, Lai L, Brian MD, et al. Pyridinic N doped graphene: synthesis, electronic structure, and electrocatalytic property. *J Mater Chem* 2011;**21**:8038−44.

38. Shao Y, Zhang S, Engelhard MH, Li G, Shao G, Wang Y, et al. Nitrogen-doped graphene and its electrochemical applications. *J Mater Chem* 2010;**20**:7491−6.

39. Lee KR, Lee KU, Lee JW, Ahn BT, Woo SI. Electrochemical oxygen reduction on nitrogen doped graphene sheets in acid media. *Electrochem Commun* 2010;**12**:1052−5.

40. Wang P, Wang Z, Jia L, Xiao Z. Origin of the catalytic activity of graphite nitride for the electrochemical reduction of oxygen: geometric factors vs. electronic factors. *Phys Chem Chem Phys* 2009;**11**:2730−40.

41. Meyer JC, Kisielowski C, Erni R, Rossell MD, Crommie MF, Zettl A. Direct imaging of lattice atoms and topological defects in graphene membranes. *Nano Lett* 2008;**8**:3582−6.

42. Boukhvalov DW, Katsnelson MI. Chemical functionalization of graphene with defects. *Nano Lett* 2008;**8**:4373−9.

43. Banhart F, Kotakoski J, Krasheninnikov AV. Structural defects in graphene. *ACS Nano* 2011;**5**:26−41.

44. Zhang L, Xia Z. Mechanisms of oxygen reduction reaction on nitrogen-doped graphene for fuel cells. *J Phys Chem C* 2011;**115**:11170−6.

45. Kim H, Lee K, Woo SI, Jung Y. On the mechanism of enhanced oxygen reduction reaction in nitrogen-doped graphene nanoribbons. *Phys Chem Chem Phys* 2011;**13**:17505−10.

46. Yu L, Pan X, Cao X, Hu P, Bao X. Oxygen reduction reaction mechanism on nitrogen-doped graphene: a density functional theory study. *J Catal* 2011;**282**:183−90.
47. Zhang L, Zhang J, Wilkinson DP, Wang H. Progress in preparation of non-noble electrocatalysts for PEM fuel cell reactions. *J Power Sources* 2006;**156**:171−82.
48. Jahnke H, Schonborn M, Zimmermann G. Organic dyestuffs as catalysts for fuel cells. *Top Curr Chem* 1976;**61**:133−81.
49. Song C, Zhang J. *PEM fuel cell electrocatalysts and catalyst layers: fundamentals and applications, chapter 2*. London: Springer; 2008.
50. Zagal J, Bindra P, Yeager E. A mechanistic study of O_2 reduction on water soluble phthalocyanines adsorbed on graphite electrodes. *J Electrochem Soc* 1980;**127**:1506−17.
51. Shi C, Anson FC. Catalytic pathways for the electroreduction of O_2 by iron tetrakis (4-*N*-methylpyridyl) porphyrin or iron tetraphenylporphyrin adsorbed on edge plane pyrolytic-graphite electrodes. *Inorg Chem* 1990;**29**:4298−305.
52. Shigehara K, Anson FC. Electrocatalytic activity of 3 iron porphyrins in the reductions of dioxygen and hydrogen-peroxide at graphite-electrodes. *J Phys Chem* 1982;**86**:2776−83.
53. Kadish KM, Fremond L, Ou Z, Shao J, Shi C, Anson FC. Cobalt(III) corroles as electrocatalysts for the reduction of dioxygen: reactivity of a monocorrole, biscorroles, and porphyrin-corrole dyads. *J Am Chem Soc* 2005;**127**:5625−31.
54. Liu H, Weaver MJ, Wang C, Chang C. Dependence of electrocatalysis for oxygen reduction by adsorbed dicobalt cofacial porphyrins upon catalyst structure. *J Electroanal Chem Interfacial Electrochem* 1983;**145**:439−47.
55. Chen Z, Higgins D, Yu A, Zhang L, Zhang J. A review on non-precious metal electrocatalysts for PEM fuel cells. *Energy Environ Sci* 2011;**4**:3167−92.
56. Bezerra CW, Zhang L, Liu H, Lee K, Marques AL, Marques EP, et al. A review of heat-treatment effects on activity and stability of PEM fuel cell catalysts for oxygen reduction reaction. *J Power Sources* 2007;**173**:891−908.
57. Li S, Zhang L, Liu H, Pan M, Zan L, Zhang J. Heat-treated cobalt−tripyridyl triazine (Co−TPTZ) electrocatalysts for oxygen reduction reaction in acidic medium. *Electrochim Acta* 2010;**55**:4403−11.
58. Taube H. Mechanisms of oxidation with oxygen. *J Gen Physiol* 1965;**49**:29−50.
59. Valentin J. Dioxygen ligand in mononuclear group VIII transition-metal complexes. *Chem Rev* 1973;**73**:235−45.
60. Griffith JS. On the magnetic properties of some haemoglobin complexes. *Proc R Soc Math Phys Sci* 1956;**235**:23−36.
61. Weiss JJ. Nature of the iron-oxygen bond in oxyhemoglobin. *Nature* 1964;**203**:182−3.
62. Tatsumi K, Hoffmann R. Metalloporphyrins with unusual geometries. 1. Mono-, di-, triatom-bridged porphyrin dimers. *J Am Chem Soc* 1981;**103**:3328−41.
63. Yeager E. Recent advances in the science of electrocatalysts. *J Electrochem Soc* 1981;**128**:C160−71.
64. Collman JP, Marrocco M, Denisevich P, Koval C, Anson FC. Potent catalysis of the electroreduction of oxygen to water by dicobalt porphyrin dimers adsorbed on graphite electrodes. *J Electroanal Chem* 1979;**101**:117−22.
65. Abel EW, Pratt JM, Whelan R. Formation of 1:1 oxygen adduct with cobalt(II)-tetrasulphophthalocyanine complex. *Chem Commun* 1971;**9**:449−50.

66. Smith TD, Pilbrow JR. Recent developments in the studies of molecular-oxygen adducts of cobalt (II) compounds and related systems. *Coord Chem Rev* 1981;**39**:295–383.
67. Vaska L. Dioxygen-metal complexes-toward a unified view. *Acc Chem Res* 1976;**9**:175–83.
68. Damjanovic A. Mechanistic analysis of oxygen electrode reactions. *Mod Asp Electrochem* 1969;**5**:369–483.
69. Damjanovic A. *Progress in the studies of oxygen reduction during the last thirty years*. New York: Plenum Press; 1992.
70. Adzic R. *Recent advances in the kinetics of oxygen reduction*. United States; 1996.
71. Anastasijević NA, Vesović V, Adžić RR. Determination of the kinetic parameters of the oxygen reduction reaction using the rotating ring-disk electrode: part I. *Theory J Electroanal Chem Interfacial Electrochem* 1987;**229**:305–16.
72. Damjanov A, Genshaw MA, Bockris JO. Distinction between intermediates produced in main and side electrodic reactions. *J Chem Phys* 1966;**45**:4057–9.
73. Wroblowa HS, Pan Y, Razumney G. Electroreduction of oxygen-new mechanistic criterion. *J Electroanal Chem* 1976;**69**:195–201.
74. Zagal JH. Metallophthalocyanines as catalysts in electrochemical reaction. *Coord Chem Rev* 1992;**119**:89–136.
75. Janik MJ, Taylor CD, Neurock M. First-principles analysis of the initial electroreduction steps of oxygen over Pt(111). *J Electrochem Soc* 2009;**156**:B126–35.
76. Kozuch S, Shaik S. Kinetic-quantum chemical model for catalytic cycles: the Haber–Bosch process and the effect of reagent concentration. *J Phys Chem A* 2008;**112**:6032–41.
77. Kozuch S, Shaik S. How to conceptualize catalytic cycles? The energetic span model. *Acc Chem Res* 2011;**44**:101–10.
78. Tripković V, Skulason E, Siahrostami S, Norskov JK, Rossmeisl J. The oxygen reduction reaction mechanism on Pt(111) from density functional theory calculations. *Electrochim Acta* 2010;**55**:7975–81.
79. Nilekar AU, Mavrikakis M. Improved oxygen reduction reactivity of platinum monolayers on transition metal surfaces. *Surf Sci* 2008;**602**:L89–94.
80. Shi Z, Zhang J. Density functional theory study of transitional metal macrocyclic complexes' dioxygen-binding abilities and their catalytic activities toward oxygen reduction reaction. *J Phys Chem C* 2007;**111**:7084–90.
81. Alt H, Binder H, Sandsted G. Mechanism of electrocatalytic oxygen reduction on metal chelates. *J Catal* 1973;**28**:8–19.

5

ROTATING DISK ELECTRODE METHOD

Chunyu Du[a], Qiang Tan[a], Geping Yin[a] and Jiujun Zhang[b]
[a]*School of Chemical Engineering and Technology, Harbin Institute of Technology, Harbin, China,* [b]*Energy, Mining and Environment, National Research Council of Canada, Vancouver, BC, Canada*

CHAPTER OUTLINE

Rotating Electrode Methods and Oxygen Reduction Electrocatalysts. http://dx.doi.org/10.1016/B978-0-444-63278-4.00005-7

5.1. Introduction

For a research on the electrode reaction mechanism and kinetics, particularly those of oxygen reduction reaction (ORR) ($O_2 + 4H^+ + 4e^- \rightarrow 2H_2O$ in acidic solution, or $O_2 + 2H_2O + 4e^- \rightarrow 4OH^-$ in alkaline solution), it is necessary to design some tools that could control and determine the reactant transportation near the electrode surface and its effect on the electron-transfer kinetics. A popular method, called the rotating disk electrode (RDE) technique has been widely used for this purpose, particularly for the ORR.

As discussed in Chapter 2, for electrolyte solution containing an excess of supporting electrolyte, the ionic migration term can be neglected, suggesting that there are only two major processes left for material transport: diffusion and convection. If there is no solution convection, the thickness of diffusion layer near the electrode surface will become thicker and thicker with prolonging the reaction time, resulting in nonsteady-state current density. However, if there is a vigorous solution convection such as stirring and electrode rotating, the thickness of diffusion layer will be fixed, leading to a steady-state current density. Therefore, the convention controls the thickness of the diffusion layer and the diffusion controls the transport rate of the reactant through the diffusion layer. Using RDE apparatus to precisely control the electrode rotating rate, the quantitatively control of diffusion layer thickness can be realized, resulting in feasible quantitative analysis of the electrode reaction kinetics. In this chapter, we will give a detailed description about the RDE theory and its associated measurement procedure.

5.2. Rotating Disk Electrode Theory[1,2]

The central part of the RDE theory and technique is the convection of electrolyte solution. Due to the solution convection, the reactant in the solution will move together with the convection at the same transport rate. Let's first consider the situation where the flow of electrolyte solution from the bottom of the electrode edge upward with a direction parallel to the electrode surface to see how the diffusion–convection layer can be formed and what is its mathematic expression.

5.2.1. Diffusion–Convection Layer Near the Electrode Surface

Here we still use the electrochemical reaction as used in Chapter 2:

$$O + ne^- \leftrightarrow R \qquad (5\text{-}I)$$

where n is the electron transfer number of the reaction. Note that in Chapter 2, we use n_α rather than n as the electron number. If Reaction (5-I) is a simple (or elementary) reaction, n_α equals n, and if Reaction (5-I) is a complicated reaction, n_α may not be equal to n, which may be larger than n in normal. In the electrolyte solution, the concentration of oxidant (O) is $C_O(x,t)$, and the reductant's concentration is $C_R(x,t)$, respectively. For simplification, let's consider the situation of oxidant reduction process. Figure 5.1(A) shows the schematic of the flow of electrolyte solution from the bottom of the electrode edge upward with a direction parallel to the electrode surface. The flow rate is marked as v_y in the figure. The flow rate distribution is not uniform along the x direction, that is, the flow rates closer to the electrode surface are smaller than those farther from the electrode surface. The flow rate distribution at y direction can be schematically expressed in Figure 5.1(B). It can be seen that at the distance of δ_O, the flow rate reaches a constant (v_y). This distance of δ_O can be called the thickness of diffusion–convection layer.

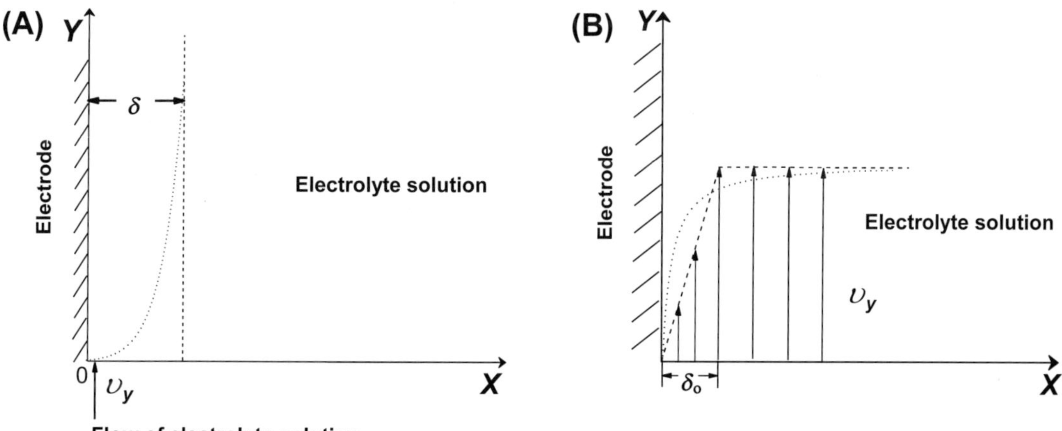

Figure 5.1 (A) Schematic of the electrolyte solution flow along the electrode surface; (B) the flow rate distribution near the electrode surface along the direction parallel to the electrode surface.

Note that this δ_O (cm) is different from that of diffusion layer thickness because within the diffusion layer, there is no solution convection, but within this diffusion–convection layer, both diffusion and convection coexist. Based on convection kinetic theory, this diffusion–convection layer thickness can be approximately expressed as Eqn (5.1):

$$\delta_O \approx D_O^{1/3} \nu^{1/6} y^{1/2} v_y^{-1/2} \tag{5.1}$$

where the thickness of diffusion–convection layer δ_O has a unit of cm, D_O is the diffusion coefficient of oxidant $(\text{cm}^2\,\text{s}^{-1})$, ν is the kinetic viscosity $(\text{cm}^2\,\text{s}^{-1})$, y is the distance at the vertical direction in Figure 5.1(A) (cm), and v_y is the solution flow rate (cm s^{-1}). It can be seen that the thickness of diffusion–convection layer is the function of both y and v_y. The larger they, the thicker the diffusion–convection layer would be, which can be seen from Figure 5.1(A); and the faster the solution flows, the thinner the diffusion–convection layer would be.

We assume that the concentration distribution within the diffusion–convection layer can be treated in the similar way to that described in Chapter 2, and then the concentration distribution of the oxidant near the electrode surface can be schematically expressed in Figure 5.2. Thus, the diffusion–convection current density $(i_{DC,O})$ can be expressed in a similar form to those Eqns (2.57) and (2.58):

$$i_{DC,O} = nFD_O \frac{C_O^o - C_O^s}{\delta_O} \tag{5.2}$$

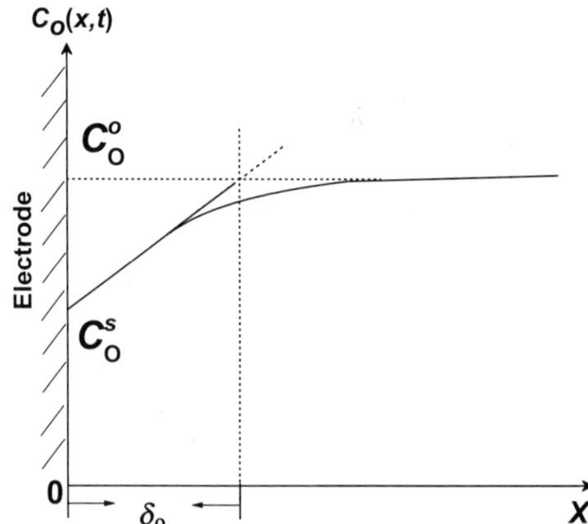

Figure 5.2 Schematic of the concentration distribution of oxidant near the electrode surface.

When C_O^s reaches to zero, the value of $i_{DC,O}$ will become the maximum $(I_{D,O})$, which is called the diffusion-limiting current density:

$$I_{DC,O} = nFD_O\frac{C_O^o}{\delta_O} \tag{5.3}$$

Substituting Eqn (5.1) into Equations (5.2) and (5.3), we can get Eqns (5.4) and (5.5), respectively:

$$i_{DC,O} = nFD_O^{2/3}v^{-1/6}y^{-1/2}v_y^{1/2}\left(C_O^o - C_O^s\right) \tag{5.4}$$

$$I_{DC,O} = nFD_O^{2/3}v^{-1/6}y^{-1/2}v_y^{1/2}C_O^o \tag{5.5}$$

Here both current densities have a unit of $A\,cm^{-2}$ if the concentration of oxidant has a unit of $mol\,cm^{-3}$. It can be seen that both Equations (5.4) and (5.5) contain y, suggesting that current density distribution at a different location over the electrode surface is not uniform, which is not a desired situation in studying the reaction kinetics and mechanism. This also suggests that only stirring a solution or flushing the solution over the electrode surface is not the favorite way for making steady-state current density. Furthermore, the current density is also strongly dependent on the solution flow—the faster the flow, the larger the current density would be. Fortunately, the RDE can make a uniform thickness of the diffusion–convection layer over the electrode surface, producing a uniform distribution of the current density.

5.2.2. Current Densities of Rotating Disk Electrode

Figure 5.3 shows a schematic structure of the RDE. The disk electrode's planner surface is made to contact with the electrolyte solution. An electronic isolator made from isolating material such as Teflon is used to cover the remaining part of the disk with only the planner surface exposed. An electrical brush is used to make the electrical connection between the electrode shaft and the wire during the electrode rotating. When the electrode is rotating, the solution will run from the bulk to the surface, and then be flushed out along the direction parallel to the disk surface, as shown in Figure 5.3(B). Using these three coordinates r, x, and ϕ, shown in Figure 5.3 through a complicated mathematic operation, the solution flow rates near the electrode surface at x and r directions can be expressed as[2,3]:

$$v_x = -0.51\omega^{3/2}v^{-1/2}x^2 \tag{5.6}$$

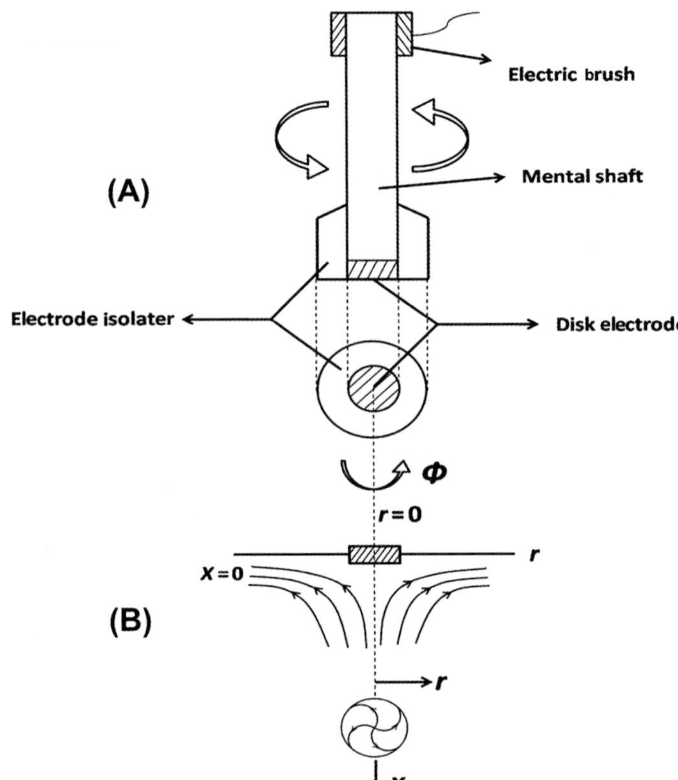

Figure 5.3 (A) Schematic of the rotating disk electrode (RDE); (B) solution flow pattern near the disk electrode surface (x is the coordinate direction perpendicular to the disk electrode surface, r is the coordinate direction parallel to the disk surface, and ϕ is the coordinate direction of the electrode rotation, respectively).

$$v_r = 0.51\omega^{3/2}\nu^{-1/2}rx \tag{5.7}$$

where v_z and v_z are the solution flow rates at x and r directions with a unit of cm s^{-1}, respectively, ω is the electrode rotation rate (s^{-1}), ν is the kinetic viscosity of the electrolyte solution (cm^2 s^{-1}), x is the distance from the electrode surface (cm), and r is the distance from the disk center at the direction parallel to the electrosurface (cm).

In order to get the current–potential relationship on the RDE, particularly the expression of limiting current density as the function of the electrode rotating rate and the reactant concentration, Fick's second law has to be used to give the equations of reactant concentration change with time at the steady-state situation of diffusion–convection. When the surface concentration of oxidant reaches zero during the reaction at the steady-state situation, the concentration distribution within the diffusion–convection layer is not changing with time anymore, meaning that the diffusion rate is equal to the convection rate. In this situation, at three directions shown in Figure 5.3(B) (r, x, and ϕ), the oxidant concentration change $\left(\left(\frac{\partial C_O}{\partial t}\right)_{\text{diffusion}}\right)$ with time induced by the diffusion can be expression as Eqn (5.8); (here the oxidant concentration is simply expressed as C_O, which should be the function of r, x, and ϕ):

$$\left(\frac{\partial C_O}{\partial t}\right)_{\text{diffusion}} = D_O\left[\frac{\partial^2 C_O}{\partial x^2} + \frac{\partial^2 C_O}{\partial r^2} + \frac{1}{r}\left(\frac{\partial C_O}{\partial r}\right) + \frac{1}{r^2}\left(\frac{\partial^2 C_O}{\partial \phi^2}\right)\right]$$

$$(5.8)$$

and the oxidant concentration change $\left(\frac{\partial C_O}{\partial t}\right)$ with time induced by the convection $\left(\left(\frac{\partial C_O}{\partial t}\right)_{\text{convection}}\right)$ can be expression as Eqn (5.9):

$$\left(\frac{\partial C_O}{\partial t}\right)_{\text{convection}} = -\left[v_r\left(\frac{\partial C_O}{\partial r}\right) + \frac{v_\phi}{r}\left(\frac{\partial C_O}{\partial \phi}\right) + v_x\left(\frac{\partial C_O}{\partial x}\right)\right] \quad (5.9)$$

At a steady-state situation, $\left(\frac{\partial C_O}{\partial t}\right)_{\text{diffusion}} - \left(\frac{\partial C_O}{\partial t}\right)_{\text{convection}} = 0$, then:

$$D_O\left[\frac{\partial^2 C_O}{\partial x^2} + \frac{\partial^2 C_O}{\partial r^2} + \frac{1}{r}\left(\frac{\partial C_O}{\partial r}\right) + \frac{1}{r^2}\left(\frac{\partial^2 C_O}{\partial \phi^2}\right)\right]$$
$$= v_r\left(\frac{\partial C_O}{\partial r}\right) + \frac{v_\phi}{r}\left(\frac{\partial C_O}{\partial \phi}\right) + v_x\left(\frac{\partial C_O}{\partial x}\right) \quad (5.10)$$

Actually, due to the originations of the three coordination directions located at the disk center, the geometric symmetry indicates that the oxidant concentration is not a function of ϕ, and therefore, $\left(\frac{\partial C_O}{\partial \phi}\right) = \left(\frac{\partial^2 C_O}{\partial \phi^2}\right) = 0$. In addition, according to Eqn (5.6), the oxidant concentration is also not a function of r, that is, $\frac{\partial^2 C_O}{\partial r^2} = \frac{\partial C_O}{\partial r} = 0$. As a result, Eqn (5.10) can be simplified as Eqn (5.11):

$$D_O\frac{\partial^2 C_O}{\partial x^2} = v_x\left(\frac{\partial C_O}{\partial x}\right) \quad (5.11)$$

Substituting Eqn (5.6) into Eqn (5.11), we can obtain:

$$\frac{\partial^2 C_O}{\partial x^2} = -\frac{0.51x^2}{D_O\omega^{-3/2}v^{1/2}}\left(\frac{\partial C_O}{\partial x}\right) \quad (5.12)$$

Note that at $x = \infty$, the oxidant concentration is equal to the bulk concentration ($C_O = C_O^0$), and when the surface concentration of oxidant drops to zero ($C_O^s = 0$ at $x = 0$), the limiting oxidant supply is reached. Resolving Eqn (5.12), the oxidant flow rate at the electrode surface ($x = 0$) can be obtained:

$$\left(\frac{\partial C_O}{\partial x}\right)_{x=0} = 1.1193 C_O^o \left(\frac{3D_O \omega^{-3/2} \nu^{1/2}}{0.51}\right)^{-1/3} \tag{5.13}$$

Combining Eqn (5.13), the limiting current density can be expressed as:

$$I_{DC,O} = nFD_O \left(\frac{\partial C_O}{\partial x}\right)_{x=0} = 0.62 nFD_O^{2/3} \nu^{-1/6} \omega^{1/2} C_O^o \tag{5.14a}$$

This is the famous Levich equation. The units of the parameters are: $I_{DC,O}$, A cm^{-2}; n, non; F, 96,487 As mol^{-1}; D_O, cm^2 s^{-1}; ν, cm^2 s^{-1}; ω, s^{-1}; and C_O^o, mol cm^{-3}, respectively. In literature, the unit of electrode rotating rate (ω) is normally taken as revolutions per minute (rpm). If using rpm as the unit of electrode rotating rate, Eqn (5.14a) can be alternatively expressed as Eqn (5.15):

$$I_{DC,O} = 0.201 nFD_O^{2/3} \nu^{-1/6} \omega^{1/2} C_O^o \tag{5.14b}$$

Comparing Eqn (5.14a) with (5.3), it can define the thickness of diffusion−convection layer for a rotating electrode:

$$\delta_O = 1.611 D_O^{1/3} \nu^{1/6} \omega^{-1/2} \tag{5.15}$$

It can be seen that this thickness of the diffusion−convection layer is not a function of the location on the electrode surface, which is different from that of Eqn (5.1), and therefore, the current density over the entire RDE surface is uniformly distributed.

The distribution of oxidant near RDE surface is schematically shown in Figure 5.4.

Regarding the situation where the oxidant surface concentration does not reach to zero, that is, $C_O^s \neq 0$, the current density on RDE can be expressed as:

$$i_{DC,O} = 0.62 nFD_O^{2/3} \nu^{-1/6} \omega^{1/2} \left(C_O^o - C_O^s\right) \tag{5.16}$$

Combining Eqn (5.14a) with (5.16), the current density can also be expressed as:

$$i_{DC,O} = I_{DC,O} \left(\frac{C_O^o - C_O^s}{C_O^o}\right) \tag{5.17}$$

In a similar way, the current density for reductant can also be expressed as:

$$i_{DC,R} = I_{l,R} \left(\frac{C_R^o - C_R^s}{C_R^o}\right) \tag{5.18}$$

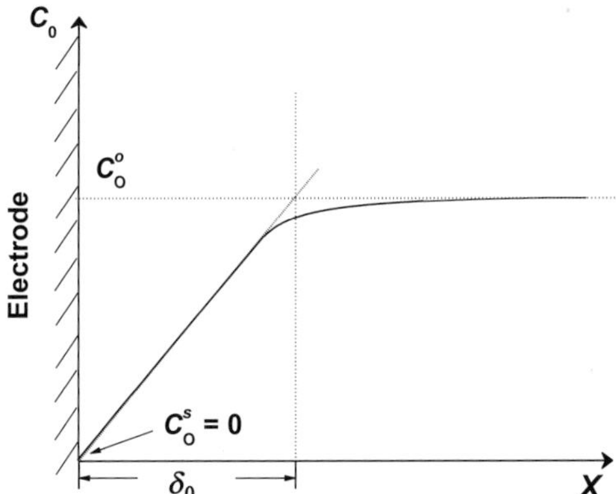

Figure 5.4 Schematic of the concentration distribution of oxidant near the RDE surface.

where $I_{DC,R}$ is the diffusion–convection limiting current density of reductant, which can be expressed as:

$$I_{DC,R} = -0.62 nF D_R^{2/3} \nu^{-1/6} \omega^{1/2} C_R^o \qquad (5.19)$$

When the electrochemical reaction is a reversible reaction that proceeds according to Reaction (5-I) from left to right, and at the beginning of the reaction, there is no reductant in the solution ($C_R^o = 0$), the Nernst electrode potential can be expressed as Eqn (5.20):

$$E = E^o + \frac{RT}{nF}\ln\left(\frac{D_R}{D_O}\right)^{2/3} + \frac{RT}{nF}\ln\left(\frac{i_{DC,O}}{I_{DC,O} - i_{DC,O}}\right) \qquad (5.20)$$

When $i_{DC,O} = \frac{1}{2} I_{DC,O}$, the electrode potential defined by Eqn (5.20) will become the half-wave potential: $E_{1/2} = E^o + \frac{RT}{nF}\ln\left(\frac{D_R}{D_O}\right)^{2/3}$, and then Eqn (5.20) can be rewritten as:

$$E = E_{1/2} + \frac{RT}{nF}\ln\left(\frac{i_{DC,O}}{I_{DC,O} - i_{DC,O}}\right) \qquad (5.21)$$

This half-wave potential is very useful in evaluating the electrochemical reaction based on the measured current–potential curves by RDE technique. Note that Eqn (5.21) is for the case of reversible reactions. As a rough estimation, this half-wave potential may be useful for those pseudoreversible reactions. However, one should be careful when using this half-wave potential to evaluation the irreversible electrochemical reactions.

5.2.3. Koutecky–Levich Equation

Normally, if the electron transfer kinetics of oxidant reduction in Reaction (5-I) is very fast so that the diffusion–convection process could not catch up the speed of this electron-transfer process, the oxidant's surface concentration is quickly exhausted to zero, and the obtained Levich plot of $I_{DC,O}$ vs $\omega^{1/2}$ according to Eqn (5.14a) will be a straight line. From the slope $(= 0.62 nFD_O^{2/3} \nu^{-1/6} C_O^o)$ of the straight line, the parameter either as n, D_O, ν, or C_O^o can be estimated if the other three are known.

However, if the electron transfer kinetics of oxidant reduction in Reaction (5-I) is much slower than that of diffusion–convection process, the oxidant's surface concentration cannot be exhausted to zero unless a very larger overpotential is controlled. In this case, the obtained Levich plot of $I_{DC,O}$ vs $\omega^{1/2}$ according to Eqn (5.14a) is not a straight line anymore, and it will gradually fall off with increasing $\omega^{1/2}$, as shown in Figure 5.5(B). In this case, for a planner smooth disk electrode, we can use the equation similar to Eqn (2.69) to take care of the effect of a slow electron-transfer kinetics:

$$\frac{1}{i_{DC,O}} = \frac{1}{i_{k,O}} + \frac{1}{I_{DC,O}} \tag{5.22a}$$

In order to use the convenient expression of current density at the disk, from now, we express the disk current densities ($i_{DC,O}$ and $I_{DC,O}$) as $i_{d,O}$ and $I_{d,O}$, respectively. The Eqn (5.22a) becomes Eqn (5.22a):

$$\frac{1}{i_{d,O}} = \frac{1}{i_{k,O}} + \frac{1}{I_{d,O}} \tag{5.22b}$$

where in Eqn (5.22b), $i_{k,O} = i^o \exp\left(-\frac{(1-\alpha)nF(E-E^{eq})}{RT}\right)$, $I_{d,O} = nFD_O\frac{C_O^o}{\delta_O} = 0.62 nFD_O^{2/3}\nu^{-1/6}\omega^{1/2}C_O^o$. Substituting these two expressions into Eqn (5.22a), we can obtain:

$$\frac{1}{i_{d,O}} = \frac{1}{i^o \exp\left(-\frac{(1-\alpha)nF(E-E^{eq})}{RT}\right)} + \frac{1}{0.62 nFD_O^{2/3}\nu^{-1/6}\omega^{1/2}C_O^o}$$

$$= \frac{1}{i^o}\exp\left(\frac{(1-\alpha)nF(E-E^{eq})}{RT}\right) + \frac{\omega^{-1/2}}{0.62 nFD_O^{2/3}\nu^{-1/6}C_O^o} \tag{5.23}$$

This Eqn (5.23) is called the Koutecky–Levich equation. The plot of $\frac{1}{i_{d,O}}$ vs $\omega^{-1/2}$ based on the experiment data allows the estimation of $i_{K,O}$ from the interception and $\frac{1}{0.62 nFD_O^{2/3}\nu^{-1/6}C_O^o}$ from the slope. If the plots can be obtained at different electrode potentials, the

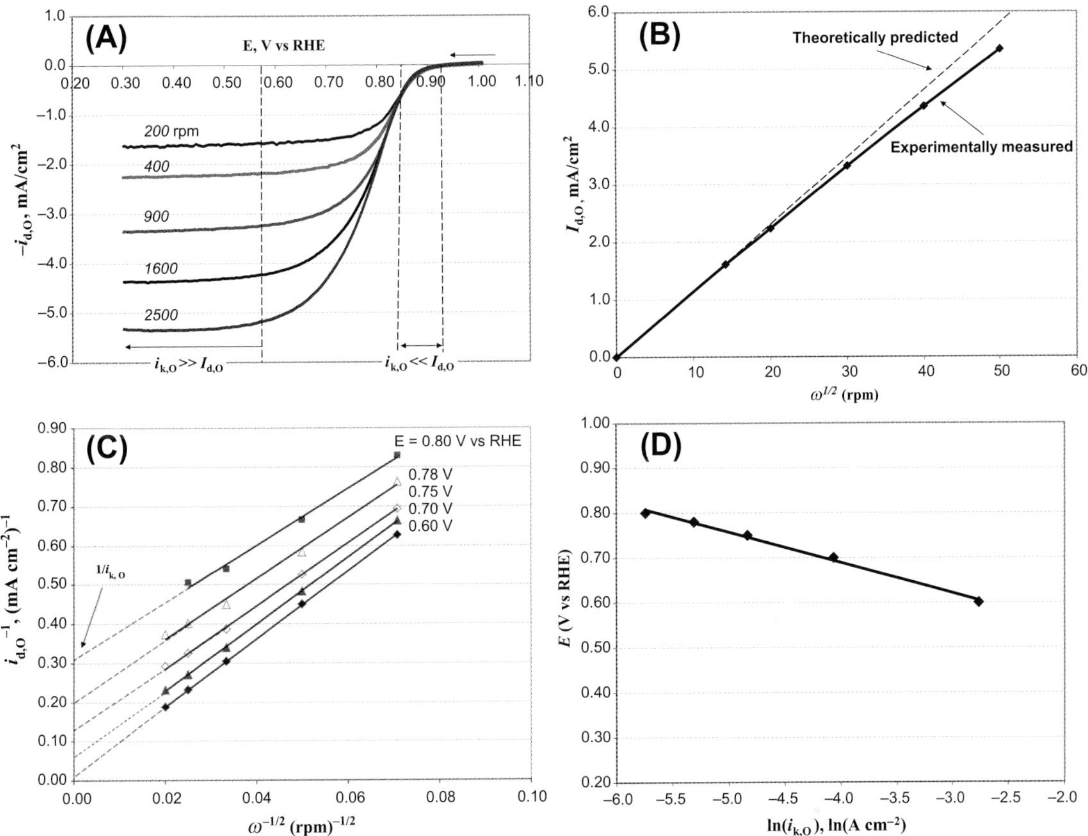

Figure 5.5 (A) Current—potential curves at different electrode rotating rates, recorded on a Pt disk electrode (0.196 cm^2) using a potential scan rate of 5 mV s^{-1} in O$_2$-saturated 0.5 M H$_2$SO$_4$ aqueous solution; (B) the Levich plot; (C) the Koutecky—Levich plots at different electrode potentials; and (D) plot of E vs ln($i_{k,O}$). The measurement was carried out at 30 °C and 1.0 oxygen pressure.[5]

values of electrode potential can be plotted against $ln(i_{k,O})$ based on the equation below:

$$E = E^{\text{eq}} + \frac{RT}{(1-\alpha)nF}\ln(i^{\text{o}}) - \frac{RT}{(1-\alpha)nF}\ln(i_{k,O}) \qquad (5.24)$$

From the slope of the E vs ln($i_{k,O}$), the electron-transfer number (n) and coefficient (α) can be obtained, and from the intercept the exchange current density (i^{o}) can be estimated if E^{eq} is known.

Therefore, using RDE measurements, the kinetic parameters of electron-transfer kinetics such as the electron-transfer number (n) and coefficient (α) and the exchange current density (i^{o}) can be estimated. At the same time, the reactant transport parameters such as the overall electron transfer number (n), reactant diffusion coefficient, and the kinetic viscosity can also be estimated. Note that n in Eqn (5.24) could be the electron transfer number in the rate determining step, which is different from the n obtained from Eqn (5.14a) if Reaction (5-I) is not an element reaction. If Reaction (5-I) is an element reaction, both electron numbers should have the same value.

For planner matrix electrode layer, the following equation will be used:

$$\frac{1}{i_{d,O}} = \frac{1}{i_{k,O}} + \frac{1}{I_{d,O}} + \frac{1}{I_{m,O}} \tag{5.25}$$

where $i_{k,O} = i^o \exp\left(-\frac{(1-\alpha)n_\alpha F(E-E^{eq})}{RT}\right)$, $I_{d,O} = 0.62nFD_O^{2/3}\nu^{-1/6}\omega^{1/2}C_O^o$, and $I_{m,O} = \frac{n_\alpha FC_m D_m}{\delta_m}$. Substituting these three expressions into Eqn (5.25), we can obtain:

$$\frac{1}{i_{d,O}} = \frac{1}{i^o}\exp\left(\frac{(1-\alpha)n_\alpha F(E-E^{eq})}{RT}\right) + \frac{\omega^{-1/2}}{0.62n_\alpha FD_O^{2/3}\nu^{-1/6}C_O^o}$$
$$+ \frac{\delta_m}{n_\alpha FC_m D_m}$$

$$\tag{5.26}$$

As an example, Figure 5.5 shows the current–potential curves at different electrode rotating rates (A), the corresponding Levich plot (B) and the Koutecky–Levich plots at different electrode potentials (C). Note that all the curves are located in the left side below the potential-axis rather than above the potential-axis. This is a traditional expression for the oxidant reduction, particularly for the ORR.

From Figure 5.5(A), it can be seen that at a very narrow potential range such as 0.85–0.93 V vs RHE as marked in the figure, the current densities are not dependent on the electrode rotating rate, indicating that, in this potential range, the current density is almost purely controlled by the electron-transfer kinetics. In this situation, the electron-transfer rate at the electrode surface is much smaller than that of the diffusion–convection limiting rate ($i_{k,O} \ll I_{d,O}$ in Eqn (5.22), then $i = i_{k,O}$), and therefore, the current density is almost equal to $i_{k,O}$. The current densities in this narrow potential range can be used to evaluate the exchange current density and the electron-transfer coefficient based on Eqn (5.24). When the potential is more negative than $\sim$0.63 V, the plateau current densities (diffusion–convection limiting current densities) appear, indicating that the reaction is controlled by the diffusion–convection process. At such a situation of $i_{d,O} \ll I_{d,O}$ in Eqn (5.22), then $i_{d,O} = I_{d,O}$. This limiting current density is increased with increasing the rotating rate of the electrode, as indicated by Eqn (5.14a).

From Figure 5.5(B), it can be seen that the Levich plot is not a straight line, which falls off with increasing the electrode rotating rate. This is because at high rotating rate the diffusion–convection rate becomes higher (or the diffusion–convection layer becomes thinner), leading to a difficulty for the electron-transfer rate at the electrode surface in catching up the diffusion–convection rate. As a result, the surface concentration of the oxidant will not be exhausted to zero, and then no limiting current density can be reached.

From Figure 5.5(C), for Koutecky–Levich plots at different electrode potentials, all lines are almost parallel to each other, suggesting that they all have the similar values of $0.62nFD_O^{2/3}\nu^{-1/6}C_O^o$, from which if D_O, ν, and C_O^o are known, the electron-transfer number, n, can be obtained. The obtained average value of $0.62nFD_O^{2/3}\nu^{-1/6}C_O^o$ from Figure 5.5(C) is 1.3×10^{-4} A s$^{-3/2}$. If we take the values of D_O, ν, and C_O^o as 1.9×10^{-5} cm^2 s^{-1}, 0.01 cm^2 s^{-1}, and 1.2×10^{-6} mol cm^{-3} at 30 °C, the calculated overall electron-transfer number is around 3.8, indicating that the ORR on Pt electrode is not a direct 4-electron transfer process from O_2 to H_2O. Some portion of the electron-transfer process is through a 2-electron transfer pathway from O_2 to H_2O_2. In Chapter 6, we will discuss how to measure the produced portion of H_2O_2 using rotating ring-disk electrode technique.

Furthermore, the Koutecky–Levich intercepts at different electrode potentials can be used to obtain the exchange current density and the electron transfer coefficient according to Eqn (5.24). Note that when $\omega^{-1/2} \to 0$, or $\omega \to \infty$, the diffusion–convection rate reaches an unlimited one , or the thickness of diffusion–convection layer is reduced to zero. In such a situation, the only limitation for the reaction is the electron-transfer process at the electrode surface. Figure 5.5(D) shows the plot of E vs $\ln(i_{k,O})$, from which the obtained $i^o = 1.2 \times 10^{-5}$ A cm^{-2}, and $(1 - \alpha)n = 0.38$ for oxygen reduction at the conditions marked in Figure 5.5. Note that in this $(1 - \alpha)n$, n is different from that of the overall ORR reaction, and it is the electron transfer number in the rate-determining step. For ORR, this n has been recognized to have a value of 2.0,[4] therefore, $\alpha = 0.81$.

In Chapter 7, we will give more examples to illustrate how these kinetic parameters can be estimated based on the RDE data.

5.3. Experimental Measurements of Rotating Disk Electrode

For electrochemical measurements using RDE technique, several steps are necessary, including fabricating the RDE instrument, electrochemical cell, electrode/electrolyte preparation, data collection and analysis. The following subsections will give some detailed description.

5.3.1. Rotating Disk Electrode Apparatus

As discussed above, the RDE is a convective electrode system for which the diffusion–convection equation can be derived and used for measuring the steady-state electrochemical reaction. The RDE includes a conductive disk embedded into an inert insulating material, as shown in Figure 5.3(A) The planner disk, like any working electrode, is generally made by glassy carbon (GC), graphite carbon, or noble metal such as Pt or Au. In principle, any conductive material can be used as the electrode for special needs. To fabricate a qualified RDE, some special attention has to be paid to ensure an exact insulating mantle and disk in order to prevent the disk edge from electrolyte leakage. The shaft is attached to a motor directly by a chuck or by a flexible rotating shaft or pulley arrangement and is rotated at a rotation rate (rpm). Actually, several kinds of RDE have been commercially available. Figure 5.6 shows three commercially available RDEs with glassy carbon, gold, and platinum disks, respectively.

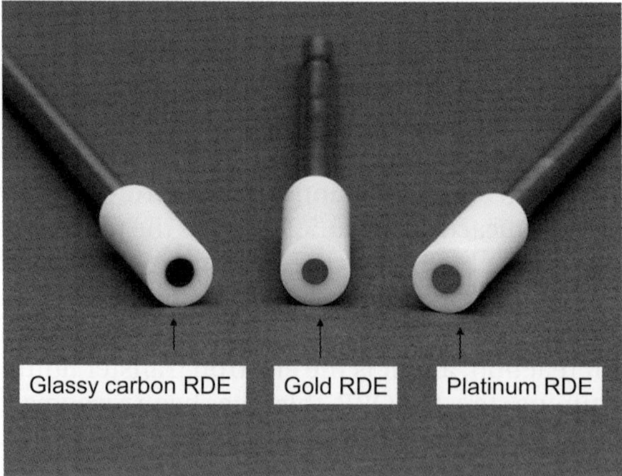

Figure 5.6 Three different rotating disk electrodes (Pine) with glassy carbon, gold, and platinum disks (from left to right), respectively. (http://www.pineinst.com/echem/viewproduct.asp?ID=45655).

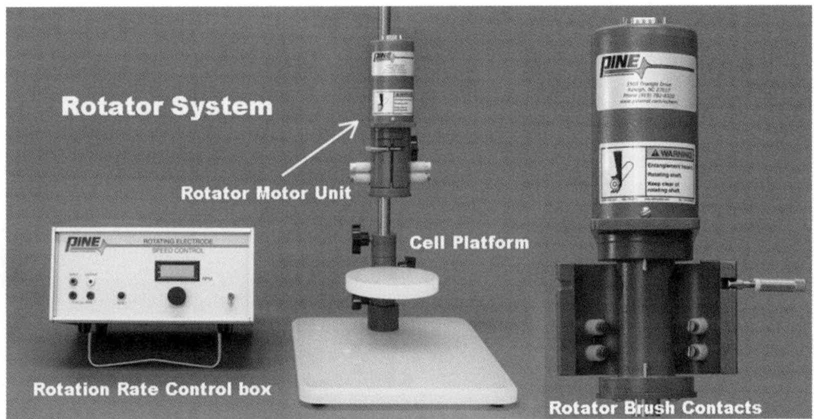

Figure 5.7 Pine electrode rotator and its associated rotation-rate control box. (http://www.pineinst.com/echem/imagebrowser.asp?productID=47123&imageID=839).

Currently, these three kinds of electrodes shown in Figure 5.6 are the most commonly used RDEs in the electrochemical measurements. For evaluating the activity of the ORR catalysts, the glass carbon or gold electrode is normally employed as the current collector on which the catalyst layer is coated. This is because both of these electrode materials are electrochemically ORR-inert over the range of electrode potentials. However, for stability evaluation of the ORR catalysts, it is better to use the gold disk electrode because GC could be oxidized at a high potential with a long period of test.

The motor (or rotator) with a rotation rate control box is also commercially available, such as from Pine company as shown in Figure 5.7. This motor can precisely control the electrode's rotating rate normally from 100 to 4900 rpm. During the electrode rotating, the electrolyte solution will be sent to the electrode surface, and then thrown out from electrode surface, as shown in Figure 5.3(B).

5.3.2. Electrochemical Cells for RDE Measurements

For evaluating catalyst performance in an ORR using RDE technique, the normally employed electrochemical cell is the conventional three-electrode cell, as shown in Figure 5.8. For RDE electrochemical cell, in which the working electrode is rotating at a high rotation rate, in order to avoid the turbulent flow inside the electrolyte solution, the electrolyte container should be a cylinder-type with the RDE located in the middle. In addition, for the same purpose, the volume of electrolyte solution should also

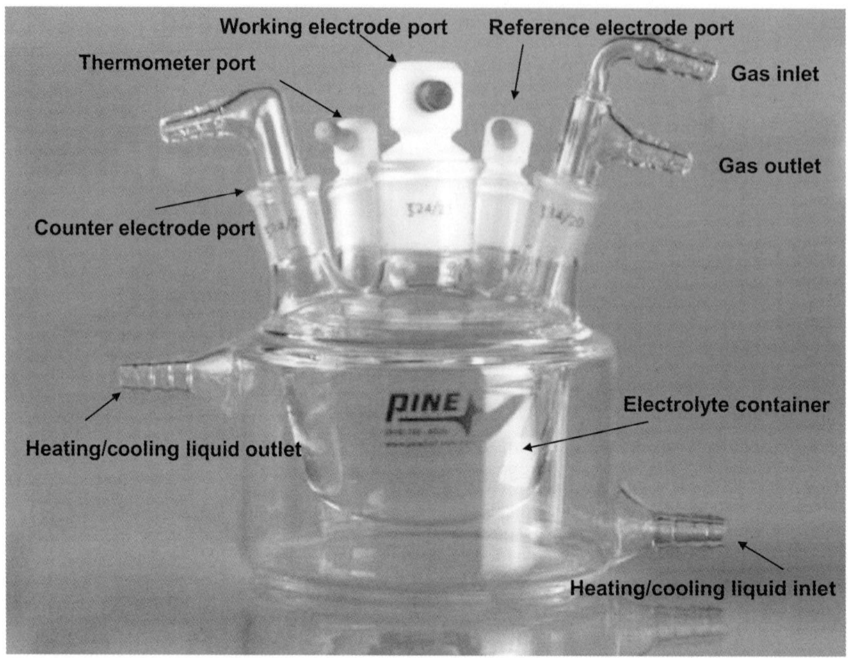

Figure 5.8 Schematic of the three-electrode cell for RDE measurement from Pine Instrument. (http://www.pineinst. com/echem/viewproduct.asp?ID=45706).

be large enough. To reduce the solution iR drop between the working electrode surface and the tip of the reference electrode, the tip should be placed as close to the electrode surface as possible, but without blocking the solution flow when the electrode is rotating.

In the cell shown in Figure 5.8, there are three electrodes: (1) the catalyst-coated disk-working electrode made of either GC or other stable metal materials such as Au, (2) the counterelectrode (e.g., Pt foil or net. For non-noble metal catalysts, the normally used counterelectrode is made of carbon or Ti material to avoid possible Pt contamination), and (3) the reference electrode such as a normal hydrogen electrode (NHE), reversible hydrogen electrode (RHE), or saturated calomel electrode (SCE). Note that the NHE uses a large surface Pt-black as the metal electrode and 1.0 M H^+ aqueous solution (such as 0.5 M H_2SO_4) as the electrolyte. Its electrode potential is defined as zero at 1.0 atm hydrogen gas, and any temperature.

The gas inlet and gas outlet shown in Figure 5.8 are used for gas (pure oxygen, nitrogen, or argon) purging. For measuring catalyst surface-cyclic voltammograms, pure N_2 or Ar gas will be used to de-aerate the electrolyte solution for 30—60 min. For an

ORR measurement, either pure oxygen or air is used to purge the electrolyte solution to make oxygen dissolved in the solution. In addition, there is a thermometer port to monitor the temperature of the electrolyte solution. Both the heating/cooling liquid inlet and outlet are used for controlling the cell's temperature at the desired level.

5.3.3. Working Electrode Preparation

5.3.3.1 Smooth Surface RDE

For preparing smooth surface RDE, the GC (or Pt, or Au) disk is polished with the 1, 0.3, and 0.05 μm γ-Al_2O_3 in succession until the mirror surface is formed. Then this electrode surface is washed using pure acetone and DI water under the ultrasonication for at least three times, and put into the electrochemical cell for measurements.

5.3.3.2 Smooth Surface RDE Adsorbed by Monolayer of Catalyst

The catalysts used for adsorption on the electrode surface are normally the organic molecules or metal-macrocyclic complexes with conjugated benzenoid structures, which can irreversibly adsorb on the electrode surface. In order to facilitate the irreversible adsorption, the commonly used electrode material is graphite carbon such as highly ordered pyrolytic graphite. Due to the similarity of the highly conjugated benzenoid surface structure of graphite electrode to those organic molecules and metal-macrocyclic complexes, the attachments between the catalyst and the electrode surface are very strong.

In this RDE preparation, the electrode material is highly ordered pyrolytic graphite with either basal or edge plane exposed to the electrolyte. Before adsorbing the catalyst onto the electrode surface, the graphite disk is polished with the 1 μm, 0.3 μm, and then 0.05 μm γ-Al_2O_3 to form a mirror surface. This mirror electrode surface is washed using pure acetone and DI water under the ultrasonication for at least three times, and put into the solution containing dissolved catalyst for a period from several seconds to several minutes to achieve the electrode surface adsorbed by the catalyst. Finally, the electrode is taken out of the soaking solution, washed using DI water, and then put into the electrochemical cell for measurements.

5.3.3.3 Catalyst Layer-Coated RDE

For catalyst layer-coated RDE, there are two major steps: one is to prepare the catalyst ink, and the other is to coat the catalyst layer onto the RDE surface. For catalyst ink preparation, a small amount of the catalyst powder such as the commercially available catalyst such as 20 wt% platinum supported on Vulcan XC-72 (E-TEK) (20 wt%Pt/C) or the non-noble metal catalyst is mixed ultrasonically with deionized water first and then added some alcohol or isopropanol (~ 1.0 ml alcohol with 5 mg catalyst) to form a well-mixed aqueous catalyst ink suspension. For coating catalyst layer onto the electrode surface, a small amount of catalyst ink is pipetted and coated onto a disk electrode surface such as GC or gold electrode surface with a geometric area of $0.2-0.5$ cm^2. The coated electrode is then left in air to dry. Normally the total catalyst loading on the electrode can be adjusted to be $0.02-0.2$ mg cm^{-2}. After the catalyst is dry, a small drop of diluted ionomer solution such as Nafion® ionomer solution is pipetted onto the catalyst-coated electrode surface to prevent the catalyst from falling off from the surface. Both catalyst and ionomer loadings on the electrode surface can be calculated according to the contents of the catalyst in the catalyst ink and the ionomer in the ionomer solution, the amount of the volumes of catalyst and ionomer solution pipetted, and the area of the electrode surface. Note that before the catalyst layer coating, the electrode surface must be polished and cleaned according to the procedure described above. As an example, the catalyst layer coating procedure on a GC can be summarized as in Figure 5.9.

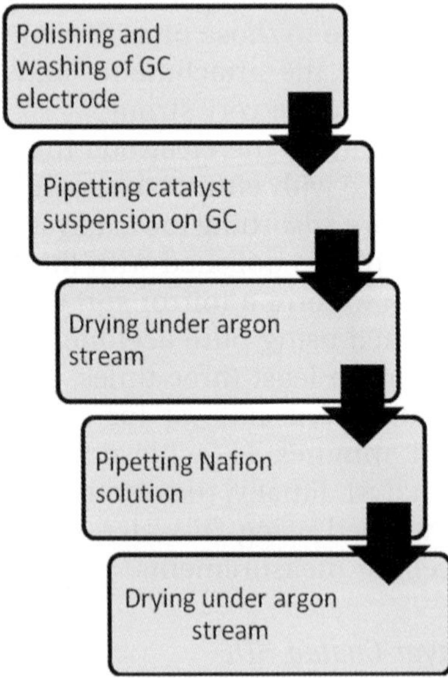

Figure 5.9 Schematic procedure for a catalyst layer coating onto the glassy carbon disk electrode surface.[6]

5.3.4. Measurement for Electrochemical Activity of the Catalysts

The electrochemical activity of the catalyst is normally measured in the electrolyte solution saturated with inert gas such as pure N_2 or Ar and the electrode is at static state (not rotated). The purpose is to obtain information about the redox activity of the catalyst itself or some information about the catalyst surface behavior, from which the catalyst electrochemical active surface area (for Pt-based catalysts) or the concentration of catalyst active center (for non-noble metal catalysts) an be obtained. To give some basic sense about these measurements, we will give two examples as follows.

5.3.4.1 Pt Catalyst-Coated Electrode Surface to Obtain the Electrochemical Pt-Surface Area (EPSA)

Here, we use the Pt Catalyst (Pt/C)-coated GC or gold electrode and the acidic electrolyte to illustrate how to obtain the *EPSA*. The electrode coated with Pt/C catalyst is immersed into the three-electrode cell containing the deaerated (high-purity N_2 or Ar bubbling for 30 min) 0.5 M H_2SO_4 electrolyte, a platinum net as the counter electrode, and a reversible hydrogen Electrode (RHE) as the reference electrode. Before taking the data, the electrode is cycled for more than 20 times between 0.05 and 0.12 V using cyclic voltammetric method at a potential scan rate of 100 mV s^{-1} for removing the possible surface contaminant. The electrochemical measurements should be conducted in a thermostated situation such as at 20–30 °C. After this cycling, the cyclic voltammograms (CVs) can be recorded from which the *EPSA* can be measured, as shown in Figure 5.10 (This figure serves

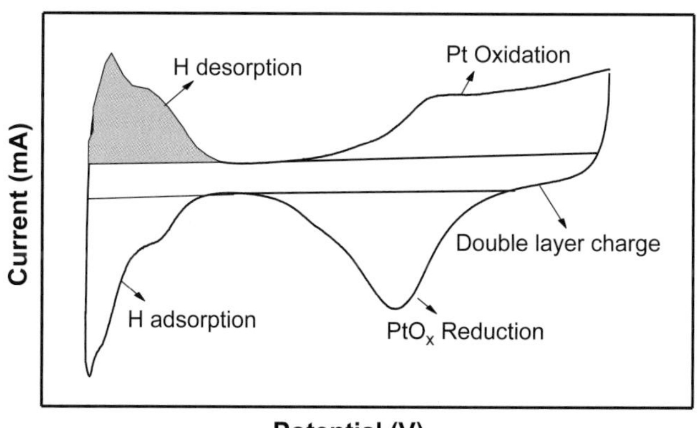

Figure 5.10 Cyclic voltammogram of Pt/C-coated glassy carbon or gold electrode in N$_2$- or Ar-saturated aqueous electrolyte solution. Potential scan rate: 50 mV s^{-1}, and electrolyte solution temperature: 25 °C. The shadow in the curve stands for H desorption.[7]

as a schematic illustration without detailed number label and experiment conditions marked). The *EPSA* is evaluated based on the electric charge of the hydrogen desorption or the average of hydrogen desorption/desorption, Q_H. It is commonly recognized that the standard charge quantity (Q_H^o) of monolayer hydrogen desorption on a smooth polycrystalline platinum is 0.21 mC cm^{-2} at 25 °C.[2,3] The *EPSA* of the catalyst layer is measured by hydrogen adsorption–desorption coulometry between 0.05 and 0.4 V vs RHE, as shown in Figure 5.10. The equation for calculating the *EPSA* can be expressed as:

$$EPSA = \frac{Q_H}{m_{Pt}Q_H^o} \tag{5.27}$$

where *EPSA* has a unit of square meter per gram of catalyst (m^2 g^{-1}), Q_H *is* the charge of adsorption–desorption area (mC), Q_H^o is the charge related to monolayer hydrogen adsorption on Pt with a value of 0.21 mC cm^{-2}, and m_{Pt} is the mass of Pt in the catalyst layer (g). The typical *EPSA* values for carbon-supported Pt-based catalysts range from 50 to 100 m^2 g^{-1} depending on the type of the catalysts and the preparation of the catalyst layer. The magnitude of *EPSA* is an indication of the Pt-active site density that is available for reactant reaction such as oxygen reduction or hydrogen oxidation. The larger the *EPSA* value, the more active the catalyst layer should be. Note that this *EPSA* is only meaningful for pure Pt catalysts. For other catalysts such as Pt alloys, one needs to be careful when using the hydrogen adsorption/desorption method to calculate the *EPSA*.

The other way to obtain the *EPSA* is to use the CO-stripping charge quantity on a Pt surface. The data obtained by CO-stripping experiment can be used to confirm the data obtained by hydrogen adsorption/desorption method. For example, Figure 5.11 shows the CO-stripping CVs recorded in CO-saturated 0.1 M HClO$_4$ solution.[8] In the measurement, the electrolyte solution was first purged with nitrogen gas for at least 30 min before preconditioning the working electrode. The Pt/C catalyst-coated working electrode was preconditioned by cyclic voltammetry in a potential range of 0.0–1.4 V vs RHE at a potential scan rate of 100 mV s^{-1} for 20 cycles. After preconditioning, CO was absorbed by purging the solution with CO gas, while holding the working electrode potential at 0.05 V vs RHE. After that, the purging gas was switched to nitrogen for 30 min to remove CO traces from the solution by keeping the potential at the same value. Then, the potential was scanned in the range of 0.05–1.2 V at a rate of 20 mV s^{-1}. Figure 5.11 shows a typical example of a CO-stripping CVs. Based on the charge of the CO peak and subtracting the

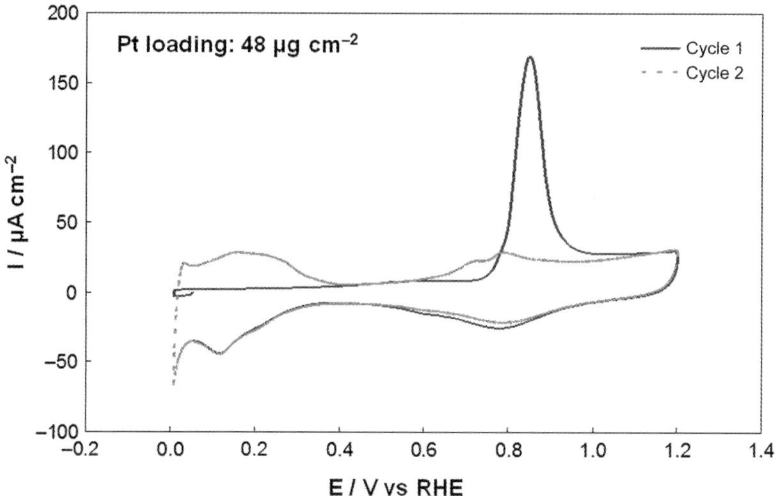

Figure 5.11 CO-stripping voltammograms on 47 wt% Pt/C catalyst-coated glassy carbon electrode, recorded in N$_2$-saturated 0.1 M HClO$_4$. Potential scan rate: 20 mV s^{-1}; Pt/C catalyst loading: 48 µg cm^{-2}.[10]

background charge arising from double layer charging and oxide formation, the *EPSA* was then calculated using the procedure reported by Vidakovic et al.[9] The obtained *EPSA* was 77 m^2 g^{-1}, which is consistent with the value obtained from the H$_2$ adsorption/desorption peaks in the following scans (75 m^2 g^{-1}).

5.3.4.2 Non-Noble Metal Catalyst-Coated Electrode Surface to Obtain the Active Center Density

For non-noble metal catalyst layer-coated electrode surface, the charges under the surface redox waves can be used to calculate the active center density. Figure 5.12 shows an example of carbon supported Fe-N catalyst-coated GC electrode.[11]

In Figure 5.12, the CV was recorded with an electrode material (heat-treated Fe-TPTZ supported on carbon, abbreviated as Fe-N$_X$/C) having a loading of 150 µg cm^{-2} in a N$_2$-purged 0.5 M H$_2$SO$_4$ solution. Both the background current and the redox current from Fe-N$_X$/C can be seen, from which the charge quantity only from Fe-N$_X$/C catalyst can be obtained. The anodic and cathodic peaks at 0.67 V (vs RHE) from the Fe(II)-N$_x$/Fe(III)-N$_x$ center have almost the same charge quantities. If the charge quantity can be expressed as Q_{Fe-N_X}, the active center density can be calculated according to the following equation:

$$C_{Fe-N_X} = \frac{Q_{Fe-N_X}}{nm_{Fe-N_X}F} \qquad (5.28)$$

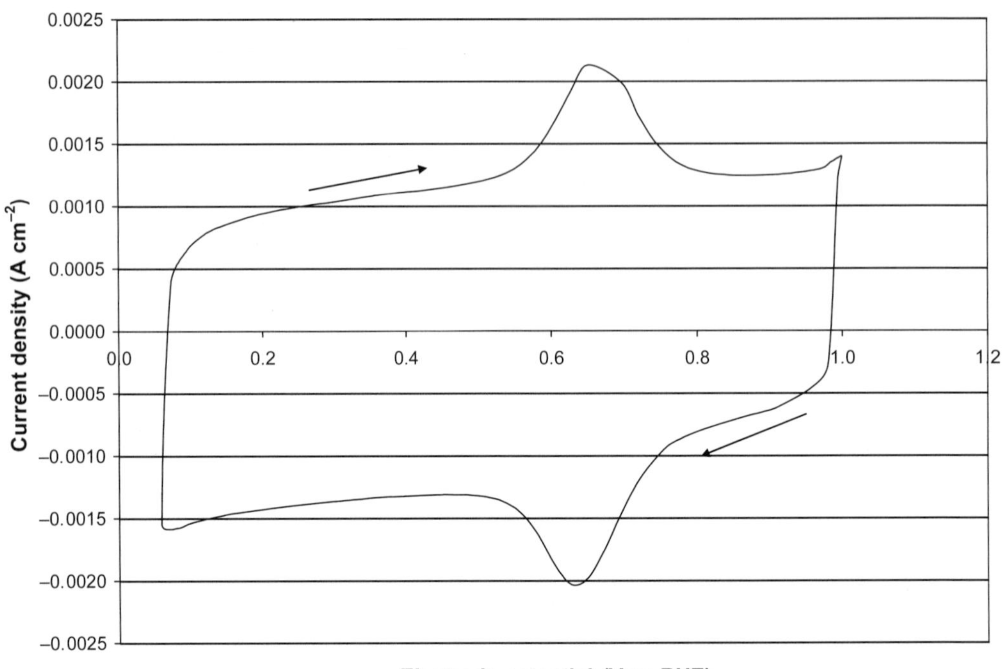

Figure 5.12 Cyclic voltammograms of Fe-N/C with 5 wt% coated on a glassy carbon electrode surface, recorded in a N₂-purged 0.5 M H₂SO₄ solution. Fe-N/C loading: 150 μg cm^{-2}; potential scan rate: 50 mV s^{-1}.[11]

where $C_{\text{Fe-N}_x}$ is the catalyst active center density with a unit of mol g^{-1}, $Q_{\text{Fe-N}_x}$ has a unit of C cm^{-2}, n is the electron transfer number of the corresponding redox process, $m_{\text{Fe-N}_x}$ is the catalyst loading (g cm^{-2}), and F is the Faraday's constant (96,487 C mol^{-1}).

5.3.5. Recording ORR Cyclic Voltammograms

For catalyst activity toward the ORR, the first step is normally to put the catalyzed electrode into the O₂- or air-saturated electrolyte solution to record the cyclic votammograms. For example, Figure 5.13 shows the cyclic voltammograms recorded on a GC disk electrode (0.28 cm^2) coated with Co-N-S/C catalyst layer in N₂-saturated and O₂-saturated 0.1 M KOH solutions, respectively.[12] It can be seen that in the presence of O₂, a large reduction wave appears with an onset potential close to that of surface Wave I/I′ in the presence of O₂, suggesting that the surface redox process expressed by Wave I/I′ is probably responsible for the ORR catalytic activity. However, there is no significant ORR

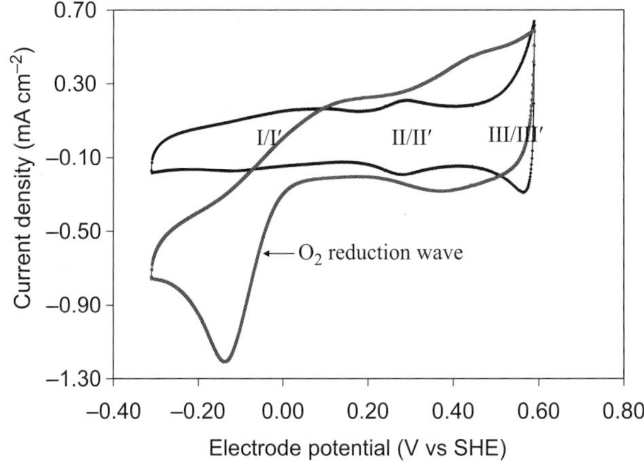

Figure 5.13 Cyclic voltammograms recorded on a glassy carbon (GC) disk electrode (0.28 cm^2) coated with Co-N-S/C catalyst in N$_2$-saturated and O$_2$-saturated 0.1 M KOH solutions, respectively. The catalyst was obtained by heat treating Co(II)-Pyridine/C precursor at 800 °C. Potential scan rate: 25 mV s^{-1}, and catalyst loading: 7.06×10^{-5} g cm^{-2}.[12]

current near both Waves II/II′ and III/III′, suggesting that these two waves may not be ORR active.

5.3.6. Recording ORR Current−Potential Curves Using RDE Technique

To study the ORR kinetics catalyzed by the catalyst or catalyst layer-coated electrode, the RDE current−potential curves are normally recorded at different electrode rotating rates when the electrolyte solution is saturated with O$_2$ or air. In order to obtain the steady-state curves, the controlled potential scan rate is normally slow such as less than 10 mV s^{-1}. In order to maintain the O$_2$ concentration inside the electrolyte solution, O$_2$ or air bubbling is maintained over the electrolyte solution during the data collection. The current−potential curves such as those shown in Figure 5.5(A) can be obtained and used to obtain the information about the ORR kinetics, as we discussed previously. In Figure 5.5, the catalyst is the Pt-based catalyst. Here, we present another example for non-noble metal catalyst to show how can the Koutecky−Levick theory be used in evaluate the ORR parameters catalyzed by non-noble metal catalyst.

5.3.7. Analysis of ORR Current−Potential Data

As discussed in Figure 5.5, Koutecky−Levick theory can be used to analyze the current−potential curves, and the equations used for this analysis are Equations (5.23) and (5.24). To obtain the ORR exchange current density (i^o), the electron-transfer coefficient (α), and the overall electron transfer number (n), the plots of the reciprocal of RDE current density ($1/i_{d,O}$) vs the reciprocal of square root of rotation rate ($\omega^{-1/2}$) can be drawn at different electrode potentials (as shown in Figure 5.14(B) and Figure 5.15 at four typical electrode potentials such as -0.10, -0.15, -0.20, and $-0.30\,\text{V}$ vs SHE). Both the slopes and the

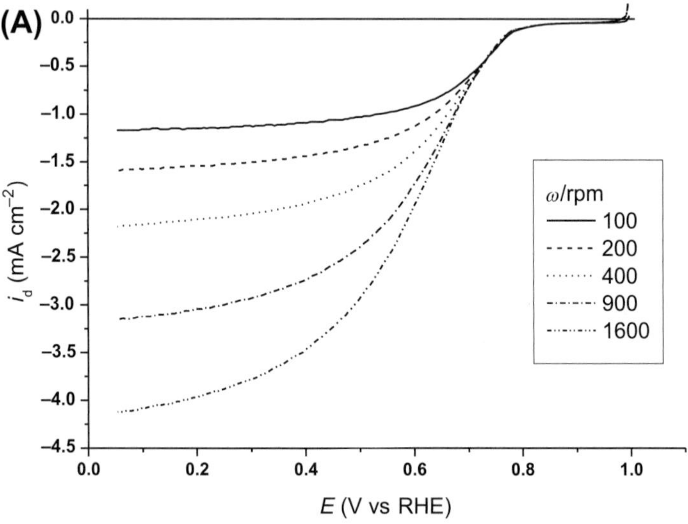

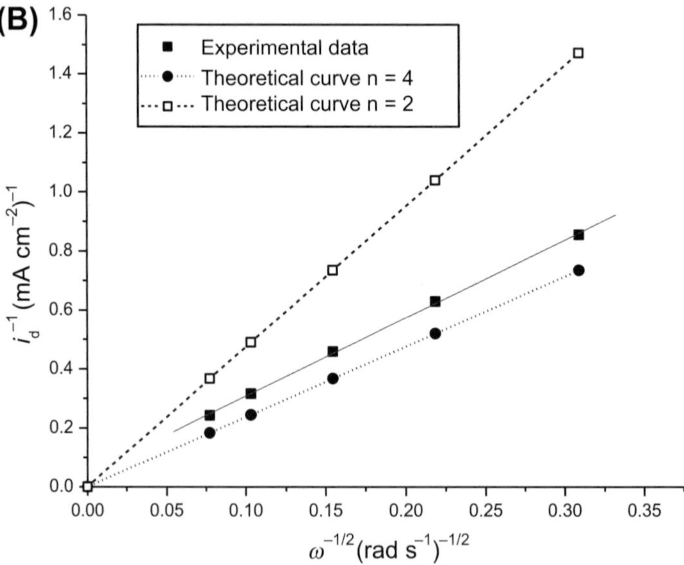

Figure 5.14 (A) Current–potential curves for ORR on a Fe-N/C-coated glassy carbon electrode. Recorded in 0.5 M H_2SO_4 solution at various electrode rotating rates as marked. Potential scan rate: 5 mV s^{-1}. Fe-N/C catalyst loading: 0.10 mg cm^{-2}; and (B) Koutecky–Levich plots for the ORR on a Fe-N/C coated glassy carbon electrode. The current data were taken at 0.1 V (vs RHE) from (A).[13]

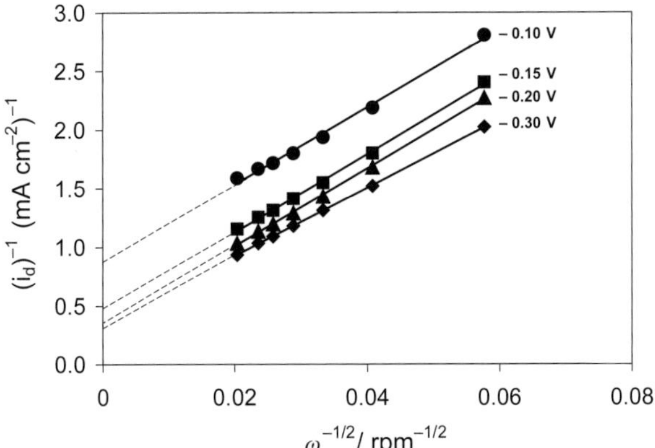

Figure 5.15 Koutecky–Levich plots at different electrode potentials. The current–potential curves recorded on the Co-N-S/C catalyst-coated GC disk electrode (0.28 cm^2) at different electrode rotation rates in O$_2$-saturated 3.0 M KOH solution. Potential scan rate: 25 mV s^{-1}, and catalyst loading: 7.06 × 10^{-5} g cm^{-2}.[12]

intercepts of these straight lines can be used to calculate the i^o, α, and n.

In Chapter 7, we will present other typical ORR examples measured by both RDE and RRDE techniques and their associated data analysis.

5.3.8. Cautions When Using RDE Technique[2]

When using RDE technique for electrochemical measurements, several cautions should be paid attention, including the iR drop between the RDE surface and the reference electrode tip, and the limitations of electrode rotating rate.

5.3.8.1 iR Drop between the RDE Surface and the Reference Electrode Tip

When the current (i) passes though the RDE surface, there will be a current distribution between the working electrode surface and the counter electrode surface, as schematically shown in Figure 5.16. Due to the electrolyte resistance (R), there will be an iR drop between the RDE surface and the reference electrode tip, which will cause an inconsistency between the controlled electrode potential ($E_{control}$ vs Reference electrode) and the actual electrode potential (E vs reference electrode):

$$E = E_{control} \pm iR \tag{5.29}$$

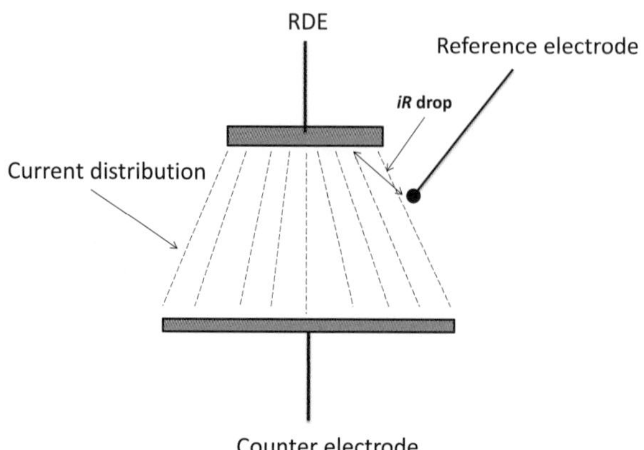

Figure 5.16 Schematic of the three-electrode system for RDE measurement, demonstrating the effect of the *iR* drop on the controlled electrode potential.

where "±" is used to take care of the current direction. This *iR* can definitely cause a measurement error. Therefore, in the experiments, this *iR* drop should be reduced or illuminated as much as possible. However, if the electrolyte solution resistance or the current density, or the distance between the RDE surface and the reference tip, or the current density is very small, the value of *iR* may be too small to affect the measurement accuracy. It is really dependent on the allowed error tolerance of the measurement. It is worthwhile to note that for a RDE measurement, if the reference tip is too close toward the electrode surface, it may change the pattern of the electrolyte convection when the electrode is rotated, leading to even larger error. It is better to compensate this *iR* drop according to the solution resistance and the current passed for a more accurate measurement. Some commercially available potentiostats have such an *iR* drop compensating function.

5.3.8.2 Limits of the Electrode Rotation Rate

As we discussed above, the RDE theory is based on the convection kinetics of the electrolyte solution. If the electrode rotating rate is too small, meaning that the solution flow rate is too slow, it will be difficult to establish the meaningful diffusion–convection layer near the electrode surface. In order to make meaningful measurement, there is a rough formula that can be used to obtain the limit of electrode rotating rate (ω_{low}):

$$\omega_{\text{low}} > \frac{10\nu}{r^2} \tag{5.30}$$

where v is the kinetic viscosity of the solution ($cm^2 s^{-1}$), and r is the radius of RDE (cm). For example, if v is 0.01 $cm^2 s^{-1}$, and r is 0.25 cm, the $\omega_{low} > 1.6$ s or >9.7 rpm.

Regarding the high end limit of the rotation rate, the major consideration is the turbulent flow of the electrolyte solution, which could make the RDE theory discussed above invalid. As we know, the RDE theory is based on the diffusion−convection kinetics of the pure solution laminar flow rather than turbulent flow. If the rate of electrode rotation is too high, the turbulent flow would happen, leading to error in data analysis using RDE theory. There is a parameter called Reynolds number (Re), which determines when the turbulent can happen. If this Re is larger than 2×10^5, the turbulent will happen. The upper limit of the electrode rotating rate is also dependent on the kinetic viscosity of the solution (v) and the radii of the RDE (r). According to this Reynolds number, the upper limit of the rotating rate (ω_{high}) can be expressed as[14]:

$$\omega_{high} < 2 \times 10^5 \frac{v}{r^2} \tag{5.31}$$

For example, if kinetic viscosity of the electrolyte solution is 0.01 $cm^2 s^{-1}$, and the RDE radius is 0.25 cm, the $\omega_{high} < 3.2 \times 10^4 s^{-1}$, or $\omega_{high} < 3.06 \times 10^5$ rpm. Therefore, the theoretically predicted range of electrode rotating rate is: 9.7 rpm $< \omega < 3.06 \times 10^5$ rpm. However, in practical RDE measurements, for safe, the normally employed RDE rotation rate range is about 100 rpm $< \omega < 10,000$ rpm.

In addition, at a very low rate of electrode rotation, the time to establish the steady-state must be very long, requiring a very low potential scan rate.

5.4. Chapter Summary

RDE is a commonly used technique for investigating the ORR in terms of both the electron transfer process on electrode surface and diffusion−convection kinetics near the electrode. To make appropriate usage in the ORR study, fundamental understanding of both the electron transfer process on electrode surface and diffusion−convection kinetics near the electrode is necessary. In this chapter, two kinds of RDE are presented, one is the smooth electrode surface, and the other is the catalyst layer-coated electrode. Based on the electrochemical reaction ($O + ne^- \leftrightarrow R$), the RDE theory, particularly those of the diffusion−convection kinetics, and its coupling with the electron-transfer process are presented. The famous Koutecky−Levich equation and its associated data analysis are deeply discussed. Using this Koutecky–Levich equation and based on the experiment data, the kinetic parameters such as exchange current density, electron-transfer coefficient and the overall electron transfer number can be obtained. For measurements using RDE technique, the apparatus of RDE and its associated devices such as rotator and electrochemical cell are described to give the readers the basic sense about this technique. The measurement procedure including RDE preparation, catalyst layer fabrication, current–potential curve recording, the data analysis, as well as the cautions in RDE measurements have also been discussed in this chapter.

We wish that the RDE knowledge and information can be useful to the reader for their continuing journey to the rest part of this book.

References

1. Zha Q. *Introduction to electrode kinetics.* Beijing: Chinese Science Press; 2005.
2. Bard AJ, Faulkner LF. *Electrochemical methods – fundamentals and applications.* New York: Wiley; 2001.
3. Levich VG. *Physicochemical hydrodynamics.* Englewood Cliffs, NJ: Prentice-Hall; 1962.
4. Song C, Tang Y, Zhang JL, Zhang J, Wang H, Shen J, et al. PEM fuel cell reaction kinetics in the temperature range of 23–120 °C. *Electrochim Acta* 2007;**52**:2552–61.
5. Fatish, K, Zhang, J, NRC-IFC, unpublished work.
6. Higuchi E, Uchida H, Watanabe M. Effect of loading level in platinum-dispersed carbon black electrocatalysts on oxygen reduction activity evaluated by rotating disk electrode. *J Electroanal Chem* 2005;**583**:69–76.
7. Pozio A, De Francesco M, Cemmi A, Cardellini F, Giorgi L. Comparison of high surface Pt/C catalysts by cyclic voltammetry. *J Power Sources* 2002;**105**:13–9.
8. Zhang J, Zhang H, Wu J, Zhang J. *PEM fuel cell testing and diagnosis.* Elsevier Press; 2013 [Chapter 12].
9. Vidaković T, Christov M, Sundmacher K. The use of CO stripping for in situ fuel cell catalyst characterization. *Electrochim Acta* 2007;**52**:5606–13.
10. Mayrhofer KJJ, Strmcnik D, Blizanac BB, Stamenkovic V, Arenz M, Markovic NM. Measurement of oxygen reduction activities via the rotating disc electrode method: from Pt model surfaces to carbon-supported high surface area catalysts. *Electrochim Acta* 2008;**53**:3181–8.
11. Yu A, Chabot V, Zhang J. *Electrochemical supercapacitors for energy storage and delivery, fundamentals and applications.* CRC Press; 2013 [Chapter 9].
12. Qiao J, Xu L, Ding L, Shi P, Zhang L, Bake R, et al. Effect of KOH concentration on the oxygen reduction kinetics catalyzed by heat-treated Co-Pyridine/C electrocatalysts. *Int J Electrochem Sci* 2013;**8**:1189–208.
13. Bezerra CWB, Zhang L, Lee K, Liu H, Zhang J, Shi Z, et al. Novel carbon-supported Fe–N electrocatalysts synthesized through heat treatment of iron tripyridyl triazine complexes for the PEM fuel cell oxygen reduction reaction. *Electrochim Acta* 2008;**53**:7703–10.
14. Newman J. *J Electrochem Soc* 1966;**113**.

6

ROTATING RING-DISK ELECTRODE METHOD

Zheng Jia[a], Geping Yin[a] and Jiujun Zhang[b]
[a]*School of Chemical Engineering and Technology, Harbin Institute of Technology, Harbin, China,* [b]*Energy, Mining and Environment, National Research Council of Canada, Vancouver, BC, Canada*

CHAPTER OUTLINE

6.1. Introduction

In Chapter 5, we presented rotating disk electrode (RDE) technique in terms of both fundamentals and applications in catalyst evaluation for oxygen reduction reaction (ORR) kinetics. In order to further understand the ORR mechanism, another commonly used technique called rotating ring-disk electrode (RRDE) method has been demonstrated to be a powerful tool, particularly in detecting the intermediate such as peroxide produced during the ORR.

In an RRDE, there are two electrodes: one is the disk electrode located in center, and the other is the ring electrode surrounding the disk electrode. During the measurement, both the ring and disk

Rotating Electrode Methods and Oxygen Reduction Electrocatalysts. http://dx.doi.org/10.1016/B978-0-444-63278-4.00006-9

electrodes rotate at the same rate. Both hydrodynamic RDE and RRDE have advantages of larger mass transport rates at the electrode surface than the rates at stationary electrodes, which thus facilitates the study of charge-transfer kinetics even in the presence of mass transport limitation. Furthermore, the hydrodynamic equations and the diffusion-convective equation of an RDE or RRDE system can be rigorously solved for the steady state. Therefore, such hydrodynamic electrode systems provide a uniform, reproducible, controllable, and calculable mass transport of electrochemically active species toward/from an electrode surface. In addition, the steady-state can be attained rather quickly for such systems, and thus measurements can be made with high precision.

The difference between an RRDE and an RDE is the addition of the second working electrode in the form of a ring around the central disk-working electrode. As the disk and the ring electrodes are electrically isolated, both the disk and ring electrodes of an RRDE function as the separate two hydrodynamic electrodes. RRDE experiments are useful in studying multielectron processes, the kinetics of a slow electron transfer, adsorption/desorption steps, and electrochemical reaction mechanisms, particularly for ORR.

In this chapter, both the fundamentals and applications of RRDE in studying electrochemical reaction kinetics such as the ORR mechanism will be presented in a detailed level.

6.2. Theory of Rotating Ring-Disk Electrode Technique[1]

A RRDE consists of a central disk electrode, and a thin insulating barrier and a ring electrode, both of which are sequentially surrounding the disk electrode in the shape of concentric circle, as shown in Figure 6.1. On the RRDE, the radius of the disk, r_1, and the inner radius, r_2, and the outer radius, r_3, of the ring are the key parameters of an RRDE.

Just like an RDE described in Chapter 5, the whole RRDE system is set into rotation during the experiment. As a result, the spinning RRDE drags the electrolyte solution to its surface and, because of a centrifugal force, flings the solution outwards from the center in a radial direction. The electrolyte solution at the RRDE surface is replenished by a flow normal to the surface. In this way, the solution is flowing in a laminar manner across the electrode surface. The convection velocity of the solution is increasing with increasing the electrode rotation rate, and as a result, the diffusion layer, which is built up near the disk/ring

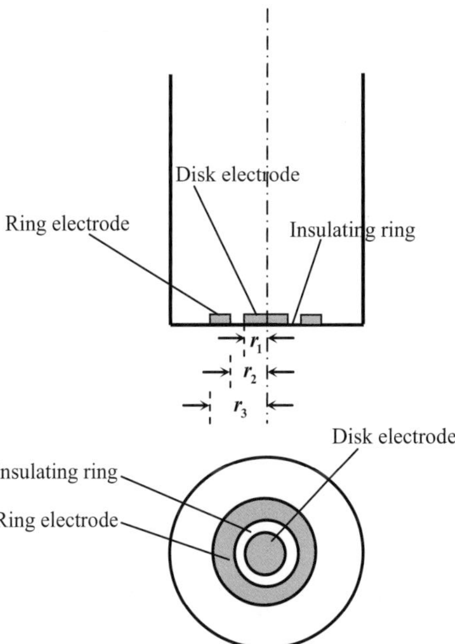

Figure 6.1 Schematic of a rotating ring-disk electrode. (For color version of this figure, the reader is referred to the online version of this book.)

electrode surface due to the electrochemical reaction, becomes thinner with increasing the diffusion–convection rate. It means that the diffusion–convection rate of reactant can be regulated by controlling the electrode rotation rate.

Due to the defined hydrodynamic conditions, the diffusion–convection process near the disk/ring electrode surface can be calculated based on the experiment conditions and the measured data, providing the information of the electrode kinetics as a function of the applied potential, which has been described in Chapter 5.

As the disk electrode is independent of the ring electrode, here we give the discussion about the electrochemical reaction occurring on the disk and the ring electrodes, respectively.

6.2.1. Electrochemical Reaction on the Disk Electrode

Let's recall the discussion in Chapter 5 for the RDE theory. Considering an electrode reaction:

$$O + ne^- \leftrightarrow R \tag{6-I}$$

where O represents the oxidant, R represents the reductant, and n is the electron-transfer number. If the laminar flow condition is fulfilled, the diffusion–convection limiting current density at the disk electrode of RRDE induced by the forward reaction in Reaction (6-I) can be described by the *Levich equation* (for detailed description, please see Chapter 5):

$$I_{d,O} = 0.62 nFD_O^{2/3} \nu^{-1/6} \omega^{1/2} C_O^o \qquad (6.1)$$

where $I_{d,O}$ is the diffusion–convection limiting current density of oxidant reduction (A cm^{-2}); F is the Faraday's constant (96,487 A s mol^{-1}); D_O is the diffusion coefficient of oxidant (cm^2 s^{-1}); ν is the solution kinetic viscosity (cm^2 s^{-1}); ω is the electrode angular rotation rate (rad s^{-1}); and C_O^o is the bulk concentration of the oxidant (mol cm^{-3}), respectively. Note this ω has a unit of revolutions per minute (rpm), and Eqn (6.1) will become Eqn (6.2) $\left(\text{rad s}^{-1} = \frac{2\pi}{60} \text{ rpm}\right)$:

$$I_{d,O} = 0.201 nFD_O^{2/3} \nu^{-1/6} \omega^{1/2} C_O^o \qquad (6.2)$$

For nonlimiting diffusion conditions, the disk electrode current density would be:

$$i_{d,O} = 0.201 nFD_O^{2/3} \nu^{-1/6} \omega^{1/2} \left(C_O^o - C_O^s\right) \qquad (6.3)$$

where, C_O^s is the disk surface concentration of the oxidant (mol cm^{-3}). For planner smooth disk electrode, the *koutecky–Levich equation* can be expressed as:

$$\frac{1}{i_{d,O}} = \frac{1}{i_{k,O}} + \frac{1}{I_{d,O}} \qquad (6.4)$$

where $i_{k,O} = i^o \exp\left(-\frac{(1-\alpha)nF(E-E^{eq})}{RT}\right)$, $I_{d,O} = 0.201 nFD_O^{2/3} \nu^{-1/6} \omega^{1/2} C_O^o$. Substituting these two expressions into Eqn (6.4), we can obtain:

$$\frac{1}{i_{d,O}} = \frac{1}{i^o} \exp\left(\frac{(1-\alpha)nF(E-E^{eq})}{RT}\right) + \frac{\omega^{-1/2}}{0.201 nFD_O^{2/3} \nu^{-1/6} C_O^o} \qquad (6.5)$$

where $i_{k,O}$ is the electron-transfer kinetic current density (A cm^{-2}), i^o is the exchange current density of Reaction (6-I) (A cm^{-2}), α is the electron-transfer coefficient in the rate determining step, E is the electrode potential (V), E^{eq} is the equilibrium electrode potential of Reaction (6-I), R is the universal gas constant (8.314 J mol^{-1} K^{-1}), and T is the absolute temperature (K). The plot of $\frac{1}{i_{d,O}}$ vs $\omega^{-1/2}$ based on the experiment data allows the estimation of $i_{k,O}$ from the interception and $\frac{1}{0.201 nFD_O^{2/3} \nu^{-1/6} C_O^o}$ from the slope. If the plots can be obtained at different electrode potentials, the values of electrode potential can be plotted against $\ln(i_{k,O})$ based on the equation below:

$$E = E^{eq} + \frac{RT}{(1-\alpha)nF}\ln(i^o) - \frac{RT}{(1-\alpha)nF}\ln(i_{k,O}) \qquad (6.6)$$

From the slope of the E vs $\ln(i_{k,O})$, the electron-transfer number (n) and coefficient (α) can be obtained, and from the intercept the exchange current density (i^o) can be estimated if E^{eq} is known.

Therefore, using RDE measurements, the kinetic parameters of electron-transfer kinetics such as the electron-transfer number (n) and coefficient (α) and the exchange current density (i^o) can be estimated. At the same time, the reactant transport parameters such as the overall electron-transfer number (n for the complex reaction), reactant diffusion coefficient, the kinetic viscosity can also be estimated.

For a planner matrix electrode layer, the following equation could be used:

$$\frac{1}{i_{d,O}} = \frac{1}{i_{k,O}} + \frac{1}{I_{d,O}} + \frac{1}{I_{m,O}} \qquad (6.7)$$

where $I_{m,O} = \frac{nFC_m D_m}{\delta_m}$ (see Chapter 2), which is the diffusion limiting current density of oxidant within the equivalent ionomer layer in the electrode layer. Substituting the $I_{m,O}$, $i_{k,O}$, and $I_{d,O}$ expressions into Eqn (6.7), we can obtain:

$$\frac{1}{i_{d,O}} = \frac{1}{i^o}\exp\left(\frac{(1-\alpha)nF(E-E^{eq})}{RT}\right) + \frac{\omega^{-1/2}}{0.201nFD_O^{2/3}\nu^{-1/6}C_O^o}$$

$$+ \frac{\delta_m}{nFC_m D_m} \qquad (6.8)$$

6.2.2. Electrochemical Reaction on the Ring Electrode

As shown in Figure 6.1, the ring electrode can be treated in the similar way to that of RDE. Then, we can set the three directions' coordinates as r, x, and ϕ, shown in Figure 5.3(B). Then, we have the Eqn (6.9) at the steady-state reaction:

$$D_O\left[\frac{\partial^2 C_O}{\partial x^2} + \frac{\partial^2 C_O}{\partial r^2} + \frac{1}{r}\left(\frac{\partial C_O}{\partial r}\right) + \frac{1}{r^2}\left(\frac{\partial^2 C_O}{\partial \phi^2}\right)\right]$$

$$= v_r\left(\frac{\partial C_O}{\partial r}\right) + \frac{v_\phi}{r}\left(\frac{\partial C_O}{\partial \phi}\right) + v_x\left(\frac{\partial C_O}{\partial x}\right) \qquad (6.9)$$

As the originations of the three-coordination direction is located at the disk center, the geometric symmetry indicates that the oxidant concentration is not a function of ϕ, and therefore, $\left(\frac{\partial C_O}{\partial \phi}\right) = \left(\frac{\partial^2 C_O}{\partial \phi^2}\right) = 0$. In addition, the term of $\frac{\partial^2 C_O}{\partial r^2} + \frac{1}{r}\frac{\partial C_O}{\partial r} \ll \frac{\partial^2 C_O}{\partial^2 x}$. As a result, Eqn (6.9) can be simplified as Eqn (6.10):

$$D_O \frac{\partial^2 C_O}{\partial x^2} = v_x \left(\frac{\partial C_O}{\partial x}\right) + v_r \left(\frac{\partial C_O}{\partial r}\right) \tag{6.10}$$

There is one more term in this equation $\left(v_r \left(\frac{\partial C_O}{\partial r}\right)\right)$ when compared to that for disk electrode in Chapter 5 (Eqn (5.11)). This is because the distribution of oxidant concentration along the r coordinate is not uniform. This $v_r \left(\frac{\partial C_O}{\partial r}\right)$ term cannot be eliminated from Eqn (6.9). Substituting Eqns (5.6) and (5.7) into Eqn (6.10), we can obtain:

$$\frac{\partial^2 C_O}{\partial x^2} = -\frac{0.51x^2}{D_O \omega^{-3/2}\nu^{1/2}}\left(\frac{\partial C_O}{\partial x}\right) + \frac{0.51rx}{D_O \omega^{-3/2}\nu^{1/2}}\left(\frac{\partial C_O}{\partial r}\right) \tag{6.11}$$

Rearranging Eqn (6.11), we can obtain:

$$\left(\frac{\partial C_O}{\partial x}\right) = \frac{r}{x}\left(\frac{\partial C_O}{\partial r}\right) - \frac{D_O \omega^{-3/2}\nu^{1/2}}{0.51x^2}\frac{\partial^2 C_O}{\partial x^2} \tag{6.11a}$$

Note that at $x = \infty$, the oxidant concentration is equal to the bulk concentration ($C_O = C_O^0$), and when the surface concentration of oxidant drops to zero ($C_O^s = 0$ at $x = 0$ in the surface range of $r_2 \leq r \leq r_3$), the limiting oxidant supply is reached. Resolving Eqn (6.11a), the oxidant flow rate $\left(\frac{\partial C_O}{\partial x}\right)_{x=0}$ at the electrode surface ($x = 0$) can be obtained, from which the ring-limiting current can be derived as described in the following.

In order to obtain the limiting current at the ring electrode, we can divide the ring thickness ($r_3 - r_2$) as shown in Figure 6.1 into numerous rings with a thickness of dr, then the small ring's area can be expressed as $\pi(r + dr)^2 - \pi r^2 \approx 2\pi r(dr)$. Then, the ring-limiting current at the small ring ($dI_{r,O}$) can be expressed as Eqn (6.12):

$$dI_{r,O} = nFD_O(2\pi r(dr))\left(\frac{\partial C_O}{\partial x}\right)_{x=0} \tag{6.12}$$

Then, the ring limiting current (note here is not current density) can be expressed by integrating all small currents:

$$I_{r,O} = \sum_{r_2}^{r_3}(dI_{R,O}) = nFD_O2\pi\sum_{r_2}^{r_3}\left(\frac{\partial C_O}{\partial x}\right)_{x=0}rdr$$

$$= nFD_O2\pi\int_{r_2}^{r_3}\left(\frac{\partial C_O}{\partial x}\right)_{x=0}rdr$$

$$= nFD_O2\pi\int_{r_2}^{r_3}\left[\frac{r}{x}\left(\frac{\partial C_O}{\partial r}\right) - \frac{D_O\omega^{-3/2}\nu^{1/2}}{0.51x^2}\left(\frac{\partial^2 C_O}{\partial x^2}\right)_{x=0}\right]rdr$$

$$(6.13)$$

Solving Eqn (6.13), the ring-limiting current can be obtained[1]:

$$I_{r,O} = 0.62nF\pi\left(r_3^3 - r_2^3\right)^{2/3}D_O^{2/3}\nu^{-1/6}\omega^{1/2}C_O^o \qquad (6.14)$$

Then, in nonlimiting situation, the ring current can be expressed as:

$$i_{r,O} = 0.62nF\pi\left(r_3^3 - r_2^3\right)^{2/3}D_O^{2/3}\nu^{-1/6}\omega^{1/2}\left(C_O^o - C_O^s\right) \qquad (6.15)$$

where $i_{R,O}$ has a unit of A rather than A cm^{-2}.

6.3. RRDE Collection Efficiency

The major advantage of RRDE is its usage in analyzing the reaction mechanism in which ring is used as a detector for disk reaction products or intermediates from which the insight of the mechanism can be obtained. The typical example is its usage in studying the ORR. The ORR happens on the disk electrode, producing intermediate such as peroxide, which can be thrown away from the disk electrode surface; one portion of the peroxide will be thrown into the bulk solution, the second portion will reach the ring electrode surface and be electrochemically oxidized or reduced, and the third portion will be chemically decomposed. The portion reacted on the ring electrode will give the ring current from which the quantity of peroxide can be detected. For the purpose of investigating the reaction mechanism, the material of disk electrode may not be necessary the same material of the ring, and also the disk and ring potential can be controlled individually at different values according to the targeted reactants.

In order to quantitatively measure the amount of the disk reaction products or intermediates reaching the ring electrode, an important parameter of the ring electrode, called the collection efficiency, has to be known. To give an expression about the collection efficiency, let us consider the following reactions on an RRDE:

$$O + ne^- \rightarrow R \qquad (6\text{-II})$$

$$R \rightarrow O + ne^- \qquad (6\text{-III})$$

As shown in Figure 6.2, when the disk electrode is controlled at a potential (E_d) to make Reaction (6-II) happen on the disk electrode to produce R, a fraction of R (xR) will be thrown onto the ring electrode, where Reaction (6-III) can happen at a limiting mass-transport rate if the ring is controlled at a sufficiently positive potential (E_r) that any R reaching the ring electrode is oxidized immediately. From both the disk and ring currents, the collection efficiency of the ring can be defined. The quantity of R (xR) detected by the ring electrode is strongly dependent on the electrode rotating rate. However, due to the limited ring area, it is impossible to detect all R coming from the disk reaction. In other words, it is impossible for the ring to collect all R produced at disk electrode. Normally, for the commercially available RRDE, the collection efficiency is in the range of 20−40% depending on the relative geometric sizes of the ring and disk electrodes. In addition, the collection efficiency is also dependent on the thickness of the isolator ring between the ring and disk electrodes (($r_2 - r_1$) in Figure 6.2).

Therefore, the collection efficiency measures what the percentage of the R produced at the disk electrode can be detected

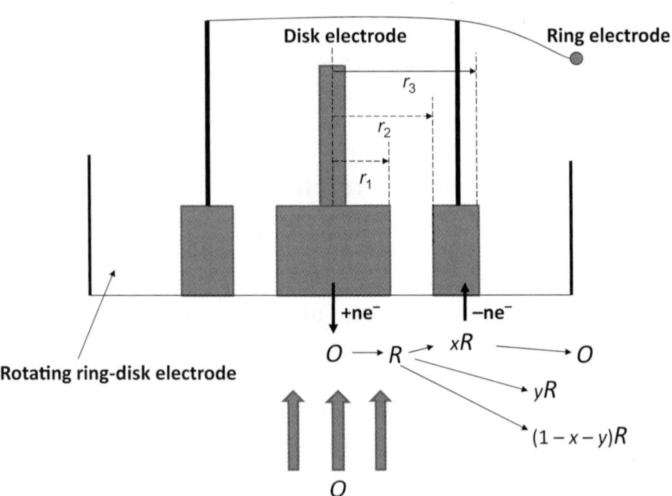

Figure 6.2 Schematic of reaction ($O + ne^- \rightarrow R \rightarrow O + ne^-$) on the rotating ring-disk electrode. xR is the portion electrochemically reacted on the ring electrode, yR is the portion thrown into the bulk solution, and $(1 - x - y)R$ is the portion chemically decomposed. (For color version of this figure, the reader is referred to the online version of this book.)

on the ring electrode, which can be obtained if both the disk current (i_d) for O reduction and the ring limiting current ($I_{r,R}$) for R oxidation are measured. By convention, when the reaction is a steady-state process, and the ring electrode current for R oxidation reaches limiting current ($I_{r,R}$), and R is stable in the solution without chemical decomposition, the collection efficiency (N) is defined as:

$$N = \left| \frac{I_{r,R}}{i_d} \right| \tag{6.16}$$

Because the RRDE itself is a universal tool for the study of any electrochemical reaction rather than a specific tool only for a specific reaction, its collection efficiency should be its intrinsic property and independent of the reaction studied. Therefore, this collection efficiency should be only the function of the geometric sizes of the disk and ring electrodes such as r_1, r_2, and r_3 in Figure 6.2. To obtain the relationship between the collection efficiency (N) and the RRDE geometric size (r_1, r_2, and r_3), that is, ($N = f(r_1, r_2, r_3)$), we still need to get the expressions of i_d and $I_{r,R}$ in Eqn (6.15).

To obtain the expression of the ring-limiting current induced by the oxidation of R (Reaction (6-III)), concentration distribution of R (C_R) near the ring surface must be known. According to Eqn (6.11a), the similar equation for R can be written as:

$$\left(\frac{\partial C_R}{\partial x} \right) = \frac{r}{x} \left(\frac{\partial C_R}{\partial r} \right) - \frac{D_R \omega^{-3/2} \nu^{1/2}}{0.51 x^2} \frac{\partial^2 C_R}{\partial x^2} \tag{6.17}$$

To solve this equation, we have to set up four boundary conditions:

1. First, we assume that there is no R in the electrolyte solution at the beginning of the reaction, and thus the bulk concentration of R is equal to zero, that is:

$$C_R = 0 \quad \text{at} \quad x = \infty \tag{6.18}$$

2. Under steady-state conditions, the rate of charge-transfer process at the disk electrode is equal to the rate of removal of R from the disk surface (According to the definition of N, there is no further reaction of R at the disk electrode and chemical decomposition of R in the electrolyte solution). Therefore, the mass balance on disk electrode surface ($x = 0$) can be expressed as:

$$\frac{i_d}{nF} = 0.62 \pi r_1^2 D_R^{2/3} \nu^{-1/6} \omega^{1/2} C_R^s \quad \text{at} \quad 0 \leq r \leq r_1 \tag{6.19}$$

where C_R^s is the disk surface concentration of R, and πr_1^2 is the surface area of the disk electrode.

3. On the surface of isolator ring of the RRDE, there is no current flowing and thus the flux of R is zero, that is:

$$\left(\frac{\partial C_R}{\partial x}\right)_{x=0} = 0 \quad \text{at} \quad r_1 < r < r_2 \tag{6.20}$$

4. On the ring electrode surface, the concentration of R is equal to zero under limiting ring current conditions, that is:

$$C_R = 0(x = 0) \quad \text{at} \quad r_2 \le r \le r_3 \tag{6.21}$$

Using these four boundary conditions, and resolving Eqn (6.17), we can obtain[2]:

$$\left(\frac{\partial C_R}{\partial x}\right)_{x=0} = \frac{0.4 i_d}{nF} \frac{r_1^2 r_2}{r^3} \frac{\left[1 - \frac{3}{4}\left(\frac{r_1}{r_2}\right)^3\right]^{1/3}}{\left[1 - \left(\frac{r_2}{r}\right)^3\right]^{1/3}\left[1 - \frac{3}{4}\left(\frac{r_1}{r}\right)^3\right]} \tag{6.22}$$

The ring limiting current can be expressed as:

$$I_{r,R} = 2\pi nF \int_{r_2}^{r_3} \left(\frac{\partial C_R}{\partial x}\right)_{x=0} r dr$$

$$= 2\pi nF \frac{0.4 i_d}{nF} \int_{r_2}^{r_3} \left(\frac{r_1^2 r_2}{r^3} \frac{\left[1 - \frac{3}{4}\left(\frac{r_1}{r_2}\right)^3\right]^{1/3}}{\left[1 - \left(\frac{r_2}{r}\right)^3\right]^{1/3}\left[1 - \frac{3}{4}\left(\frac{r_1}{r}\right)^3\right]}\right) r dr$$

$$\tag{6.23}$$

Rearranging Eqn (6.23), we can get:

$$\frac{I_{r,R}}{i_d} = 0.8\pi \int_{r_2}^{r_3} \left(\frac{r_1^2 r_2}{r^3} \frac{\left[1 - \frac{3}{4}\left(\frac{r_1}{r_2}\right)^3\right]^{1/3}}{\left[1 - \left(\frac{r_2}{r}\right)^3\right]^{1/3}\left[1 - \frac{3}{4}\left(\frac{r_1}{r}\right)^3\right]}\right) r dr \tag{6.24}$$

Equation (6.24) shows that the collection efficiency is only the function of RRDE geometric size such as r_1, r_2, and r_3. It is nothing related to the reaction's parameters such as the electrode rotating rate (ω), and solution viscosity (ν), reactant and product diffusion coefficients (D_O and D_R), as well as the electron-transfer number (n).

A result of Eqn (6.24) for collection efficiency has been summarized by Bard and Faulkner[1]:

$$N = 1 - F\left(\frac{r_2^3 - r_1^3}{r_3^3 - r_2^3}\right) + \left(\frac{r_3^3 - r_2^3}{r_1^3}\right)^{2/3}\left[1 - F\left(\left(\frac{r_2}{r_1}\right)^3 - 1\right)\right]$$
$$- \left(\frac{r_3}{r_1}\right)^2\left\{1 - F\left[\left(\frac{r_2^3 - r_1^3}{r_3^3 - r_2^3}\right)\left(\frac{r_3}{r_1}\right)^3\right]\right\}$$

$$(6.25)$$

If $\beta_1 = \frac{r_2^3 - r_1^3}{r_3^3 - r_2^3}$, and $\beta_2 = \left(\frac{r_2^3 - r_1^3}{r_3^3 - r_2^3}\right)\left(\frac{r_3}{r_1}\right)^3$, $F\left(\frac{r_2^3 - r_1^3}{r_3^3 - r_2^3}\right) = F(\beta_1)$ and $F\left[\left(\frac{r_2^3 - r_1^3}{r_3^3 - r_2^3}\right)\left(\frac{r_3}{r_1}\right)^3\right] = F(\beta_2)$ in Eqn (6.25) can be expressed as Eqns (6.26) and (6.27), respectively:

$$F(\beta_1) = \left(\frac{\sqrt{3}}{4\pi}\right)\ln\left[\frac{\left(1 + \beta_1^{1/3}\right)^3}{1 + \beta_1}\right] + \frac{3}{2\pi}\arctan\left(\frac{2\beta_1^{1/3} - 1}{\sqrt{3}}\right) + \frac{1}{4}$$

$$(6.26)$$

$$F(\beta_2) = \left(\frac{\sqrt{3}}{4\pi}\right)\ln\left[\frac{\left(1 + \beta_2^{1/3}\right)^3}{1 + \beta_2}\right] + \frac{3}{2\pi}\arctan\left(\frac{2\beta_2^{1/3} - 1}{\sqrt{3}}\right) + \frac{1}{4}$$

$$(6.27)$$

Equation (6.25) shows that when the area of ring electrode increases (($r_3 - r_2$) increases), the collection efficiency will increase; and when the area of isolator ring is increased (($r_2 - r_1$) increases), the collection efficiency will decrease.

According to Eqn (6.25), the collection efficiency can be calculated if r_1, r_2, and r_3 are known. Table 6.1 lists the calculated collection efficiencies of commercially available RRDEs from Pine Instruments.

Besides the calculation using Eqn (6.25), this collection efficiency can also be experimentally measured using a simple reaction such as $[Fe(III)(CN)_6]^{3-} + e^- \leftrightarrow [Fe(II)(CN)_6]^{4-}$. The detailed calibration of the RRDE collection efficiency will be given in the following section.

It is also worthwhile to note that the expression of the collection efficiency is derived based on the reaction with a simple mechanism as expressed by Reactions (6-II) and (6-III) ($O + ne^- \rightarrow R \rightarrow O + ne^-$). However, for a complicated reaction such as ORR, this collection efficiency is also very useful, from which the quantitative estimation of the intermediates

Table 6.1. RRDE Geometric Parameters and Their Corresponding Collection Efficiencies (Commercially Available RRDEs from Pine Instruments)[3]

| | Metric Radii | | | Metric Areas (cm^2) | | |
| | Disk | Ring | Ring | | | Collection Efficiency (%) |
	Outer Radius	Inner Radius	Outer Radius	Disk	Ring	
E6 series	2.50	3.25	3.75	0.1963	0.1100	25.6%
E6HT series	2.75	3.25	4.25	0.2376	0.2356	38.3%
E7R8 series	2.29	2.47	2.69	0.1640	0.0364	21.8%
E7R9 series	2.81	3.13	3.96	0.2472	0.1859	37.0%

can be made, providing an insight into the reaction mechanism. This will be discussed in a late section of this chapter.

6.4. RRDE Instrumentation

In Chapter 5, the instrumentation of RDE technique including electrode design and electrochemical cell construction have been presented, which are also applicable to RRDE technique.

Regarding RRDE technique, a variety of RRDEs with different materials, configuration, and size can be constructed for targeted researches. Due to the relative stability of platinum, gold, and glassy carbon (GC), they are commonly used as the disk or ring electrode materials. Figure 6.3 shows several commercially available RRDEs from ALS. However, different combinations between the disk and ring materials are not limited to those three shown in Figure 6.3. For example, in total, there are nine different combinations among three materials of Pt, Au, and GC for a ring-disk electrode arrangement. For special researches, some in-house RRDEs can also be constructed using different materials from those three and the geometric sizes in Figure 6.3 by the researchers to fit the requirements of their study.

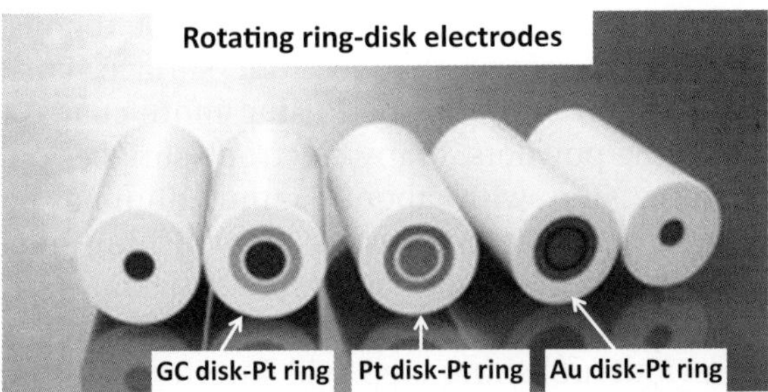

Figure 6.3 Commercially available rotating ring-disk electrode from ALS. Here GC represents the glassy carbon.[4] (For color version of this figure, the reader is referred to the online version of this book.)

It has been recognized that the RRDE is an important tool for characterizing the fundamental properties of ORR electrocatalysts. For example, in a proton exchange membrane fuel cell, ORR at the cathode is often enhanced by an electrocatalyst. When an ORR is catalyzed by an electrocatalyst, either a noble metal catalyst or a non-noble metal catalyst, a by-product, or intermediate such as peroxide may be produced. To detect how much peroxide is produced, an RRDE collection experiment can be used to probe the peroxide-generating tendencies of an electrocatalyst. For evaluating the activity of ORR electrocatalysts, GC or gold material is normally used as the disk electrode substrate on which the catalyst layer is fabricated. This is because they do not have a significant ORR catalytic activity and also have reasonable affinity to the catalyst layer. For detecting the ORR intermediates, Pt is commonly used as the ring material.

The insulating gap, made from solvent-resistant organic materials such as PTFE or PEEK, between the disk and ring electrodes is desirable to be designed small to increase the collection efficiency, which usually ranges from 0.1 to 0.5 mm. This is because the narrow insulating gap could reduce the "transit time" necessary for an intermediate species generated at the disk electrode to successfully reach the ring electrode and to be detected, which is especially important for "short-lives" (unstable) intermediate products.

In order to maintain a uniform and controllable hydrodynamic condition, the RRDE surface must be retained quite plain and regular. Great cautions for electrode polishing are necessary. The electrode polishing must be operated on a very flat surface, such as a glass plate. During polishing, the electrode surface must be kept as parallel to the surface of the glass plate as possible, which is especially important for ensuring that the surrounding plastic is not worn unevenly.

The used RRDE electrode usually needs to be repolished after each experiment to refresh the electrode surface. For this purpose, the polish can be performed using more coarse than fine alumina polishing powders. Many electrode surfaces only need a single polishing step with about 50 nm alumina powder to physically remove the contaminants. All polishing steps require extensive rinsing of the electrode before moving onto the next stage. Without extensive rinsing, the contaminated materials from the previous polishing step will interfere the progression toward a finely polished surface.

Electric connection between a rotating disk and ring parts of an RRDE and external circuit is performed through carbon brushes contacting a stainless steel rotation shaft. One end of the rotation shaft is fixed to a DC servo motor to achieve rotation, while the other end of the rotation shaft consists of a disk tip and a ring tip, which are electrically insulated by an organic resin. An RRDE electrode head can be fabricated on or removed from this end of the rotation shaft easily by twisting with fingers. The disk tip contacts with the disk electrode, so the disk current goes through the shaft center and flows to the top of the shaft. The ring tip contacts with the ring electrode and the ring current flows to the middle of the shaft. By using two pairs of the carbon brushes contacting to the upper and middle parts of the shaft, the ring and disk currents can flow to the external circuit respectively during electrode rotation.

In order to obtain the mass transport rate in good reproducibility, it is very important to keep the electrode rotating in a stable speed during the measurement, which requires high rotation accuracy of the motor and its associated controller.

As discussed in Chapter 5, all quantitative hydrodynamic equations are derived based on the laminar flow kinetics of the electrolyte solution, and therefore, the maintenance of laminar flow conditions near the RRDE surface in the measurements is necessary. In addition, the electrode rotation rate should be modulated at a proper range. Commonly, a rotation rate higher than 100 rpm is necessary to eliminate the impact of natural convection. However, if the rate is too high, the turbulent flow may occur, causing significant measurement error. Furthermore, if the rotation rate is too high, the noise may be caused, and the service life of RRDE electrode, motor and carbon brushes may be shortened. Normally the high end of electrode rotation rate is controlled at below 5000 rpm although theoretically it allows to be as high as 10,000 rpm without producing turbulent flow conditions.

In order to maintain a certain atmosphere in the electrolytic cell, a gas purge equipment should be provided. For example, inert gas such as nitrogen or argon gas can be bubbled in the

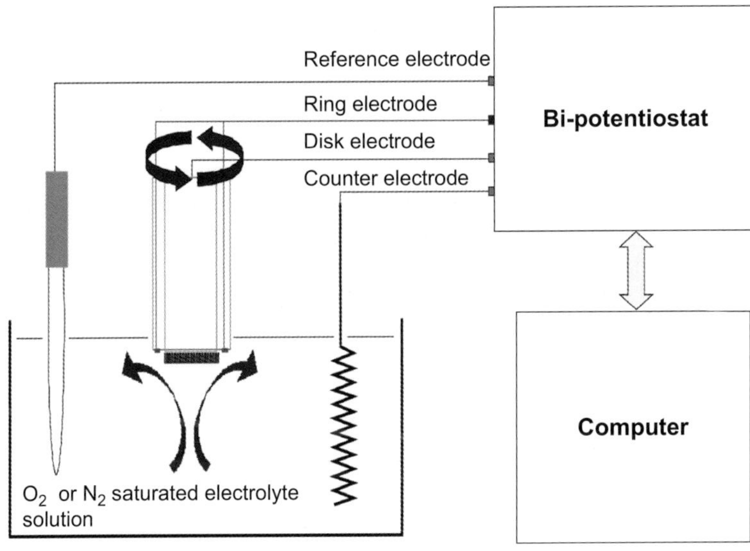

Figure 6.4 Connection of the electrodes with the bi-potentiostat. (For color version of this figure, the reader is referred to the online version of this book.). Modified from Ref. 5.

solution for 10–30 min before the experiment to expel the dissolved oxygen in the solution. If the oxygen-reduction reaction is the research target, oxygen gas should be purged in the solution to obtain an oxygen-saturated solution.

Regarding the control of electrode potential, a bi-potentiostat is normally used to control the disk and ring electrodes at independent electrode potentials. This bi-potentiostat can give the separate control and adjustment of the disk electrode potential, E_d, and the ring electrode potential, E_r, with a common reference electrode and counter electrode. This means that the bi-potentiostat has four electrode connectors—the first one is connected to the disk electrode, the second to the ring electrode, the third to the reference electrode, and the fourth to the counter electrode. Figure 6.4 shows the electrical connection of the disk electrode, the ring electrode, the reference electrode, and the counterelectrode with the corresponding leads of the bi-potentiostat.

In Figure 6.4, the disk and ring electrode potentials can be independently controlled to linearly sweep or be constant adopting linear sweep voltammetry or chronoamperometry. The most frequently used is the method that the disk potential is scanned linearly and the ring electrode is fixed at a constant potential.

6.5. RRDE Measurements

For RRDE measurement, the first step is to calibrate its collection efficiency. As mentioned previously, the collection efficiency is one of the most important parameters in measurement and data analysis. Although the collection efficiency can be calculated according to Eqn (6.25) based on the geometric sizes of the radius of the disk electrode, the thicknesses of the ring electrode, and the isolating ring, the actual electrode surface of an RRDE may be somewhat uneven, and sometimes the real collection efficiency is hardly to be coinciding with the theoretical calculation value. To calibrate the collection efficiency, in general, reversible and simple systems such as $[Fe(CN)_6]^{4-}/[Fe(CN)_6]^{3-}$, hydroquinone/benzoquinone, ferrocene$^{0/+}$, Br^-/Br_3^- and so on can be used for RRDE electrode collection efficiency measurement.

As an example, let's take the well-known reversible reaction of ferricyanide/ferrocyanide on an RRDE with Pt as ring and Pt as disk electrodes:

$$[Fe(III)(CN)_6]^{3-} + e^- \leftrightarrow [Fe(II)(CN)_6]^{4-}$$

$$E^o = 0.361 \text{ V vs SHE}(25\,^\circ\text{C}, 1.0 \text{ atm}) \qquad (6\text{-IV})$$

Due to the high reversibility of this Reaction (6-IV), when the electrode potential is more negative than 0.361 V (vs SHE), the forward 1-electron-transfer reaction will happen from $[Fe(III)(CN)_6]^{3-}$ to $[Fe(II)(CN)_6]^{4-}$, when the potential is more positive than this value, the backward reaction from $[Fe(II)(CN)_6]^{4-}$ to $[Fe(III)(CN)_6]^{3-}$ will happen.

At the beginning of the reaction, the electrolyte solution contains only $[Fe(III)(CN)_6]^{3-}$. Initially, both the ring and the disk electrodes are held at a more positive potential than 0.361 V vs SHE (E^o), for instance, 0.2 V more positive than E^o, at which no reaction occurs. Then, the potential of the disk electrode is slowly swept (~ 50 mV s^{-1}) toward more negative potentials till a limiting disk current ($I_{d,[Fe(III)(CN)]^{3-}}$) can be observed, which corresponds to the reduction of ferricyanide to ferrocyanide at the disk surface, as shown in Figure 6.5. As ferricyanide is reduced at the disk electrode, the ferrocyanide generated by this process is radially swept outward away from the disk electrode and toward the ring electrode. The ring electrode is held constant at 0.2 V more positive than E^o throughout the experiment. Then, some (but not all) of the ferrocyanide generated at the Pt-disk travels close enough to the ring electrode where it is oxidized back to ferricyanide. Thus, a ring current ($-I_{r,[Fe(II)(CN)]^{4-}}$) can be observed at the ring electrode due to the oxidation of ferrocyanide to ferricyanide at the ring. Generally, due to the limited ring electrode surface, only a fraction of the $[Fe(II)(CN)_6]^{4-}$ can react on the ring electrode unless the

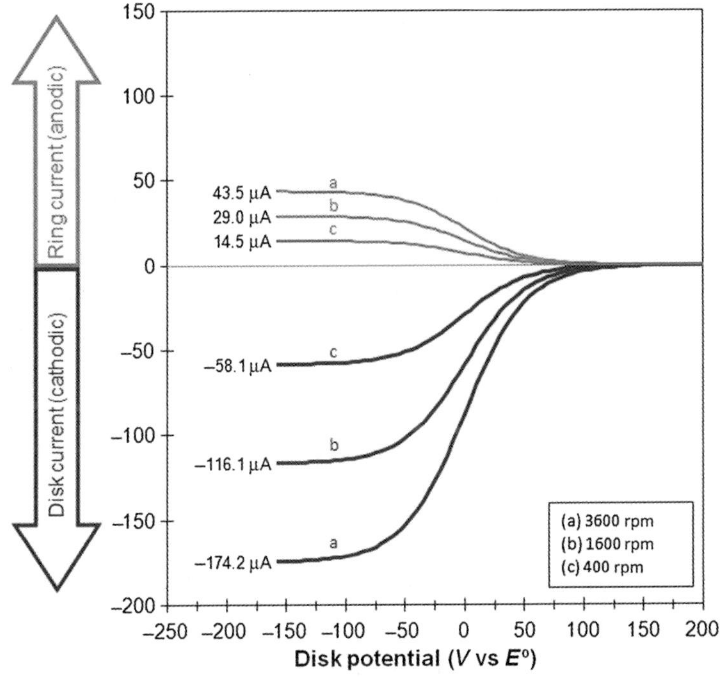

Figure 6.5 Current–potential curves recorded on an RRDE electrode in O_2-free 1.0 M KNO_3 + 0.01 M $[Fe(III)(CN)_6]^{3-}$ aqueous solution with a disk potential scan rate of 50 mV s^{-1}, where E^o is 0.361 vs SHE.[6] (For color version of this figure, the reader is referred to the online version of this book.)

ring area is large enough. In other words, only a fraction of the produced $[Fe(II)(CN)_6]^{4-}$ can be collected on the ring electrode. Then the RRDE's collection efficiency (N) can be experimentally calculated using Eqn (6.28):

$$N = \left| \frac{I_{r,[Fe(II)(CN)]^{4-}}}{-I_{d,[Fe(III)(CN)]^{3-}}} \right| \tag{6.28}$$

Because this N is measured by an experiment, it is called the empirical collection efficiency. From the limiting disk and ring currents marked beside each curve in Figure 6.5, the calculated empirical collection efficiencies for the RRDE used at three different rotating rates are all 25%, indicating that the RRDE used has a collection efficiency of 25%. Note that this percentage of collection efficiency is mole percentage rather than weight percentage.

6.6. RRDE Data Analysis for ORR

ORR is probably one of the most widely studied processes due to its important applications. In literature, the ORR mechanisms have been proposed and extensively studied using RRDE technique.[7–9] Here, we take the RRDE measurements of ORR as examples to illustrate the RRDE data analysis methods.

6.6.1. Proposed ORR Mechanisms and Their Analysis Based on the RRDE Data

Generally to say, ORR has a rather complex mechanism, usually containing a series of electrochemical and/or chemical steps. The ORR mechanism is strongly dependent on many specific operating conditions, such as electrocatalyst materials, solution composition, pH, and applied potential range, etc. In general, there are two distinct parallel reaction pathways that may occur simultaneously when oxygen is reduced. The first one is a direct 4-electron-transfer pathway from O_2 to H_2O (acidic solution) or OH^- (basic solution); and the other one is to reduce oxygen to H_2O_2 (acidic solution) or HO_2^- (basic solution) with receipt of two electrons and then H_2O_2 or HO_2^- may be further reduced with receipt of another two electrons to produce H_2O or OH^-; or H_2O_2 or HO_2^- is decomposed chemically. In practical applications such as in fuel cells, the direct 4-electron-transfer pathway (from O_2 to H_2O or OH^-) is desirable because the formation of H_2O_2 can reduce the number of electrons transferred per oxygen molecule reacted. Furthermore, H_2O_2 is also believed to be responsible for the degradation of electrode performance over time, through its attack upon the carbon support, the catalyst and the membrane. Therefore, the determination of the detailed reaction mechanism of oxygen reduction and the role of H_2O_2 in oxygen reduction is of great importance in both the fundamental understanding and applications.

To study the ORR mechanism, RRDE technique has been recognized as one of the most useful techniques. If hydrogen peroxide is formed during an ORR at the disk electrode with a given potential, it will be carried by the forced convection to the ring electrode. If the potential of the ring electrode is kept high enough, all H_2O_2, which is transported to the ring surface, will be oxidized immediately.

Assuming that the intermediate H_2O_2 is stable without chemical decomposition such as disproportionation, the ORR at and near RRDE surface can be schematically represented by Figure 6.6[9,10]:

For the convenience of mathematical treatment below, we may express O_2 as A and H_2O_2 or HO_2^- as B. In Figure 6.6, the diffusion−convection constants for O_2 can be expressed as $B_A = 0.201D_A^{2/3}\nu^{-1/6}$ and for H_2O_2 or HO_2^- as $B_B = 0.201D_B^{2/3}\nu^{-1/6}$, respectively, and k_1, k_2, and k_3 are the reaction rate constants for their corresponding reactions, all of which are the function of electrode potential. If all the reactions are at the steady states

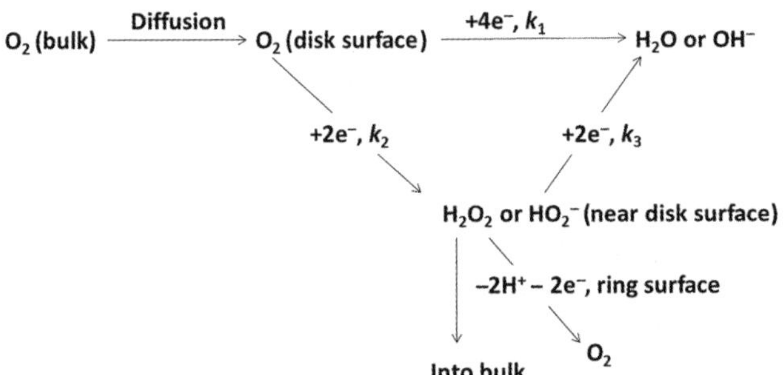

Figure 6.6 Schematic of the processes of oxygen reduction reaction on RRDE (Mechanism 1).[10]

on the disk electrode, O_2 mass balance can be expressed as Eqn (6.29):

$$B_A\omega^{1/2}\left(C_A^o - C_A^s\right) = (k_1 + k_2)C_A^s \tag{6.29}$$

and for H_2O_2,

$$k_2 C_A^s = B_B\omega^{1/2}C_B^s + k_3 C_B^s \tag{6.30}$$

or

$$\frac{C_A^s}{C_B^s} = \frac{B_B\omega^{1/2} + k_3}{k_2} \tag{6.30a}$$

The total current of the disk electrode (I_d) can be expressed as:

$$I_d = 4F\pi r_1^2 k_1 C_A^s + 2F\pi r_1^2 k_2 C_A^s + 2F\pi r_1^2 k_3 C_B^s \tag{6.31}$$

Note that I_d in Eqn (6.31) is not necessarily the limiting current. If at the beginning of the reaction there is no H_2O_2 in the solution, which means the bulk concentration of H_2O_2 is zero, the diffusion current for H_2O_2 removal from the disk electrode should be expressed as:

$$I_{d,B} = 2F\pi r_1^2 B_B\omega^{1/2}C_B^s \tag{6.32}$$

According to the definition of collection efficiency (N), the current at the ring electrode (I_r) can be expressed as:

$$I_r = I_{d,B}N = 2F\pi r_1^2 B_B\omega^{1/2}C_B^s N \tag{6.33}$$

Combining Eqns (6.36), (6.31) and (6.33), the following equation can be obtained:

$$\frac{I_d}{I_r} = \frac{1}{N}\left(1 + \frac{2k_1}{k_2}\right) + \frac{2k_3}{B_B N}\left(1 + \frac{k_1}{k_2}\right)\omega^{-1/2} \tag{6.34}$$

Based on this Eqn (6.34), we can discuss three situations as follows:

1. If k_3 is much small, the produced H_2O_2 or HO_2^- cannot be further reduced to H_2O or OH^- on the disk electrode. Based

on Eqn (6.34), the slope will be equal to zero, leading to a rotating rate independent plot of $\frac{I_d}{I_r}$ vs $\omega^{-1/2}$. As an example, Figure 6.7(A) shows the experiment data recorded at different potentials in O_2-saturated 0.5 M H_2SO_4 solution. The potential dependent horizontal lines can be observed, indicating that $k_3 \to 0$ according to Eqn (6.34). However, the intercept of the line is dependent on the electrode potential, from which the potential dependences of $\frac{k_1}{k_2}$ can be obtained.

2. If the plot of $\frac{I_d}{I_r}$ vs $\omega^{-1/2}$ can give a straight line showing both slope of $\frac{2k_3}{B_{H_2O_2}N}\left(1 + \frac{k_1}{k_2}\right)$ and intercept of $\frac{1}{N}\left(1 + \frac{2k_1}{k_2}\right)$, from both of which k_3 can be obtained. Figure 6.7(B) shows an example measured in O_2-saturated 0.1 M KOH aqueous solution. It can be seen that both the slope and intercept are dependent on the electrode potential, from which the potential dependences of k_3 and $\frac{k_1}{k_2}$ can be obtained.

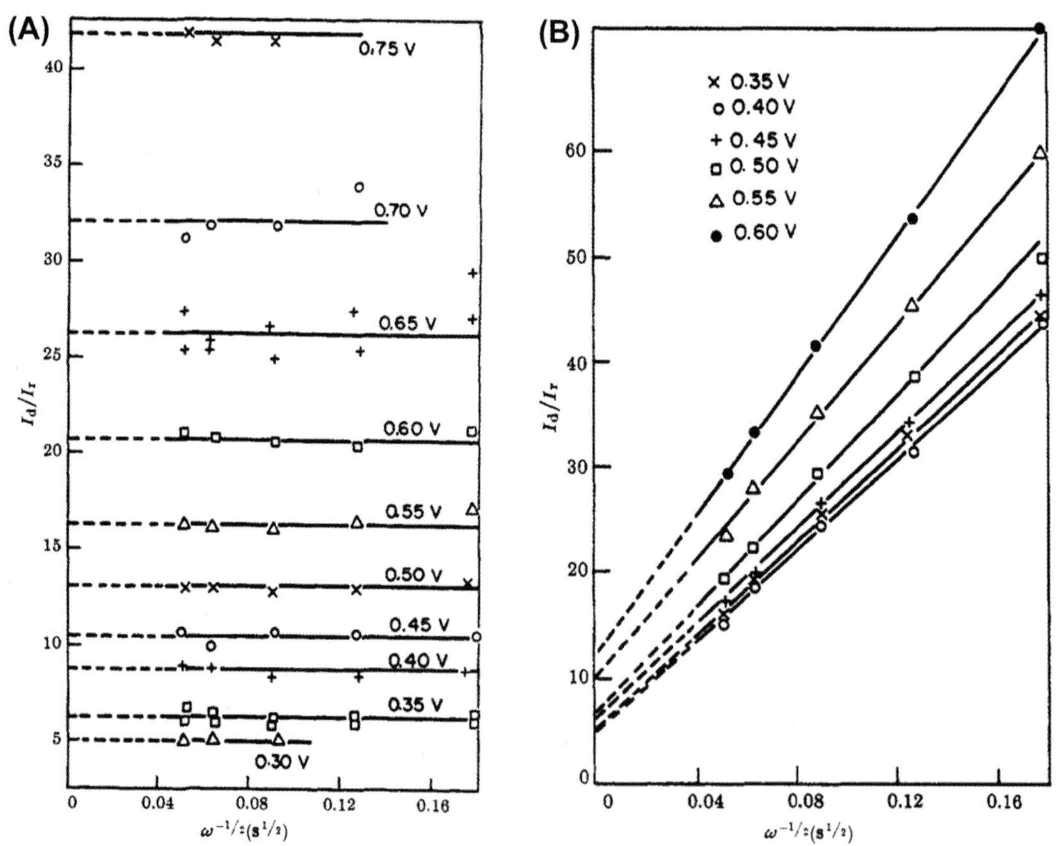

Figure 6.7 (A) Electrode potential-dependent plots of $\frac{I_d}{I_r}$ vs $\omega^{-1/2}$ recorded on an RRDE with a rhodium disk electrode and a platinum ring electrode (the collection efficiency: 38%) in O_2-saturated 0.5 M H_2SO_4 aqueous solution; (B) the same condition as (A) except the electrolyte solution is O_2-saturated 0.1 M KOH aqueous solution.[7]

3. If the ORR is a dominating 2-electron-transfer process from O_2 to H_2O_2 or HO_2^-, meaning that $k_2 \gg k_1$, according to Eqn (6.34), the obtained intercept from the plot of $\frac{I_d}{I_r}$ vs $\omega^{-1/2}$ at any electrode potential would be close to a single value of $\frac{1}{N}$.

4. If the ORR is a direct 4-electron-transfer process from O_2 to H_2O or OH^-, $k_1 \gg k_2$, meaning that Eqn (6.34) is not applicable anymore because $I_r \to 0$. In this case, the limiting current at the disk electrode would be:

$$I_{d,A} = 4F\pi r_1^2 B_A \omega^{1/2} C_A^o \tag{6.35}$$

At any situations, the transported O_2 from the bulk solution to the disk surface should be equal to that consumed by the electrochemical Reactions of 4-electron-transfer reduction and 2-electron-transfer reduction:

$$B_A \omega^{1/2}\left(C_A^o - C_A^s\right) = (k_1 + k_2)C_A^s \tag{6.36}$$

Combining Eqns (6.35) and (6.36), we can get:

$$I_{d,A} = 4F\pi r_1^2\left(k_1 + k_2 + B_A \omega^{1/2}\right)C_A^s \tag{6.37}$$

Combining Eqns (6.30a), (6.33), (6.34) and (6.37), Eqn (6.38) can be obtained:

$$\frac{I_{d,A} - I_d}{I_r} = \frac{1}{N}\left(1 + \frac{2k_3}{k_2}\frac{B_A}{B_B}\right) + \frac{2B_A}{k_2 N}\omega^{1/2} \tag{6.38}$$

For a mixed 2-electron and 4-electron-transfer process, the I_d and I_r can be measured and can be calculated using Eqn (6.35) at different electrode rotating rates. Then the plot of $\frac{I_{d,A}-I_d}{I_r}$ vs $\omega^{-1/2}$ will give a slope of $\frac{2B_A}{k_2 N}$, from which k_2 can be obtained. Substituting this k_2 into the intercept $\left(\frac{1}{N}\left(1 + \frac{2k_1}{k_2}\right)\right)$ obtained by Eqn (6.34), the k_1 can be calculated. In this way, all three reaction-rate constants k_1, k_2, and k_3 can be obtained. Note again that all of these three constants are the functions of the disk electrode potential.

Note that if $I_d \gg I_r$, the surface concentration of O_2 (C_A^s) can be simplified as:

$$C_A^s = C_A^o\left(1 - \frac{I_d}{I_{d,A}}\right) \tag{6.39}$$

Combining Eqns (6.30) and (6.39), the following Equation can be obtained[9]:

$$\frac{I_{d,A}}{I_{d,A} - I_d} = 1 + \frac{k_1 + k_2}{B_A}\omega^{-1/2} \tag{6.40}$$

According Eqn (6.40), plot of $\frac{I_{d,A}}{I_{d,A}-I_d}$ vs $\omega^{-1/2}$ will give a straight line with a slope of $\frac{k_1+k_2}{B_A}$, from which $k_1 + k_2$ can be obtained.

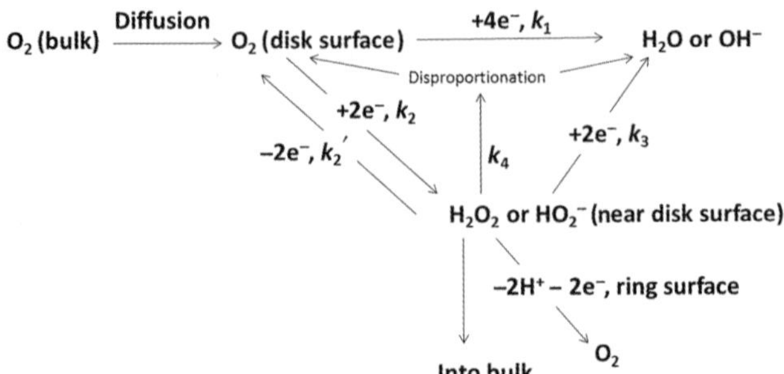

Figure 6.8 Schematic of the processes of oxygen-reduction reaction on RRDE (Mechanism 2).[10]

In the situation where the formed H_2O_2 or HO_2^- on the disk electrode surface is disproportionated with a reaction rate constant of k_4, and the reaction from O_2 to H_2O_2 or HO_2^- is reversible with a backward reaction rate constant of k_2', the similar analysis as for that shown in Figure 6.6 can be carried out. The mechanism is schematically shown in Figure 6.8. The obtained relationships between $\frac{I_d}{I_r}$ and $\omega^{-1/2}$ are as follows:

$$\frac{I_d}{I_r} = \frac{1}{N}\left(1 + \frac{2k_1}{k_2}\right) + \frac{1}{B_B N}\left[\left(1 + \frac{2k_1}{k_2}\right)(k_2' + k_3 + k_4)\right.$$
$$\left. + k_3 - k_2'\right]\omega^{-1/2} \tag{6.41}$$

$$\frac{I_{d,A} - I_d}{I_r} = \frac{1}{N}\left(1 + 2\frac{k_2' + k_3 + k_4}{k_2}\frac{B_A}{B_B}\right) + \frac{2B_A}{k_2 N}\omega^{1/2} \tag{6.42}$$

Using experiment data and plotting $\frac{I_d}{I_r}$ and $\frac{I_{d,A} - I_d}{I_r}$ against $\omega^{-1/2}$ straight lines, the obtained slopes and intercepts can be used to obtained the all values of k_1, k_2, k_2', k_3, and k_4.

In the situation where the formed H_2O_2 or HO_2^- on the disk electrode surface is adsorbed on the disk electrode, and the reaction from O_2 to H_2O_2 or HO_2^- is reversible with a backward reaction rate constant of k_2', the similar analysis as for that shown in Figure 6.6 can be carried out. The reaction mechanism is schematically shown in Figure 6.9, where two reaction rate constants, k_5 and k_5', can be used to describe the adsorption and desorption of H_2O_2 or HO_2^-.

The obtained relationships between $\frac{I_d}{I_r}$ and $\omega^{-1/2}$ are as follows[11]:

$$\frac{I_d}{I_r} = \frac{1}{N}\left(1 + \frac{2k_1}{k_2} + \frac{2k_1}{k_2 k_5}(k_2' + k_3 + k_4) + \frac{2k_3 + k_4}{k_5}\right)$$
$$+ \frac{k_5'}{B_B N}\left(\frac{2k_1}{k_2 k_5}(k_2' + k_3 + k_4) + \frac{2k_3 + k_4}{k_5}\right)\omega^{-1/2} \tag{6.43}$$

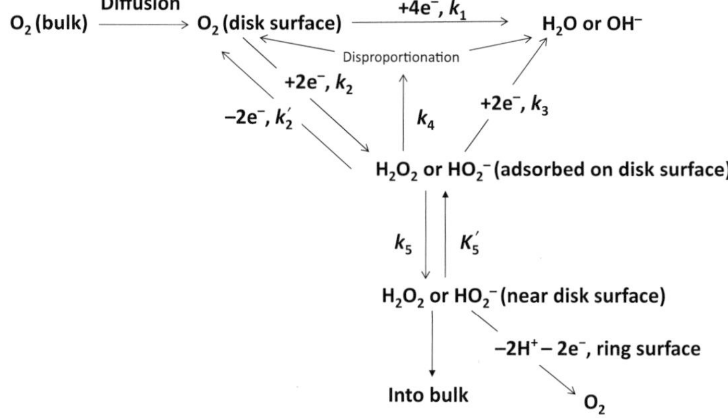

Figure 6.9 Schematic of the processes of oxygen reduction reaction on RRDE (Mechanism 3).[10]

$$
\frac{I_{d,A} - I_d}{I_r} = \frac{1}{N}\left(1 + 2\frac{B_A}{B_B}\frac{k_5'}{k_5}\frac{k_2' + k_3 + k_4}{k_2}\right) \\
+ \frac{2B_A}{k_2 N}\left(1 + \frac{k_2' + k_3 + k_4}{k_5}\right)\omega^{1/2}
$$

(6.44)

Unfortunately, there are six reaction rate constants in Eqns (6.43) and (6.44), and using experimental data and plotting $\frac{I_d}{I_r}$ and $\frac{I_{d,A} - I_d}{I_r}$ against $\omega^{-1/2}$ straight lines cannot resolve all of these constants. More additional experiments are needed in order to obtain all of these parameters.

6.6.2. ORR Apparent Electron-Transfer Number and the Formed Percentage of Peroxide Measured by RRDE Technique

Due to the production of H_2O_2 or HO_2^- through a 2-electron-transfer pathway, the overall electron-transfer number of the ORR process is always less than 4. This electron-transfer number is normally called the apparent number of electrons transferred per O_2 molecule. Actually, this apparent number of electron transfer can be measured by the RRDE technique, from which the percentage of H_2O_2 formation in the ORR can also be calculated. Generally to say, the apparent number of electron transfer and the percentage of peroxide produced in the ORR process are two important pieces of information in evaluating the ORR catalyst's catalytic activity.

As discussed in Chapter 5 for RDE technique, this apparent electron-transfer number of ORR or the overall ORR electron-transfer number can be obtained using the slope of the

Koutecky–Levich plot, measured at different electrode rotating rates, if the O_2 concentration, the O_2 diffusion coefficient, and the solution kinetic viscosity are known. Here, we give the expression of apparent electron-transfer number of ORR as functions of both disk and ring currents, measured using the RRDE technique.

To express the relationship among the apparent electron number (n), the disk current (I_d), and the ring current (I_r), we use the ORR mechanism shown in Figure 6.6 for our analysis. Based on Eqn (6.29), the disk current can be expressed as Eqn (6.45):

$$I_d = nF\pi r_1^2 B_A \omega^{1/2}\left(C_A^o - C_A^s\right) = nF\pi r_1^2(k_1 + k_2)C_A^s$$
$$\Rightarrow \frac{4}{n}I_d = 4FAk_1 C_A^s + 4FAk_2 C_A^s \tag{6.45}$$

According to Figure 6.6, the disk current can also be expressed as:

$$I_d = 4F\pi r_1^2 k_1 C_A^s + 2F\pi r_1^2 k_2 C_A^s + 2F\pi r_1^2 k_3 C_B^s \tag{6.46}$$

and the limiting current at the ring electrode (I_r) can be expressed as:

$$I_r = 2F\pi r_1^2 B_B \omega^{1/2} C_B^s N$$
$$\Rightarrow C_B^s = \frac{I_r}{N}\frac{1}{2F\pi r_1^2 B_B \omega^{1/2}} \tag{6.47}$$

Combining Eqns (6.45)–(6.47), the following equation can be obtained:

$$\frac{4I_d}{n} = I_d - \frac{I_r k_3}{B_B \omega^{1/2} N} + 2FAk_2 C_A^s \tag{6.48}$$

According to Eqns (6.30a) and (6.47), C_A^s can be expressed as:

$$C_A^s = C_B^s \frac{k_3 + B_B \omega^{1/2}}{k_2} = \frac{I_r}{N}\frac{1}{2F\pi r_1^2 B_B \omega^{1/2}}\frac{k_3 + B_B \omega^{1/2}}{k_2} \tag{6.49}$$

Substituting Eqn (6.49) into Eqn (6.48), the expression of apparent electron-transfer number of ORR can be obtained:

$$n = \frac{4I_d N}{I_d N + I_r} \tag{6.50}$$

When the ORR is a total 4-electron-transfer process, $I_r = 0$, Eqn (6.50) indicates that the apparent number is 4, and when the ORR is a pure 2-electron-transfer process, $I_r = I_d N$, Eqn (6.50) will give an apparent number of 2.

Regarding the percentage of H_2O_2 or HO_2^- produced in the ORR (%H_2O_2), we can use the mass flows to make calculation based on the mechanism shown in Figure 6.6:

$$\%H_2O_2 = 100\frac{\text{total mole of proxide produced}}{\text{total mole of oxygen reacted}}$$

$$= 100\frac{k_2C_A^s - k_3C_B^s}{k_1C_A^s + k_2C_A^s} = 100\frac{2F\pi r_1^2 k_2C_A^s - 2F\pi r_1^2 k_3C_B^s}{\dfrac{2}{n}\left(nF\pi r_1^2 k_1C_A^s + nF\pi r_1^2 k_2C_A^s\right)} \quad (6.51)$$

Substituting Eqns (6.49) and (6.50) into Eqn (6.51), we can obtain:

$$\%H_2O_2 = 100\frac{nI_r}{2I_dN} = 100\frac{2I_r}{I_dN + I_r} = 100\frac{4 - n}{2} \quad (6.52)$$

With the known value of N and experimental results of disk current, I_d, and ring current, I_r, Eqns (6.41) and (6.52) are often used to calculate the apparent number of electrons transferred per O_2 molecule, n, and the percentage of peroxide formation, % H_2O_2 to determine the ORR mechanism.

As an example, Figure 6.10 shows the ORR current–potential curves at the disk electrode and their corresponding currents at the ring electrode, recorded at different electrode rotating rates, where the horizontal abscissa is the disk potential, and the upper vertical ordinate is the ring current and the lower vertical ordinate is the disk current.[12]

The curves presented in Figure 6.10 can provide much electrochemical information about the mechanism of the disk reaction. First of all, it can be seen that these four catalysts all have catalytic ORR activities. However, their ORR activities are different, which can be seen from their diffusion limiting currents and the onset potentials. The electrocatalytic capability of Ag-W_2C/C catalyst is comparable to that of Pt/C catalyst, and much superior to the other two catalysts.

Secondly, the ORR mechanism at these catalysts can be deduced from the ring-response current curves. In Figure 6.10(A), the kinetic region of the ring curve is the same as the disk curve and the ring current increases synchronically. This suggests that the reduction of oxygen on W_2C/C is via a two-step pathway (the formation of intermediate of HO_2^-, which was oxidized immediately at the ring electrode). In contrast, for other catalysts, the current at the ring electrode is hardly observed in the kinetic region. This suggests the catalyzed ORR having no formation of HO_2^-, indicating that the ORR is a four-electron reduction mechanism. Using the disk and ring current data presented in

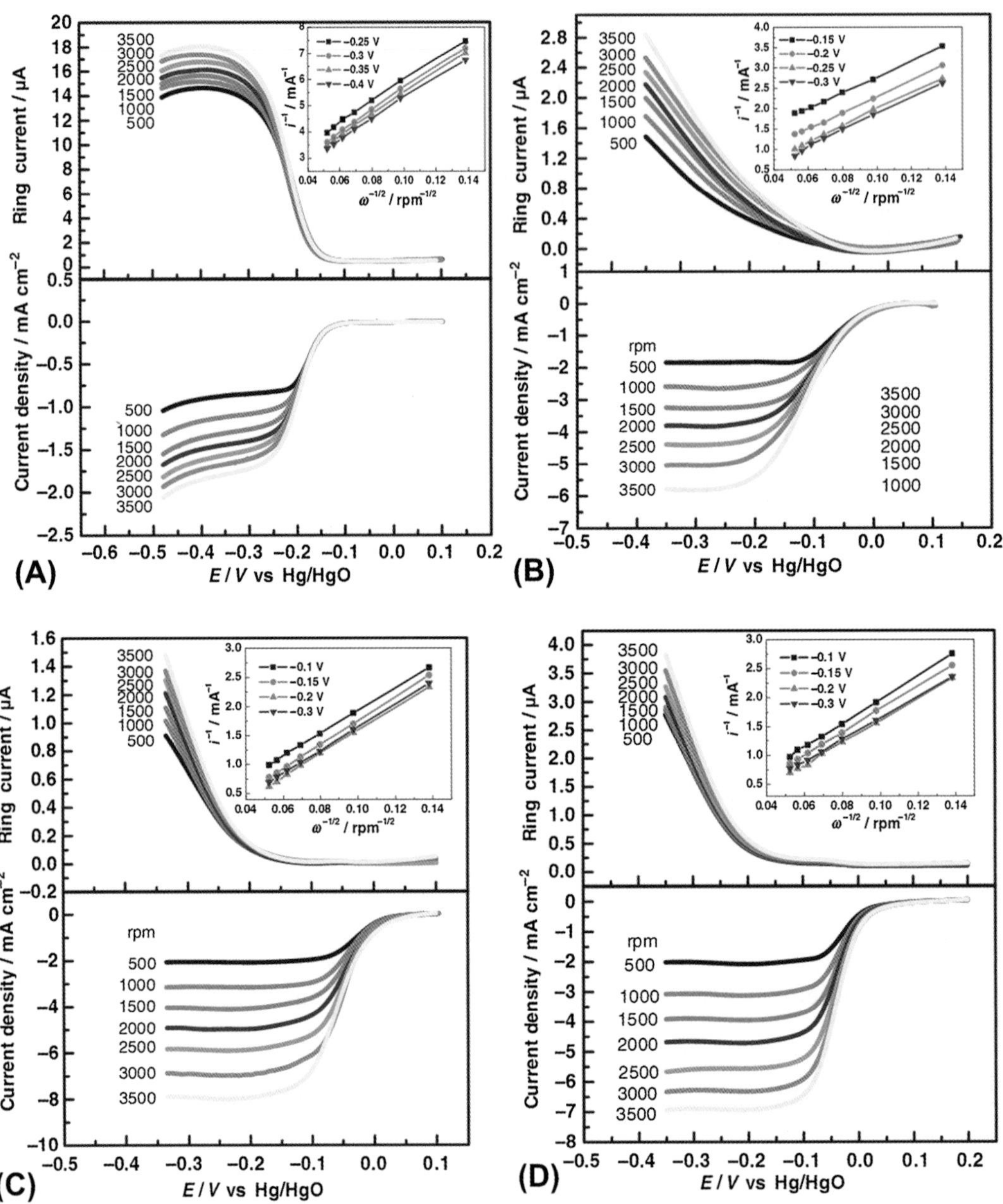

Figure 6.10 Current−potential curves at the disk electrode (below the *x*-axis in each figure) and the current at the ring electrode (Pt) (above the *x*-axis in each figure), recorded in O_2-saturated 1.0 mol dm^{-3} KOH aqueous solution at different electrode-rotating rates. Disk electrode surface coated with a layer of (A) W_2C/C, (B) Ag/C, (C) Ag-W_2C/C, or (D) Pt/C. Potential scan rate: 5 mV s^{-1}, ring potential fixed at 0.474 V vs Hg/HgO, and ring collection efficiency: 20%. The insets present the *Koutecky−Levich* plots of the disk electrode at different potentials.[12] (For color version of this figure, the reader is referred to the online version of this book.)

Figure 6.10, and according to the discussion on Figures 6.6, 6.8 and 6.9 and their associated theory, the reaction kinetic constants, the apparent electron-transfer number, as well as the percentage of peroxide produced can be determined.

Actually, the disk curves in Figure 6.10 can be transformed into *Koutecky–Levich* plots at different potentials. The plots show a linear dependence at all potentials, as inserted in each figure. The linearity and the parallelism of these plots are usually taken to indicate first-order kinetics with respect to oxygen. The *Koutecky–Levich* slope in the insets for W_2C/C is very close to the theoretical value of two-electron reaction, and other three are close to that of four-electron reaction, indicating that W_2C/C can only catalyze a 2-electron-transfer ORR from O_2 to peroxide, and Ag/C, Ag-W_2C/C, and Pt/C can catalyze a 4-electron-transfer ORR from O_2 to OH^-.

Figure 6.11 presents the RRDE experimental results of ORR at various modified GC electrodes.[13] At the bare GC-disk electrode (Figure 6.11a), there are two reduction waves in the O_2-saturated solution, both of which are ascribed to 2-electron reduction of O_2 to HO_2^-. The first is mediated by the active native

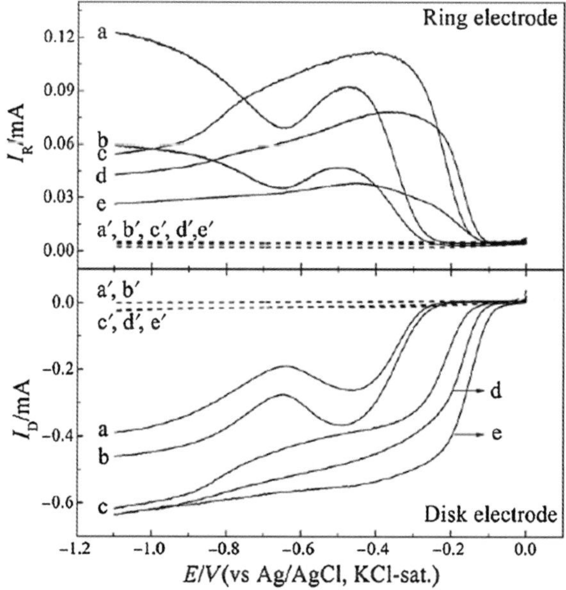

Figure 6.11 RRDE voltammograms with a rotation rate of 400 rpm at the bare glassy carbon (GC) (a and a′), Co(OH)$_2$/GC (b and b′), graphene/GC (c and c′), NaNO$_3$-treated graphene/GC (d and d′), and Co(OH)$_2$/graphene/GC (e and e′) disk electrodes and Pt ring electrode in N$_2$- (a′, b′, c′, d′, and e′) and O$_2$-saturated (a, b, c, d, and e) 0.1 M KOH aqueous solutions. Potential scan rate: 10 mV s^{-1}.[13]

surface quinone-like groups with superoxide anion ($O_2^{\bullet-}$) as the intermediate, while the second is a direct 2-electron reduction at the GC surface. When the GC electrode surface was electro-deposited by a layer of $Co(OH)_2$, a larger disk current for O_2 reduction and a smaller ring current for HO_2^- oxidation without obvious potential shifts could be observed as shown in Figure 6.11b, confirming that $Co(OH)_2$ possesses a good cata-lytic activity for the disproportionation of $O_2^{\bullet-}$ (provides addi-tional O_2 leading to the enlarged disk current) and HO_2^- (consumes HO_2^- causing the reduced ring current).

Two reduction waves can also be observed at the graphene/GC disk electrode (Figure 6.11c). However, this modification of graphene can lead to a significant increase in disk current and also a 130 mV positive shift of the half-wave potential, indicating the better catalytic capability of graphene for O_2 reduction to HO_2^-, possibly due to the excellent redox-mediated effect of quinone-like groups on graphene surfaces. Furthermore, very different from bare GC electrode, the second reduction process on the graphene/GC disk electrode is attributed to the reduction of HO_2^-, which is supported by the decrease in ring current with increasing disk overpotential.

After electrochemical treatment in 20 mM $NaNO_3$ solution, the second reduction process on graphene-modified GC elec-trode almost disappears as shown in Figure 6.11d, indicating the enhanced disproportionation of HO_2^- induced by activated gra-phene. The disproportionation of HO_2^- provides additional O_2 for the first reduction step, and decreases the amount of HO_2^- for further reduction at the second step. Consequently, the current interval between these two stages is much smaller than that at the graphene/GC disk electrode. The deposited $Co(OH)_2$ can further enhance the disproportionation of HO_2^-, and as a result, only one reduction step is observed at the $Co(OH)_2$/graphene/GC disk electrode, as seen in Figure 6.11e. In comparison with bare GC electrode, the half-wave potential for O_2 reduction at the $Co(OH)_2$/graphene/GC electrode is shifted c. 190 mV in the positive direction, and also the reduction current increases obviously. These results demonstrate that the graphene–$Co(OH)_2$ composite possesses an enhanced catalytic activity toward ORR.

On the basis of RRDE voltammograms, Eqns (6.50) and (6.52) can be used to calculate the apparent number of electrons transferred (n) and percentage of HO_2^- (%HO_2^-) for ORR, as shown in Figure 6.12. It can be seen from Figure 6.12(A) that the value of n at the bare GC electrode remains at around 2 in the potential window employed (Curve a), and the introduction of

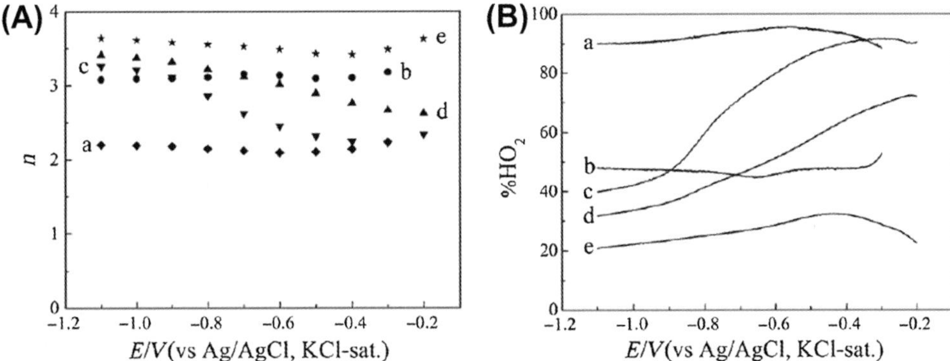

Figure 6.12 Dependence of n (A) and %HO_2^- (B) on electrode potential at the bare GC (a), Co(OH)$_2$/GC (b), graphene/GC (c), NaNO$_3$-treated graphene/GC (d), and Co(OH)$_2$/graphene/GC (e) disk electrodes. Experiment conditions are the same as that in Figure 6.11.[13]

Co(OH)$_2$ results in its increase to greater than 3 (Curve b), which is in accordance with the consumption of HO_2^- in the disproportionation mentioned above. For the graphene/GC electrode (Curve c), the n value is close to 2 at low overpotentials ($E > -0.4$ V) and gradually increases ($n > 3$ at -0.9 V), suggesting HO_2^- produced reduces to OH^- at more negative potentials. The electrochemical activation of the graphene/GC electrode in 20 mM NaNO$_3$ causes a higher n value, and the value can reach to 3 at -0.6 V (Curve d), which confirms that the activated graphene does catalyze the disproportionation of HO_2^- effectively. The value of n at the Co(OH)$_2$/graphene/GC reaches 3.5 at quite low overpotentials due to the synergistic effect of activated graphene and Co(OH)$_2$ (Curve e), indicating the prepared graphene– Co(OH)$_2$ composite has high catalytic activity toward the reduction of O$_2$ to OH^-.

As shown in Figure 6.12(B), the value of %HO_2^- at the bare GC and Co(OH)$_2$/GC electrodes changes minimally with potential (Curves a and b), and the introduction of Co(OH)$_2$ induces a decrease of the value from greater than 90% to c. 45%. Different from the bare GC electrode, the %HO_2^- value at the graphene/GC electrode is dependent on potential (Curve c), in which it is higher than 80% at potentials for the first reduction step and drops to c. 40% at potentials more negative than -0.9 V. This difference is attributed to the occurrence of HO_2^- reduction rather than the direct 2-electron reduction of O$_2$ at the second step at the graphene/GC electrode. After electrochemical treatment in 20 mM NaNO$_3$, the value of %HO_2^- decreases obviously (Curve d), and further precipitation of Co(OH)$_2$ results in a value of c. 20% (Curve e), which is in accordance with the prediction on the catalytic activity of activated graphene and $Co(OH)_2$ toward HO_2^- disproportionation.

6.7. Chapter Summary

RRDE is a commonly used technique for investigating ORR in terms of both the electron-transfer process on an electrode surface, diffusion—convection kinetics near the electrode, and the ORR mechanism, particularly the detection of intermediate such as peroxide. To make appropriate usage in an ORR study, fundamental understanding of both the RRDE fundamentals and practical usage is necessary. In this chapter, one of most important parameters of RRDE, the collection efficiency, is extensively described including its concept, theoretical expression, as well as experiment calibration. Its usage in evaluating the ORR kinetic parameters, the apparent electron transfer, and percentage of peroxide formation is also presented. In addition, the measurement procedure including RRDE preparation, current—potential curve recording, and the data analysis are also discussed in this chapter.

We wish the RRDE knowledge and information can be useful to the readers for their study.

References

1. Bard AJ, Faulkner LF. *Electrochemical methods — fundamentals and applications.* New York: John Wiley; 2001.
2. Tian Z. *Methods for electrochemical research.* Beijing: Chinese Science Press; 1984.
3. http://www.pineinst.com/ECHEM/files/RRDE_EFFICIENCY.XLS.
4. http://www.als-japan.com/1205.html.
5. http://www.helmholtz-berlin.de/forschung/enma/solare-brennstoffe/ analytische-methoden/rotierende-scheibenelektrode_de.html.
6. http://wiki.voltammetry.net/pine/rotating_electrode/theory.
7. Damjanovic A, Genshaw MA, Bockris JO'M. The role of hydrogen peroxide in the reduction of oxygen at platinum electrodes. *J Phys Chem* 1966;**70**(11):3761—2.
8. Genshaw MA, Damjanovic A, Bockris JO'M. The role of hydrogen peroxide in oxygen reduction at rhodium electrodes. *J Phys Chem* 1967;**71**(12):3722—31.
9. Hsueh KL, Chin DT, Srinivasan S. Electrode kinetics of oxygen reduction: a theoretical and experimental analysis of the rotating ring-disk electrode method. *J Electroanal Chem* 1983;**153**:79—95.
10. Cha Q. *An introduction to electrode process kinetics.* Beijing: Chinese Science Press; 2002.
11. Wroblowa HS, Yen CP, Razumney G. Electroreduction of oxygen: a new mechanistic criterion. *J Electroanal Chem Interfacial Electrochem* 1976;**69**(2):195—201.

12. Meng H, Shen PK. Novel Pt-free catalyst for oxygen electroreduction. *Electrochem Commun* 2006;**8**(4):588—94.
13. Wu J, Zhang D, Wang Y, Wan Y, Hou B. Catalytic activity of graphene—cobalt hydroxide composite for oxygen reduction reaction in alkaline media. *J Power Sources* 2012;**198**:122—6.

APPLICATIONS OF RDE AND RRDE METHODS IN OXYGEN REDUCTION REACTION

Chunyu Du[a], Yongrong Sun[a], Tiantian Shen[a], Geping Yin[a] and Jiujun Zhang[b]

[a]*School of Chemical Engineering and Technology, Harbin Institute of Technology, Harbin, China,* [b]*Energy, Mining and Environment, National Research Council of Canada, Vancouver, BC, Canada*

CHAPTER OUTLINE

Rotating Electrode Methods and Oxygen Reduction Electrocatalysts. http://dx.doi.org/10.1016/B978-0-444-63278-4.00007-0

7.1. Introduction

As we discussed previously, the oxygen reduction reaction (ORR) is considered one of the most important reactions in several electrochemical energy devices such as fuel cells and metal-air batteries. However, ORR continues to be a challenge because of its complex kinetics and the need for better electrocatalysts. Recently, a great deal of researches has focused on the development of ORR catalysts in order to improve the performance of fuel cells. The most notable efforts are to further improve both the activity and the stability of the existing Pt-based ORR catalysts, and at the same time to develop non-noble metal electrocatalysts. In these approaches, both rotating disk electrode (RDE) and rotating ring-disk electrode (RRDE) methods have been recognized as the commonly employed methods in evaluating the catalysts' activity and stability as well as performance optimization. Using RDE and RRDE techniques, the ORR has been studied on planar Pt, carbon electrode, monolayer metal catalyst, Pt-based catalyst and non-noble metal-based catalysts. To give more detailed information, this chapter will review on the applications of RDE and RRDE techniques in the ORR research and its associated catalyst evaluation.

7.2. RDE/RRDE Study for ORR on Pt-based Electrode Surfaces

Pt-based rotating electrode surfaces include pure Pt and Pt-alloy surfaces. This kind of electrode surfaces are the mostly used electrode surfaces for ORR by RDE/RRDE techniques in literature. In this section, we will give some typical examples in the following subsections.

7.2.1. Oxygen Reduction Reaction on Crystal Pt-Facet Electrode Surfaces

Platinum is still the most widely used ORR electrocatalyst due to both their relatively high catalytic activity and stability when compared to other catalyst materials. Normally, Pt crystal has three interesting facets, namely Pt(111), Pt(110), and Pt(100), respectively. Both the density functional theory (DFT) and RRDE experiments have confirmed that these three facets have different ORR activities.

Figure 7.1 shows the patterns of RRDE measurements recorded on different facets of platinum low-index single crystal (expressed as Pt(hlk)) in O_2-saturated 0.1 M KOH aqueous solution. The insets

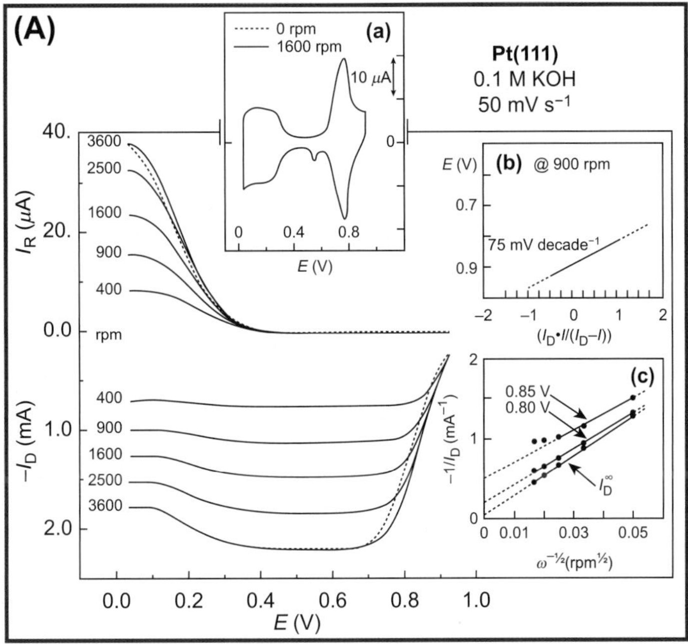

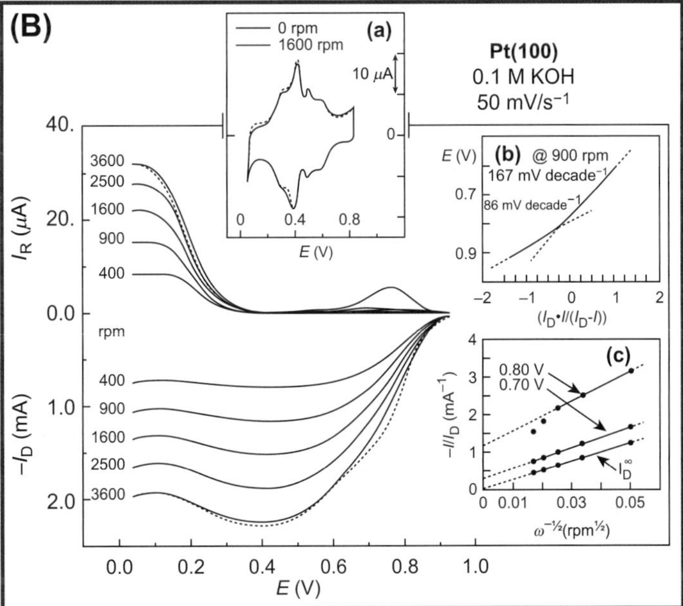

Figure 7.1 (A) Disk and ring currents during oxygen reduction on (A) Pt(111), (B) Pt(100), and (C) Pt(110) facets, respectively, in 0.1 M O_2-saturated KOH aqueous solution at a potential sweep rate of 50 mV s^{-1} (ring potential = 1.15 V (RHE)). Cyclic voltammograms in the insets (a) are recorded in KOH aqueous solution without dissolved oxygen. Reprinted with permission from Ref. 2.

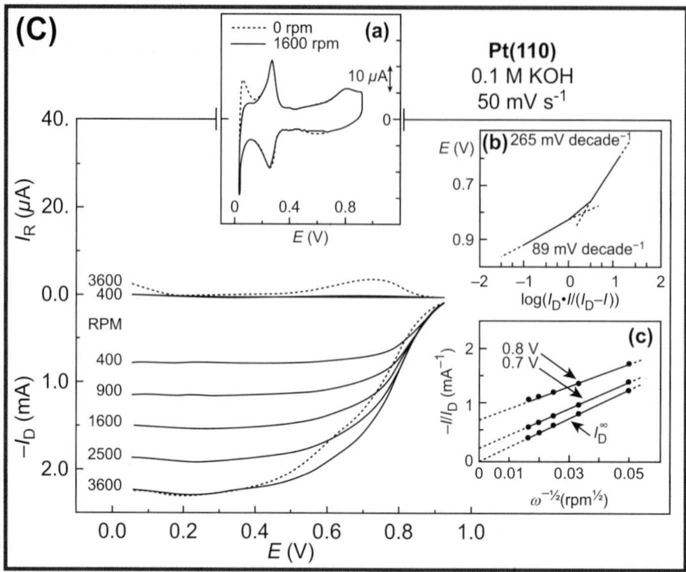

Figure 7.1 (*continued*).

in each figure show the surface cyclic voltammogram (CV) of the corresponding Pt facet, recorded in the same electrolyte solution but saturated by inert gas such as argon. For Pt(111) facet shown in Figure 7.1(A), in the potential region between 0.8 and 0.925 V, the corresponding ring currents are insignificant, implying that the oxygen reduction proceeds almost entirely through the 4-electron-transfer pathway. However, the activity of Pt(111) for O_2 reduction becomes suppressed when potential is more native than 0.2 V, which parallels with the onset of hydrogen adsorption on Pt(111). Then the formation of peroxide can be seen from the increase of ring current, and then the ORR starts to convert from a 4-electron-transfer pathway to a 2-electron-transfer one. In the potential range of a double layer, well-defined diffusion limiting currents can be observed, indicating that the O_2 reduction is purely controlled by a mass transport. For Pt(100) facet shown in Figure 7.1(B), the potential region for combined kinetic-diffusion control for O_2 reduction is markedly, and the true diffusion limiting currents for O_2 reduction can only observed in a very narrow potential range. The polarization curves for ORR on Pt(110) facet as shown in Figure 7.1(C) exhibit a quite similar behavior to those reported in the literature for polycrystalline platinum electrodes.[1] Koutecky–Levich (K–L) plots at various potentials display intercepts corresponding to the kinetic currents, i_k—at the same potential (e.g., 0.8 V), $i_k(111) > i_k(110) > i_k(100)$, giving the order of absolute kinetic activity of the three surfaces. The slope of the straight lines

$(= (0.201nFD_{O_2}^{2/3} \nu^{-1/6} C_{O_2})^{-1})$ allows one to assess the number of apparent or overall electron-transfer number involved in the ORR. The experimental value of this slope, $4.00 \times 10^{-2}(\pm 1\%)$ mA rpm$^{-1/2}$ agrees well with the theoretical value, 3.99×10^{-2} mA rpm$^{1/2}$, for a $4e^-$ reduction. On Pt(111) surface, a single linear Tafel slope (≈ 75 mV decade^{-1}) is obtained in the entire potential region, which appears to be related to the potential-dependent surface charge of adsorbed OH (Q_{OH} ($= current \times time = it$) vs E) shown in Figure 7.1(A) (a). However, on Pt(100) or Pt(110), there are two Tafel slopes: one is 86 or 89 mV decade^{-1}, and the other is 167 or 265 mV decade^{-1}. The transition in the Tafel slopes on both Pt(110) and Pt(100) appears to be related to the change in Q_{OH}.

In ORR, due to the oxygen atoms competing with species in a solution for adsorption sites, some adsorbed anions such as OH$^-$, HSO$_4^-$, SO$_4^{2-}$, Cl$^-$, H$_2$PO$_4^-$, or Br$^-$ are detrimental for the kinetics of O$_2$ reduction. Therefore, the ORR activity of Pt(hlk) facets are strongly dependent on the relative adsorption strengthens of O$_2$ and electrolyte anion, which have been extensively studied using both the DFT theoretical analysis and experimental observation in literature.[3-18] Normally, the combination of surface CVs with the ORR current−potential curves measured by RRDE can give some useful information about the correlation between the surface adsorption and the ORR activity of Pt facets.

For example, in acidic solution, bisulfate adsorption onto Pt(hkl) surfaces could inhibit the reduction of molecular O$_2$, most likely by blocking the initial adsorption of O$_2$. But it did not affect the pathway of the reaction, since no H$_2$O$_2$ is detected on the ring electrode for any of the surfaces in the kinetically controlled potential region, as reviewed by Markovic et al.[19] Regarding the ORR activity sequence of Pt(hlk) facets in H$_2$SO$_4$ solution, activity order of Pt(110) > Pt(100) > Pt(111) was observed. However, this ORR activity order is also dependent on the type of acid used.[20] Another example is the relation between Pt(hkl) and Cl$^-$ in acidic solution. It was found that the ORR activity in a solution containing Cl$^-$ was several orders of magnitude lower than in 0.05 M H$_2$SO$_4$, which was due to the strong adsorption of Cl$^-$ on Pt(111), which blocked the adsorption of O$_2$.[19] This result indicates that the Pt(111)-Cl$_{ad}$ system has the same effect on the ORR as bisulfate anions.

7.2.2. Oxygen Reduction Reaction on Polycrystalline Pt Surface

Since polycrystalline Pt electrodes show the ideal ORR activity and proceed mainly via the 4-electron step,[21] they are the most commonly used and practical electrode surfaces for ORR in

RDE/RRDE measurements. Actually, the polycrystalline Pt surface is composed of many Pt facets, and therefore, its ORR activity should be different from those of individual Pt facets, and be considered all the contributions from different facets.[21] Figure 7.2 shows the CV of a polycrystalline Pt surface on which the hydrogen adsorption/desorption peaks at the lower potential range and the Pt oxidation and its reduction waves at the higher potential range can be observed. Normally, only a bare Pt surface has a high ORR activity, while that portion of surface covered by Pt oxides such as PtO should have low/non-ORR activity. Therefore, when the bare Pt surface is covered by a layer of Pt oxide, the ORR activity will be significantly reduced or eliminated due to the surface blocking effect. In Figure 7.2(A), when the electrode potential is positively scanned, starting from 0.7 V, the Pt surface starts to be oxidized, forming a Pt oxide-covered surface, and when the potential is negatively scanned back from 1.2 V, starting from 1.05 V, the Pt oxide starts to be reduced back to a bare Pt surface. The RDE measurement for an ORR activity can be used to support this point. As shown in Figure 7.2(B), when electrode potential is negatively scanned from 1.2 V where the surface is probably totally coved by Pt oxide, there is no significant ORR current can be observed. However, when the potential is near the onset potential of the Pt oxide reduction to form a bare Pt surface in Figure 7.2(A), an ORR current appears, probably indicating that only the bare surface Pt is more active for the ORR. While the potential is positively scanned back, there is a delay of the ORR current due to the delayed conversion of the bare Pt surface to the Pt oxide-covered surface.

In measuring the ORR activity, obviously, the ORR current–potential curve (I–V curve) obtained by scanning the potential at the direction from negative to positive will give larger value than that obtained by scanning the potential at the direction from positive to negative. In estimating the ORR activity of Pt surface, particularly the ORR activity of the Pt catalyst, the RDE I–V curve is normally recorded by scanning the electrode potential at the direction from negative to positive.

7.2.3. Oxygen Reduction Reaction on Pt Monolayer Coated on Other Metal Electrode Surfaces

A Pt monolayer on other metal surface such as Pd(111) could be prepared by the galvanic displacement and the ORR activity on such a Pt monolayer was studied using RDE/RRDE techniques.[3] The RRDE results revealed that no peroxide could be detected at

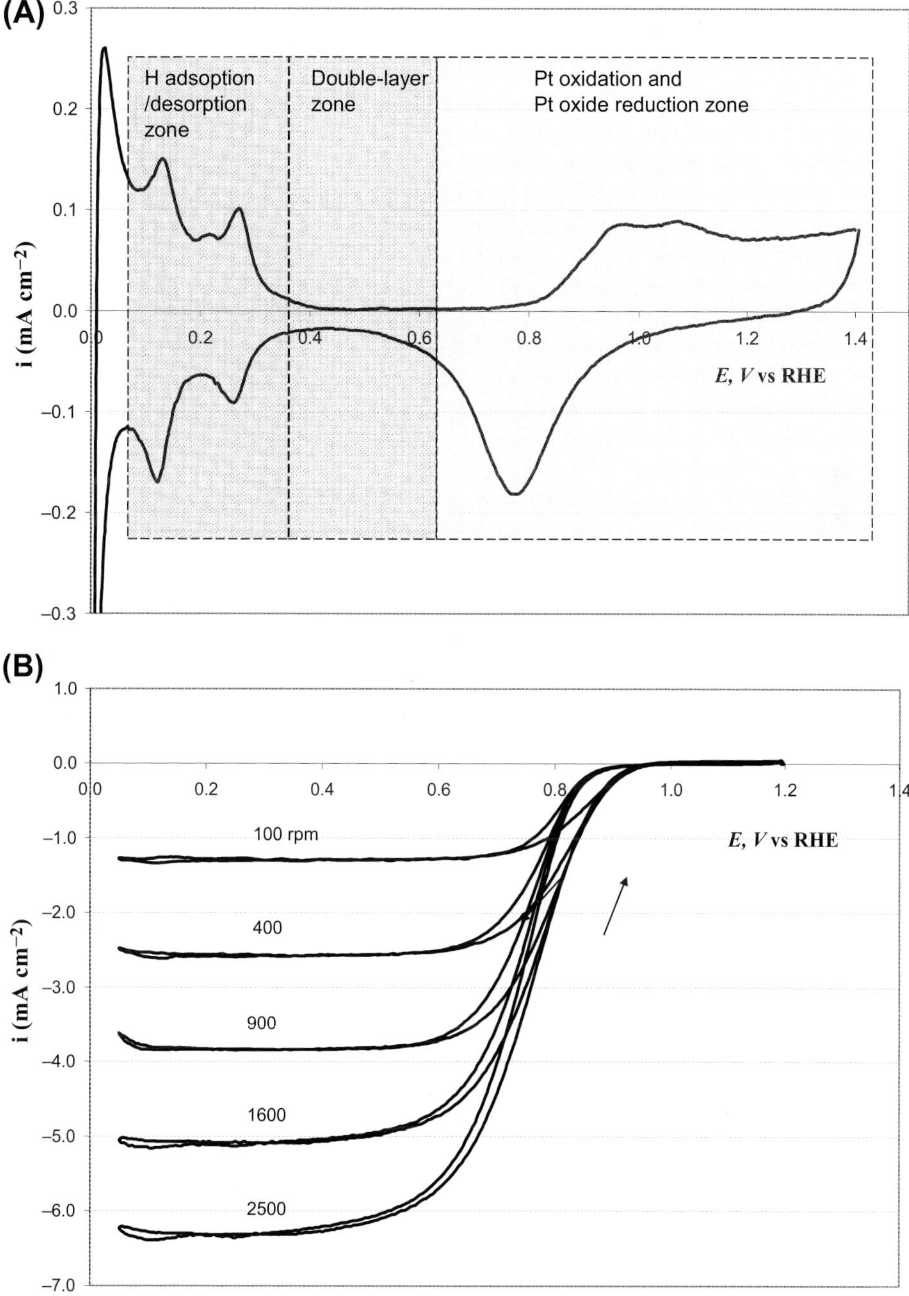

Figure 7.2 (A) Cyclic voltammogram of RDE with polycrystalline Pt electrode surface in N_2-saturated 0.5 M H_2SO_4 aqueous solution, potential scan rate: 50 mV s^{-1}; (B) current—potential curves of RDE with polycrystalline Pt electrode surface in O_2-saturated 0.5 M H_2SO_4 aqueous solution, potential scan rate: 10 mV s^{-1}. Authors' own work.

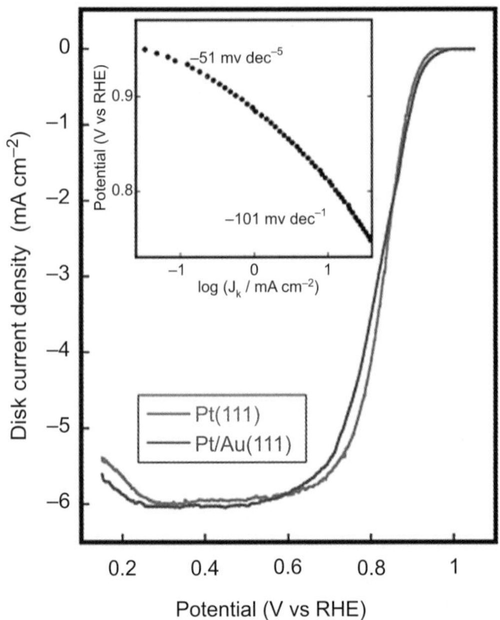

Figure 7.3 Linear sweep voltammetry for ORR of Pt/Au(111) (red line) and Pt(111) (blue line) electrodes measured in oxygen-saturated 0.1 M $HClO_4$ at a scan rate of 10 mV s^{-1} with a rotation rate of 1600 rpm. Inset: diffusion-corrected Tafel plot of the Pt(111) electrode obtained with a rotation rate of 1600 rpm. (For interpretation of the references to color in this figure legend, the reader is referred to the online version of this book.) Reprinted with permission from 4.

the ring electrode in the kinetic region in 0.1 M $HClO_4$ solution. Also, the ORR activity of the Pt/Pd(111) surface is larger than that of Pt(111). The mechanism of the enhanced ORR activity for Pt monolayer catalyst is widely accepted to be the reduced Pt-OH coverage by the lateral repulsion between the OH adsorbed on Pt and the OH or O adsorbed on the neighboring metal atom.[1] According to DFT calculations, the experimentally measured electrocatalytic ORR activity of Pt monolayers showed a volcano-type dependence on the center of their d-bands. While a higher lying d band center tends to facilitate O—O bond breaking, a lower lying one tends to facilitate bond formation (e.g., hydrogen addition).

The ORR catalytic activity of the electrochemically constructed Pt monolayer on the Au(111) was also investigated using the RRDE technique,[4] as shown in Figure 7.3. Two Tafel slopes could be obtained on Pt(111) surface to be −51 mV decade^{-1} in the potential range of 0.89–0.93 V and −101 mV decade^{-1} in the potential range of 0.76–0.83 V, respectively. The Tafel slope of

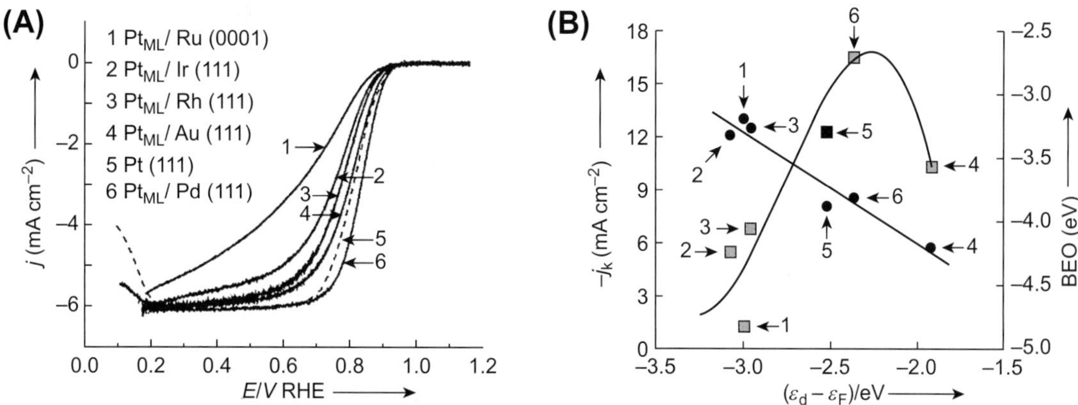

Figure 7.4 (A) Polarization curves for O_2 reduction on platinum monolayers (Pt_{ML}) on Ru(0001), Ir(111), Rh(111), Au(111), and Pd(111) surfaces in O_2-saturated 0.1 M $HClO_4$ solution on a disk electrode. The rotation rate is 1600 rpm, and the sweep rate is 20 mV s^{-1} (50 mV s^{-1} for Pt(111)); j = current density, RHE = reversible hydrogen electrode. (Reprinted with permission from Ref. 6); (B) kinetic currents (j_K; square symbols) at 0.8 V for O_2 reduction on the platinum monolayers supported on different single-crystal surfaces in O_2-saturated 0.1 M $HClO_4$ solution and calculated binding energies of atomic oxygen (BEO; filled circles) as functions of calculated d-band center ($\varepsilon_d - \varepsilon_F$; relative to the Fermi level) of the respective clean Pt monolayers. Labels: (1) Pt_{ML}/Ru(0001), (2) Pt_{ML}/Ir(111), (3) Pt_{ML}/Rh(111), (4) Pt_{ML}/Au(111), (5) Pt(111), (6) Pt_{ML}/Pd(111). Reprinted with permission from Ref. 22.

Pt/Au(111) was different from that of Pt(111), because the coverage of the adsorbed oxygen species in these potential ranges should be different from each other on the Pt/Au(111) surface. Thus, it was believed that the change in the Tafel slope was reflected by potential-dependent changes in coverage of adsorbed OH on the Pt surfaces. The reason that the ORR catalytic activity of the Pt monolayer on the Au(111) is different from Pt was ascribable to an electronic effect from the underlying Au. The underlying noble Au is expected to shift the d-band center of the Pt overlayer to a higher state. The shift of the d-band center to a higher state should lower the 5d vacancy in the Pt_{ML}, thus leading to a contracted 2π-electron donation from the oxygen molecule to Pt. As a result, oxygen adsorption on the surface decreased, lowering the kinetic of oxygen reduction.[5]

The electrocatalytic ORR activity of platinum monolayers supported on Au(111), Rh(111), Pd(111), Ru(0001), and Ir(111) surfaces in a 0.1 M $HClO_4$ were also studied,[6] as shown in Figure 7.4(A). There was a volcano-type dependence of monolayer catalytic activity on the substrate, and the most active was Pt_{ML}/Pd(111), as shown in Figure 7.4(B). The weighted center of the d-band (ε_d) was believed to play a decisive role in determining a surface reactivity. The positions of the ε_d of the Pt monolayers depend both on the strain (geometric effects) and on the electronic interaction between the Pt monolayer and its substrate (ligand effect).[23] Pt_{ML}/Au(111) lies at the high ε_d end of the plot and binds oxygen much more strongly than Pt(111), whereas Pt_{ML}/Ru(0001), Pt_{ML}/Rh(111), and Pt_{ML}/Ir(111) lie at the other end and bind oxygen considerably less strongly than Pt(111). The superior ORR catalytic activity of Pt_{ML}/Pd(111) was probably associated with a reduced OH coverage, which proved that OH hydrogenation rates enhance for Pt_{ML}/Pd(111) in view of DFT again.

7.3. RDE/RRDE Study for ORR on Carbon-Based Electrode Surfaces

The most commonly used carbon electrode materials for ORR are pyrolytic graphite (PG)[24] and glassy carbon (GC).[23] Normally, the solution pH has a strong effect on the ORR activity. For example, the ORR activity in alkaline solution is much higher than that in acidic solution. For both bare PG and GC electrodes, the ORR activity in acidic solution is insignificant. This may be the reason why they can be used as the electrode substrates on which the catalyst is coated for evaluating the catalyst's ORR activity in acidic solutions.

The ORR behaviors observed by RRDE technique on these bare carbon surfaces in alkaline solution are slightly different although there are two distinguishable waves on their current–potential curves, as shown in Figure 7.5(A and B), respectively.

Regarding the PG electrode, Figure 7.5(A) shows the RRDE curves that display two waves. However, using the slopes Kouctecky–Levich plot at different potential from −0.2−0.8 V vs 1.0 M NaOH//Hg/HgO, the electron-transfer numbers were calculated, which were close to 2 in the entire potential range, indicating the ORR on PG surface was a 2-electron-transfer process from O_2 to HO_2^- in an alkaline solution. The feature of two waves was believed to be caused by the change of surface active sites. However, when the electrode was oxidized at 2 V in the same solution for several minutes to produce some surface active sites such as quinine-like groups, the second wave at more negative potential range became a 4-electron-transfer process from O_2 to OH^-. In addition, after this oxidation treatment, the ORR activity was enhanced significantly, which can be seen from the onset potential of the ORR current–potential curves in Figure 7.5(B) when compared to that in Figure 7.5(A). Therefore, the oxidation treatment of the electrode surface is effective in improving the ORR activity of the carbon electrode surface.

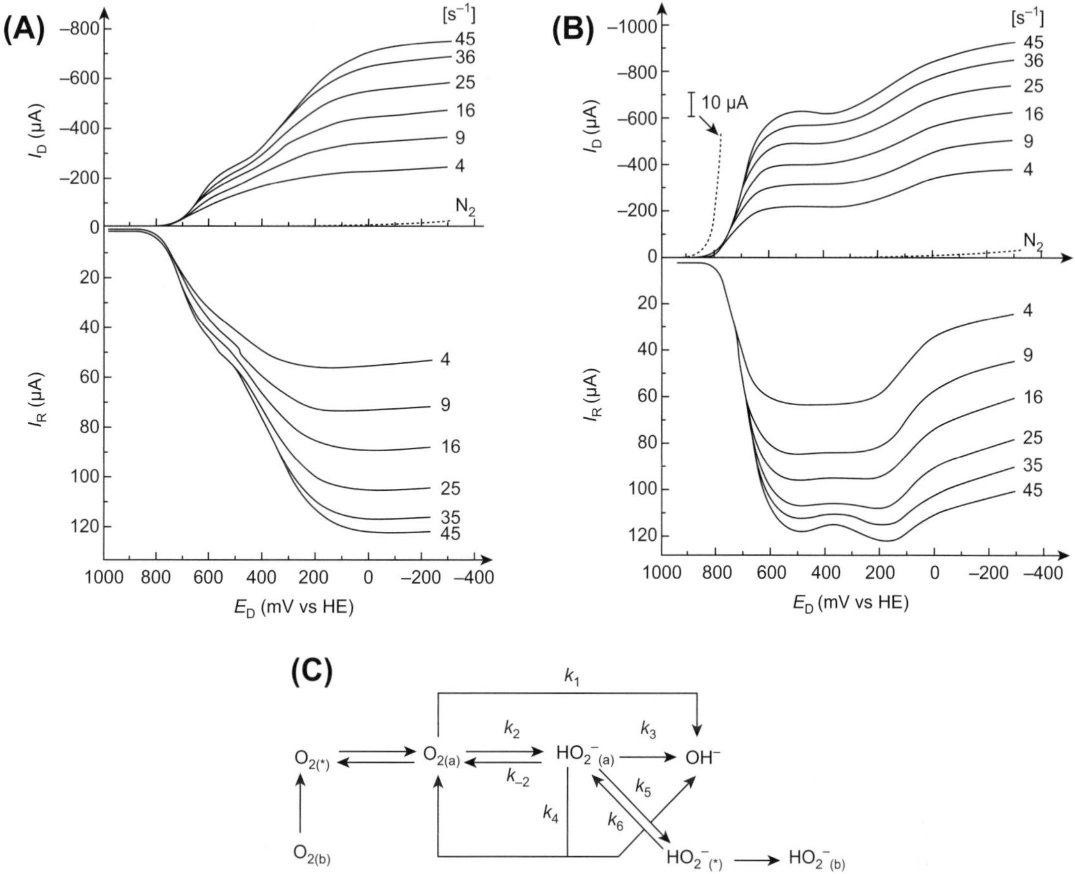

Figure 7.5 (A) Oxygen reduction and hydrogen peroxide oxidation on a bare pyrolytic graphite disk-Pt ring electrode; (B) Oxygen reduction and hydrogen peroxide oxidation on an oxidized pyrolytic graphite disk-Pt ring electrode in O_2-saturated 0.1 M KOH; and (C) Proposed ORR mechanism. Potential scan rate: 5 mV s⁻¹, ring potential: 1.34 V vs 1 M NaOH//Hg/HgO. Reprinted with permission from Ref. 24.

An ORR mechanism was proposed as shown in Figure 7.5(C), based on which and the approach discussed in Chapter 6 of this book, the relationships of $N\frac{I_D}{I_R}$ vs $\omega^{-1/2}$ and $N\frac{I_L-I_D}{I_R}$ vs $\omega^{-1/2}$ were formulated and the corresponding plots, as examples, are shown in Figure 7.6 (note that the expression of current densities in this figure are different from the expressions we used in Chapter 6). From the slopes and the intercepts of the plots in Figure 7.6, the ORR constants can be obtained. For example, the obtained values for k_2 at different potentials were in the range of 1.0×10^{-1} to 1.1×10^{-2} cm s⁻¹ when potential was changed from 0.2 to 0.7 V.

Regarding the bare GC electrode, Figure 7.7 shows the RRDE current–potential curves. There are also two waves. From the

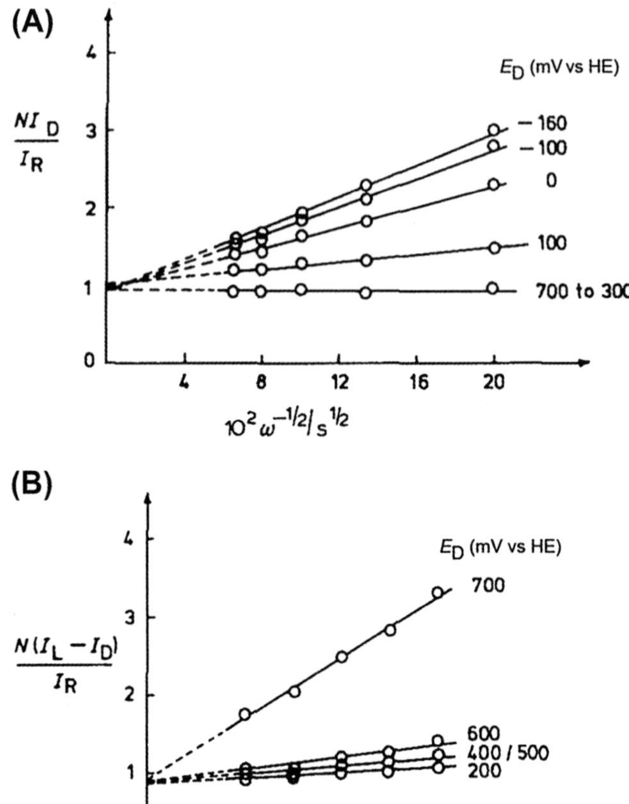

Figure 7.6 (A) $N\frac{I_D}{I_R}$ vs $\omega^{-1/2}$ plots obtained from the RRDE data of Figure 7.5(A), (B) $N\frac{I_L - I_D}{I_R}$ vs $\omega^{-1/2}$ plots obtained from Figure 7.6(B). Reprinted with permission from Ref. 24.

slopes and intercepts of K—L plots, the electron-transfer number and the electrode kinetic constant can be estimated. The obtained electron-transfer number was 2 in the entire potential range, indicating that the ORR process on a GC surface is similar to that on a PG surface. A similar reaction mechanism was also proposed, as shown in Figure 7.7(C). Based on this mechanism and the data shown in Figure 7.7(A and B), the similar approach as that in Figure 7.6 was also carried out to estimate the reaction kinetics constants. Therefore, the ORR on both bare PG and GC electrodes is a 2-electron-transfer process from O_2 to peroxide in basic solution.

It is worthwhile to note that the ORR activity on carbon-based electrode surface is strongly dependent on the pretreatment such as polishing, dipping a carbon electrode into chromic acid, exposing to radio frequency plasma in oxygen atmosphere, and

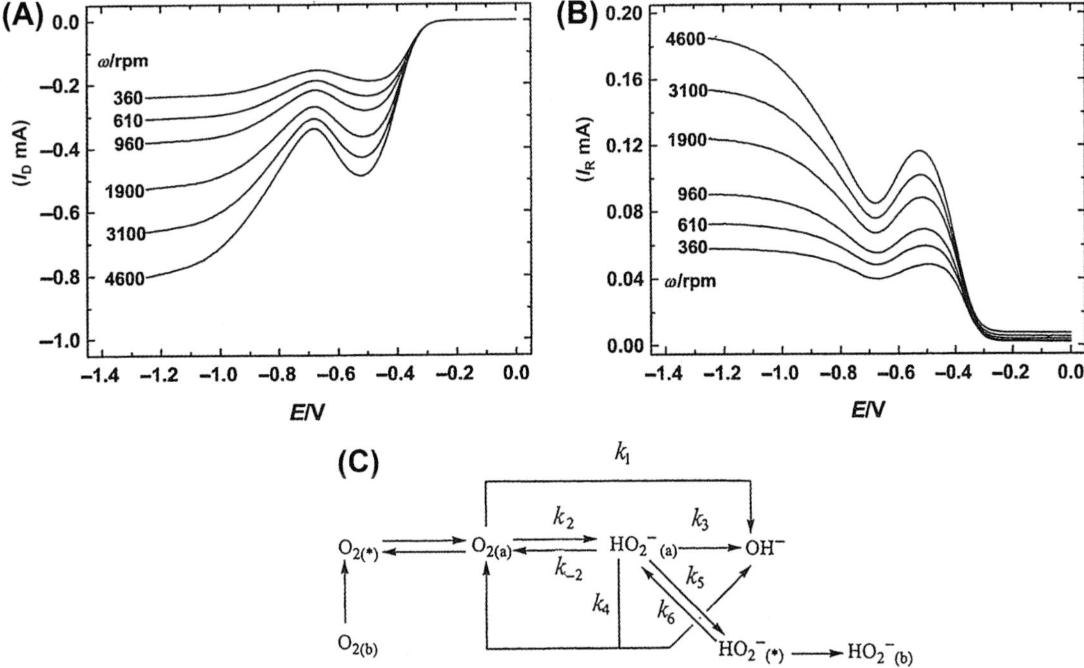

Figure 7.7 (A) Oxygen reduction on a bare GC disk electrode, and (B) peroxide oxidation on an Au ring electrode of RRDE at different electrode rotating rates in O_2-saturated 0.1 M KOH; potential scan rate: 10 mV s^{-1}, ring potential: 0.05 V vs SCE; and (C) proposed ORR mechanism.[23]

anodizing in an electrolyte solution.[8] It was reported that the pretreatment can create the oxygen-containing functional groups such as quinine on the carbon surface, which might serve as mediators to affect the ORR activity.[25,26] At the same time, the oxidized carbon surfaces could also catalyze the further reduction of hydrogen peroxide in alkaline solutions.

In general, the ORR mechanism on carbon electrode surface in basic solution can be expressed as follows[27]:

$$O_2 + e^- \rightarrow O_2^- \, (ads) \tag{7-I}$$

followed by a disproportionation:

$$2O_2^- \, (ads) + H_2O \rightarrow HO_2^- + O_2 + OH^- \tag{7-II}$$

or an electrochemical reduction:

$$HO_2(ads) + e^- \rightarrow + HO_2^- \tag{7-III}$$

Besides, the GC and PG electrode materials, other carbon materials, have also been used as the electrode substrates, including high-oriented pyrolytic graphite (HOPG)[12] and carbon nanotubes (CNTs).[28–30]

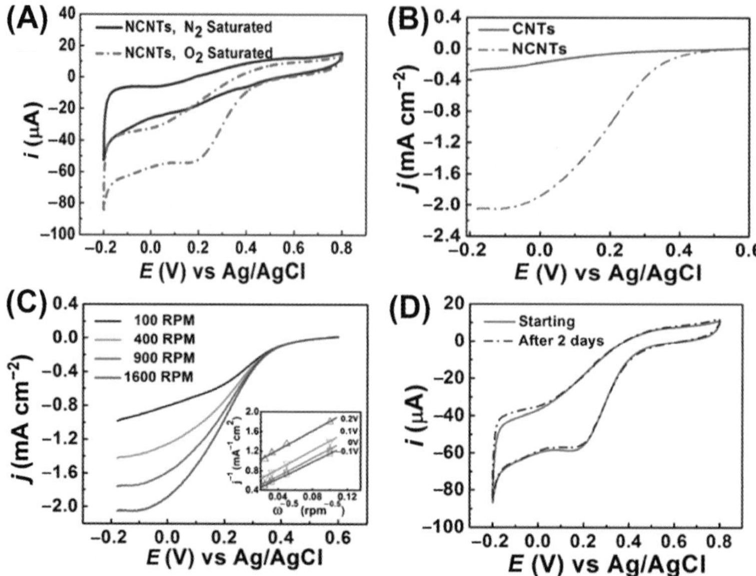

Figure 7.8 (A) Cyclic voltammograms of the nitrogen-doped CNTs (NCNTs) in N_2- and O_2-saturated 0.5 M H_2SO_4 solutions; (B) rotating-disk-electrode (RDE) linear sweep voltammograms of the NCNTs and undoped CNTs in O_2-saturated 0.5 M H_2SO_4 at a scan rate of 10 mV s^{-1}; (C) RDE curves of the NCNT in O_2-saturated 0.5 M H_2SO_4 with different rotating rates of 100, 400, 900, and 1600 rpm at a potential scan rate of 10 mV s^{-1} (the inset showing the Koutecky–Levich plots of the NCNT derived from RDE measurements); and (D) electrochemical stability measurements of the NCNT by using continuous cyclic voltammetry in O_2-saturated 0.5 M H_2SO_4 at a scan rate of 50 mV s^{-1}. Reprinted with permission from Ref. 29.

The ORR activity of undoped and/or nitrogen-doped single-walled CNTs in acidic solution by RDE technique are as shown in Figure 7.8.[29] It seems that nitrogen-doped CNTs have a relatively good electrocatalytic activity and also a long-term stability toward the ORR. Using the slopes of the K–L plots at different electrode potentials, the overall electron-transfer number for the ORR were calculated to be in the range of 3.5–3.9, indicating that the ORR on nitrogen-doped CNTs is dominated by a 4-electron-transfer process from O_2 to H_2O.

N-doped graphene was also studied as the ORR electrode material.[30,31] Poly (diallyldimethylammonium chloride) PDDA, as nitrogen source, was introduced into a graphene material. It was observed that PDDA as a p-type dopant could cause the partial electron transfer from the electron-rich graphene substrate to the positively charged N$^+$ centers of PDDA.[31] Thus, carbon atoms in the graphene sheets with the adsorbed PDDA could create somewhat delocalized positive charges on the conjugated carbon surface of graphene. The ORR results revealed that the electron-transfer number (3.5–4) at the PDDA-graphene electrode was always higher than that on the pure graphene electrode. Furthermore, the amount of hydrogen peroxide ions generated on the PDDA-graphene electrode was significantly less than that on the pure graphene, indicating that PDDA-graphene is a more efficient ORR electrocatalyst. It was also found that the bonding state of the N atom has a significant effect on the selectivity and catalytic activity for ORR. And the electrocatalytic activity of the catalyst was also found to be dependent on the graphitic N content, which determined the limiting current density, while the pyridinic N content improved the onset potential for ORR[32] and pyridinic N species, might convert the ORR reaction mechanism from a 2-electron-dominated process to a 4-electron-dominated process. Thus, pyridinic N at the carbon edge plane helps the reduction of oxygen. This is due to pyridinic N, which possesses one lone pair of electrons in addition to the electron donated to the conjugated π bond, facilitates reductive O_2 adsorption.

Boron atoms, with a strong electron-withdrawing capability, are also doped into graphene frameworks forming boron-doped graphene. Due to its particular structure and unique electronic properties, the resultant B-doped graphene exhibits a Pt-like electrocatalytic activity toward ORR in alkaline electrolytes.[9,33–35]

7.4. Oxygen Reduction Reaction on Monolayer Substances-Modified Carbon Electrode Surfaces

As mentioned above, the pretreatment of the carbon electrode surface can significantly affect its ORR activity, and even alter the ORR mechanism. Actually, the carbon surface can also be modified by some monolayer substances such as anthraquinones (AQ), and transition metal macrocycle complexes. Due to the conjugate structures of these molecules, they can irreversibly adsorb on the carbon surface to form monolayers that could serve as an electrocatalyst for ORR. This kind of electrode structure can be used to evaluate the electrocatalyst's ORR activity, particularly in acidic solution.

7.4.1. ORR on Anthraquinone-Modified Carbon Electrodes

It has been proposed that quinone groups (Q) on the carbon surface are responsible for the enhanced activity toward

2-electron-transfer ORR since the presence of semiquinone radicals $Q^{\cdot-}$ can promote the first step of 1-electron transfer to form superoxide, which is considered the rate-determining step. The electrochemical–chemical (EC) mechanism can be expressed as follows[27]:

$$Q + e^- \rightarrow Q^- \tag{7-IV}$$

$$Q^- + O_2 \rightarrow O_2^-(ads) + Q \tag{7-V}$$

Reaction (7-V) has been proposed as the rate-determining step. After the formation of $O_2^-(ads)$, it will go further reactions as expressed as Reactions (7-II) and (7-III).

The surface CV of anthraquinone- or its derivative-adsorbed electrode shows a reversible 2-electron-transfer redox couple. For example, Figure 7.9(A) shows the surface CV of anthraquinone-carboxlyic-allyl ester (ACAE) adsorbed on a basal plane graphite electrode surface.[36] The pH dependent of the peak potentials gives a slope of 57 mV pH^{-1}, suggesting that the redox process involves two electrons and two protons. The surface reaction reactions on the electrode surface can be assigned as Reaction (7-VI):

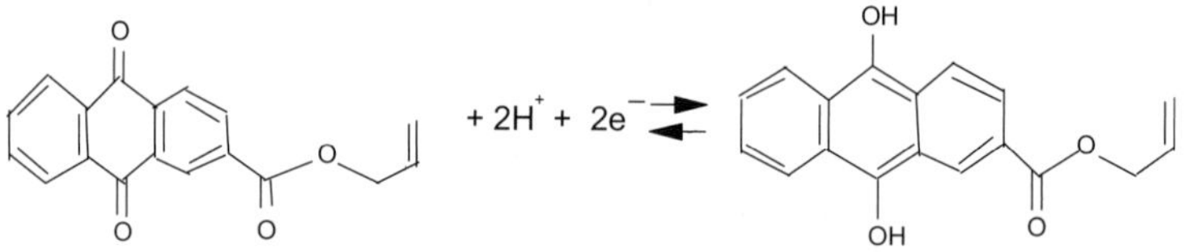

$$\tag{7-VI}$$

When the ACAE-coated RDE was exposed to O_2-saturated solution, some strong ORR activity was observed as shown in Figure 7.9(B). The slope of K–L plot indicates a 2-electron-transfer ORR (Figure 7.9(C)), confirming that ACAE can catalyze a 2-electron-transfer ORR process to produce H_2O_2. The intercept can give a kinetic current of $(i_k)_{O_2}$ with a unit of A, which was expressed as:

$$(i_k)_{O_2} = nFAk_{O_2}\Gamma_{ACAE}C_{O_2} \tag{7.2}$$

where n is the electron-transfer number ($=2$), F is the Faraday's constant (96,487 C mol^{-1}), A is the electrode surface area (cm^2), k_{O_2} is the kinetic constant of ORR (mol^{-1} cm^3 s^{-1}), Γ_{ACAE} is the surface concentration of ACAE (mol cm^{-2}), and C_{O_2} is the concentration of the dissolved oxygen (mol cm^{-2}). From the experimental data and Eqn (7.1), the catalyzed ORR kinetic constant was obtained.

In literature, different anthraquinones were employed as ORR catalysts, as listed in Ref. 12 together with their redox potentials of

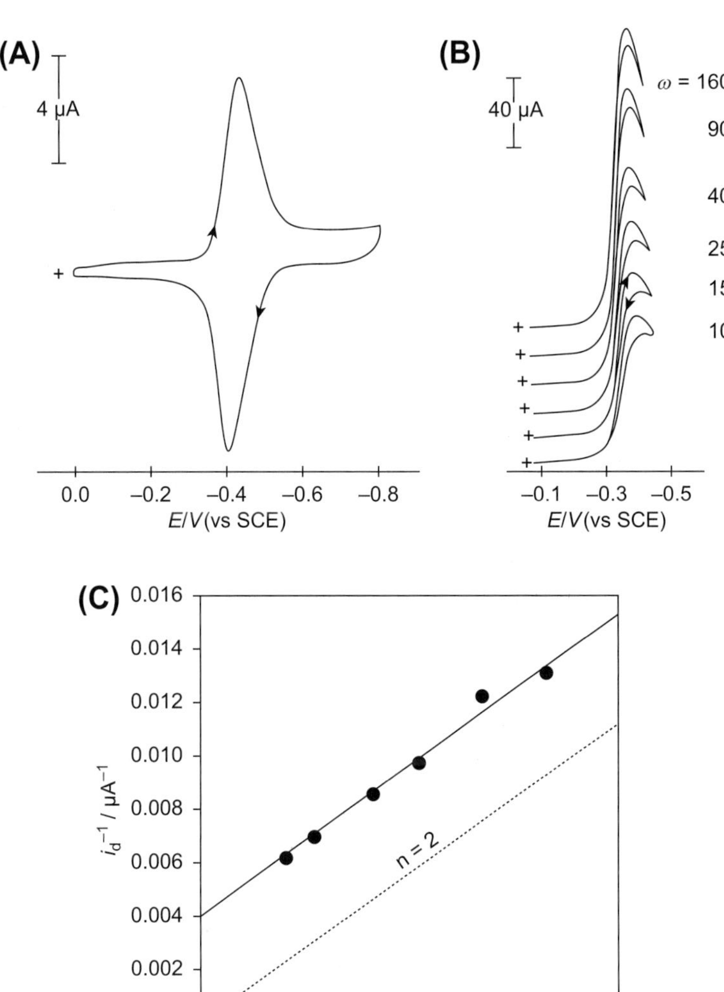

Figure 7.9 (A) Cyclic voltammogram of the basal plane graphite (BPG) coated by anthraquinone-carboxlyic-allyl ester (ACAE with a loading of 8.9×10^{-10} mol cm^{-2}) in N$_2$-saturated 0.1 M Na$_2$SO$_4$ + 0.1 M CH$_3$COOH/CH$_3$COONa (pH = 5.8). Potential scan rate: 100 mV s^{-1}. (B) ORR current–potential curves of ACAE (1.0×10^{-10} mol cm^{-2})-coated BPG electrode recorded in O$_2$-saturated 0.1 M Na$_2$SO$_4$ + 0.1 M CH$_3$COOH/CH$_3$COONa (pH = 5.6) at different electrode rotating rates as marked beside the curves. (C) Koutecky–Levich plot at electrode potential of −0.4 V vs SCE, data from (B). Reprinted with permission from Ref. 36.

the surface quinone/hydroquinone couple (E_f) and the ORR peak potentials (E_p) on high-orientated pyrilytic graphite electrode surface. It can be seen that AQ derivatives, electron-withdrawing substituents (halogens and substituted sulphonyl groups), can shift the redox potential to more positive values. Therefore, the redox potentials are strongly dependent on anthraquinone's structures.

9,10-phenanthrenequinone (PQ)[37] has the most positive E_p, suggesting that it probably has the best ORR activity when compared to others. A GC electrode was modified by PQ was studied by Vaik et al. using RRDE technique.[12] The GC was modified by adsorption, covalent binding, and electrochemical reduction of appropriate diazonium salts. The half-wave potential for the PQ-modified GC electrode was more positive than that of the bare GC, implying that its ORR catalytic activity was enhanced. It was revealed that the reactive intermediate semiquinone radical anion was important for enhancing the ORR activity. It was found that the ORR rate of was dependent on the surface concentration of the quinone radical anion.

7.4.2. ORR on Metal Macrocycle Complex-Coated Carbon Electrodes

Monolayers of transition metal (e.g., Fe, Co) macrocyclic compounds (e.g., porphyrins, phthalocyanines, tetraazannulenes), strongly adsorbed on the graphite electrode surface, have been studied for ORR catalysts for many years.[38] These electrocatalysts' coated graphite surfaces are capable of selectively catalyzing the 2-electron or 4-electron-transfer reduction of oxygen to produce H_2O_2 or water. However, compared to Pt-based catalysts, both their ORR activity and stability are insufficient. In the effort to overcome the challenges of activity and stability as well as to obtain the fundamental understanding of oxygen reduction electrocatalysis, RDE and RRDE are the mostly commonly used techniques for evaluation. Among the metal macrocyclic complexes, such as metal porphyrins have been chosen as the typical systems for reaction mechanism exploration.[39–45]

For example, three iron porphyrin complexes' adsorbed graphite electrode surface were studied by Anson's group using both RDE and RRDE techniques.[39] The ORR activities of these three complexes were compared using their Levich plots and their K–L plots, as shown in Figure 7.10. From Figure 7.10(A and B), it can be seen that the ORR activity is in the order of iron(III) protoporphyrin IX (FePPIX) > iron(III) meso-tetraphenylporphine (FeTPP) > iron(III) meso-tetra(3-pyridyl)porphine (FeTPyP), which can be seen from the intercepts of their K–L plots. Although all three complexes could catalyze ORR through the pathways dominated by the 4-electron-transfer process because the K–L plots of three complexes are all closely parallel to that theoretically calculated, the ring currents in RRDE test displayed in Figure 7.10(C) shows that there are still a considerable amount of H_2O_2 produced during the catalyzed ORR.

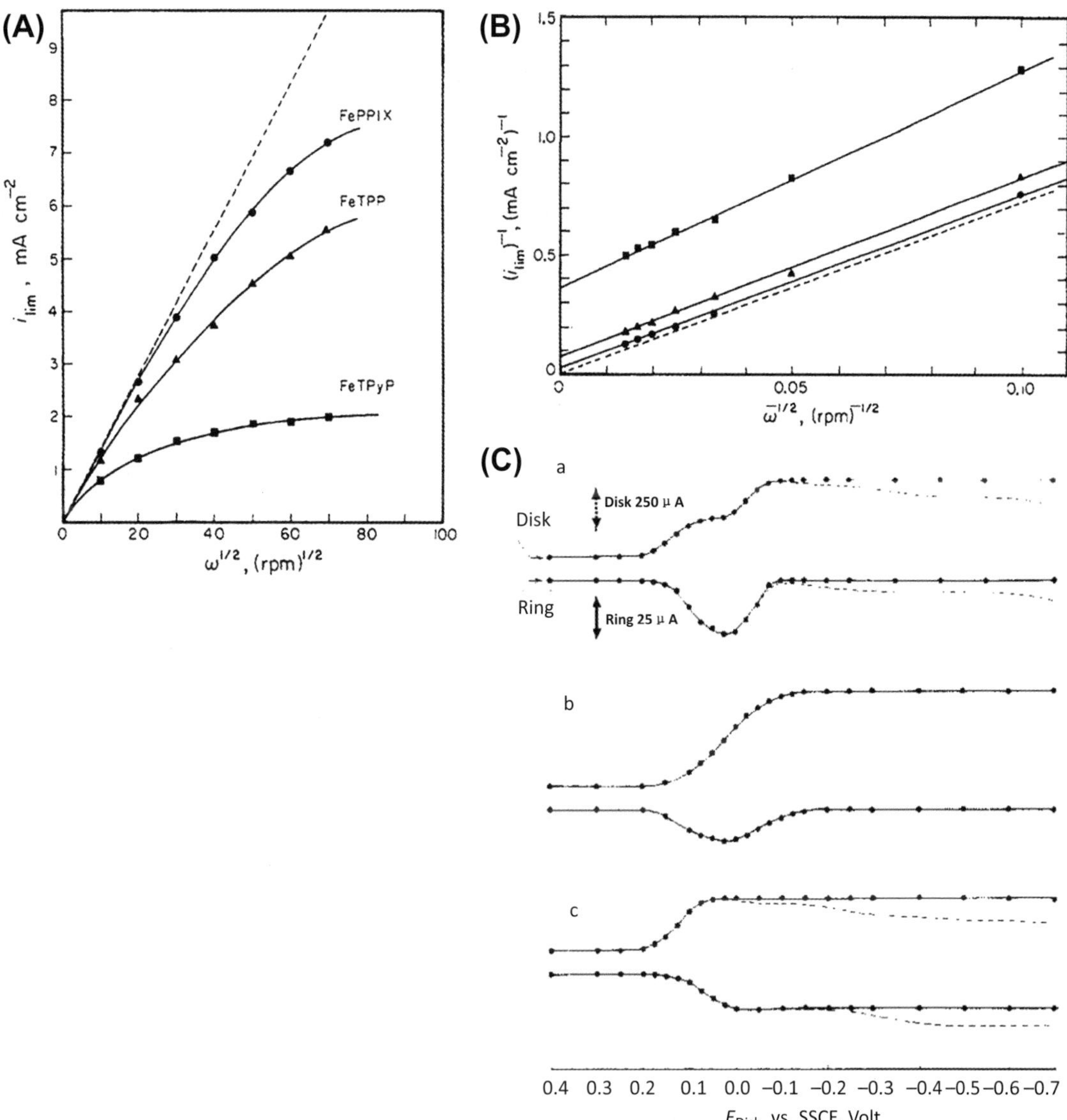

Figure 7.10 (A) Levick plots and (B) Koutecky–Levich plots of the limiting current for ORR at rotating graphite disk-electrode coated with three iron porphyrin catalysts, measured in O_2-saturated 0.1 M HClO$_4$ + 0.1 M NaClO$_4$ solution. The dashed line gives the calculated response for 4-electron, mass transfer-limiting reduction; and (C) current–potential curves during ORR at a rotated graphite disk-Pt ring electrode. The graphite disk (0.34 cm^2) was coated with 6×10^{-8} mol cm^{-2} of (a) FeTPP, (b) FePPIX, and (c) FeTPyP. Ring electrode potential: 1.2 V. Potential scan rate: 10 mV s^{-1}. Rotation rate: 100 rpm. Electrolyte is the same as (A) and (B). Reprinted with permission from Ref. 39.

The temperature dependence of the catalyst activity of an iron fluoro-porphyrin-coated graphite electrode was studied by RDE coupled with the surface cyclic voltammetry.[46] The purpose was to investigate the surface adsorption and reaction, O_2 reduction catalysis kinetics, and especially the temperature effect on the catalyst activity. Figure 7.11(A) shows the surface CVs of 5,10,15,20-Tetrakis(pentafluorophenyl)-21H,23H-porphine iron (III) chloride (abbreviated as $Fe^{III}TPFPP$)-coated graphite electrode, recorded in a pH ~ 1.0 Ar-saturated solution at different potential scan rates. The 1-electron reversible redox peak of approximately 0.35 V can be seen, which has a peak current increased linearly with increasing the potential scan rate, indicating the electrochemical behavior of this peak follows the feature of a reversible redox reaction of an adsorbed species on the electrode surface.

Regarding the catalyzed ORR catalyzed by $Fe^{III}TPFPP$, Figure 7.12(A) shows the CVs recorded at different conditions. The bare graphite electrode does not show ORR activity in the presence of oxygen (a). If the surface was coated with $Fe^{III}TPFPP$, a large oxygen reduction current onset from 0.45 V (vs NHE) can be observed as in (b), indicating that $Fe^{III}TPFPP$ has strong electrocatalytic activity toward oxygen reduction. The CV for the $Fe^{III}TPFPP$-adsorbed electrode in the absence of oxygen is shown in (c) where a redox wave near 0.35 V with an onset potential of ~0.45 V can be observed. The oxygen reduction onset potential (b) of 0.45 V is consistent with the onset potential of the surface couple of $Fe^{III}TPFPP/Fe^{II}TPFPP$, suggesting that the oxygen reduction is catalyzed by surface-adsorbed $Fe^{II}TPFPP$. The CV of H_2O_2 reduction on the same electrode coated with $Fe^{III}TPFPP$ also shows the reduction wave with an onset potential near the potential of O_2 reduction, suggesting that $Fe^{II}TPFPP$ is not only active toward O_2 reduction, but also catalyzes the reduction of H_2O_2.

ORR kinetics evaluated by RDE experiments at different temperatures was also carried out. The current–potential curves at different rotating rates were recorded using a rotating disk-graphite electrode coated with $Fe^{III}TPFPP$, as shown in Figure 7.12(B). The reciprocal of the limiting current (peak current in Figure 7.12(B)) was plotted against the reciprocal of the square root of the rotation rate as in Figure 7.12(C), together with a calculated plot according to the Levich theory for a 4-electron O_2 reduction process. The experimental line is approximately parallel to the calculated line, indicating that the catalyzed O_2 reduction is a 4-electron process, producing water.

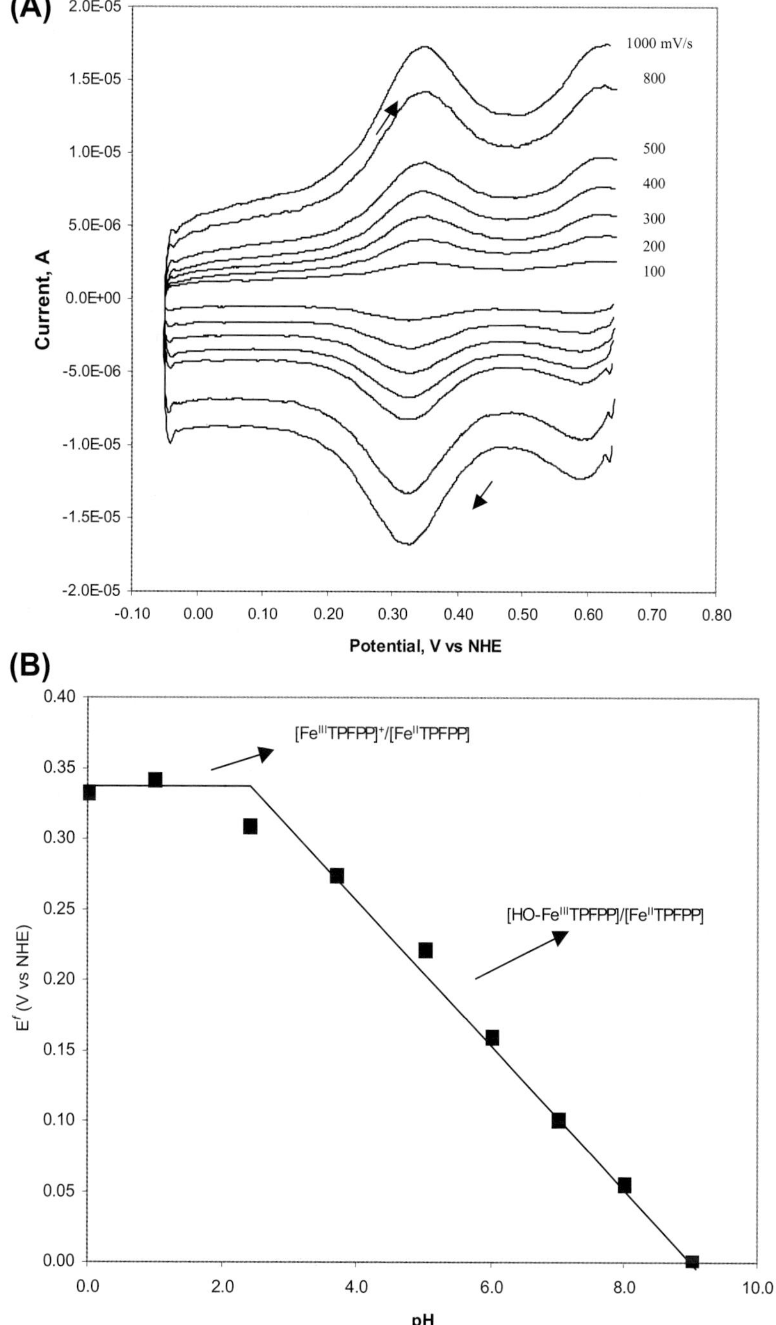

Figure 7.11 (A) Cyclic voltammograms of Fe^{III}TPFPP adsorbed on a graphite electrode at room temperature. Ar-saturated solution containing 0.5 M H_2SO_4 as a supporting electrolyte. Potential scan rates as marked beside each traces; (B) pH dependence of formal potential E^f for the Fe^{III}TPFPP adsorbed on a graphite electrode at 35 °C. Potential scan rate: 300 mV s^{-1}. Reprinted with permission from Ref. 46.

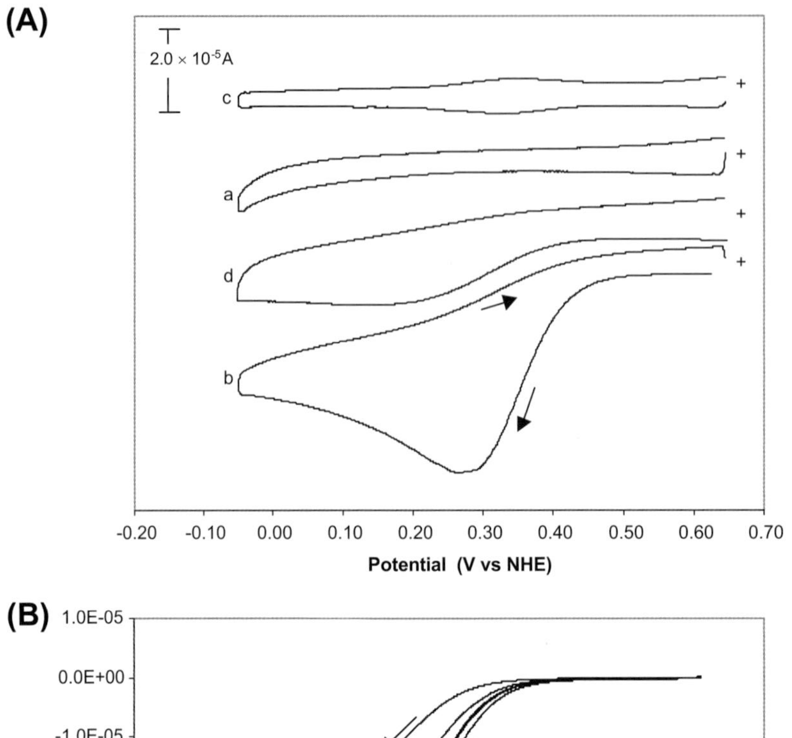

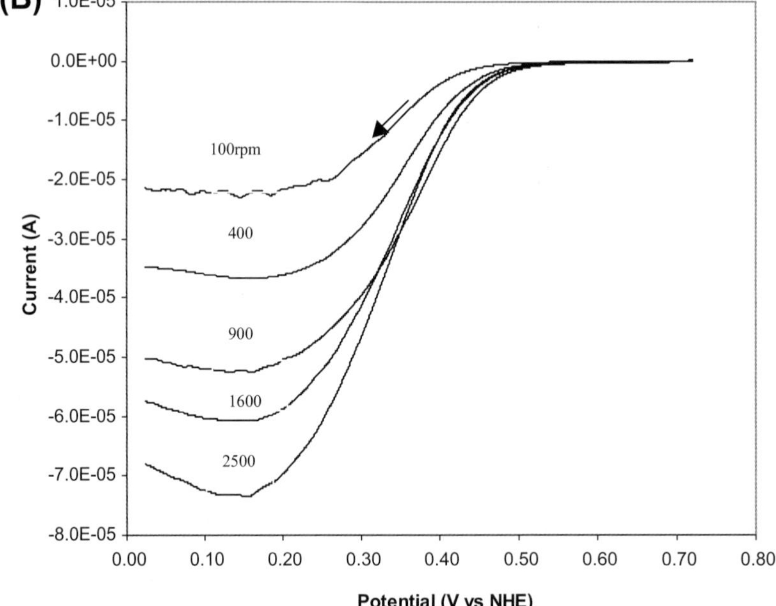

Figure 7.12 (A) Cyclic voltammetry of Fe^{III}TPFPP adsorbed on a graphite electrode: (a) background current at the bare graphite electrode in an air-saturated solution; (b) after Fe^{III}TPFPP was adsorbed on the electrode in the solution of (a); (c) after Fe^{III}TPFPP was adsorbed on the electrode in an argon-saturated solution; (d) after Fe^{III}TPFPP was adsorbed on the electrode in 0.28 mM H_2O_2 solution. Supporting electrolyte: 1 M H_2SO_4. Potential scan rate: 300 mV s^{-1}; (B) current—potential curves for Fe^{III}TPFPP adsorbed on a rotating graphite disk electrode with different rotating rates as marked on each trace. Recorded in O_2-saturated 0.5 M H_2SO_4 solution at 55 °C; (C) Koutecky—Levich plot, data from (B), the solid line is calculated according to the Levich theory for a 4-electron O_2 reduction process. The solid line with points on it is experimental. Reprinted with permission from Ref. 46.

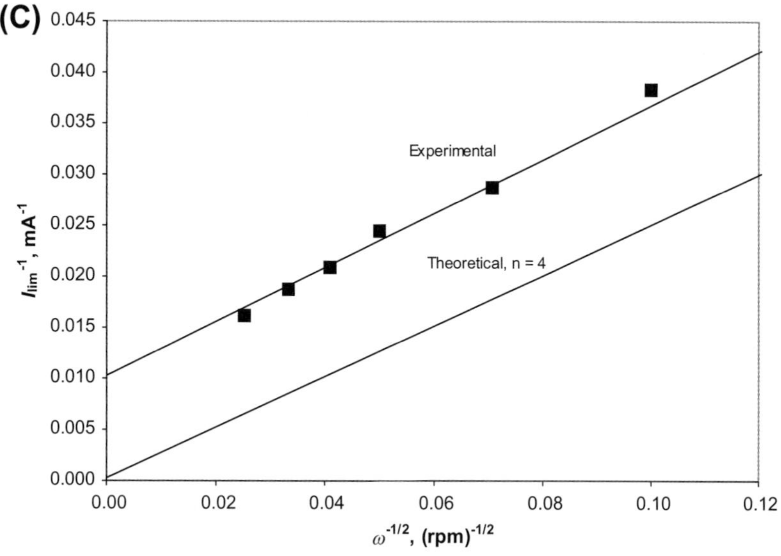

Figure 7.12 (*continued*).

The intercept of the K−L plot can give a kinetic current, which can be expressed as the same form as that of Eqn (7.1):

$$(i_k)_{O_2} = nFAk_{O_2}\Gamma_{Fe^{III}TPFPP}C_{O_2} \tag{7.2}$$

The obtained kinetic constants are summarized in Table 7.1.

The temperature effect on the catalyzed ORR was also studied using RDE at several temperatures listed in Table 7.2. Based on the Arrhenius theory, the reaction rate can be expressed as a function of reaction temperature:

$$\ln(k_{O_2}) = \ln(k_{O_2}^A) - \frac{E_a}{RT} \tag{7.3}$$

Table 7.1. Summary of Kinetic Parameters for O_2 Reduction Catalyzed by Fe^{III}TPFPP

T (K)		293	308	328	343
k_{O_2} mol^{-1} cm^3 s^{-1}	pH = 0	2.7×10^8	3.5×10^8	3.8×10^8	5.0×10^8
	pH = 5.0	1.1×10^8	1.7×10^8	2.8×10^8	3.8×10^8

Reprinted with permission from Ref. 46.

Table 7.2. $Fe^{II}TPFPP$ Electrocatalyzed O_2 Reduction Arrhenius Activation Energies and Absolute Reduction Rates at Two pH Levels

Solution pH	0.0	5.0
E_a, KJ mol^{-1}	9.9	21.3
$k_{O_2}^A$, mol^{-1} cm^3 s^{-1}	1.6×10^{10}	6.4×10^{11}

Reprinted with permission from Ref. 46.

where $k_{O_2}^A$ is the absolute reaction rate constant at $E_a = 0$, E_a is the reaction activation energy. Figure 7.13 shows the plots of $\ln(k_{O_2})$ vs $1/T$ at pH $\sim$ 0.0 and pH $= 5.0$, respectively, from which $k_{O_2}^A$ and E_a can be estimated as listed in Table 7.2.

Based on the experiments, the mechanism of oxygen reduction catalyzed by iron–porphyrin complexes was proposed as follows:

In acidic solutions (pH < 2.7):

$$\left[Fe^{III}TPFPP\right]^+ + e^- \leftrightarrow \left[Fe^{II}TPFPP\right] \tag{7-VII}$$

$$\left[Fe^{II}TPFPP\right] + O_2 \rightarrow \left[O_2 - Fe^{II}TPFPP\right] \tag{7-VIII}$$

$$\left[O_2 - Fe^{II}TPFPP\right] + 4H^+ + 3e^- \rightarrow \left[Fe^{III}TPFPP\right]^+ + 2H_2O \tag{7-IX}$$

In solutions with pH > 2.7:

$$\left[HO - Fe^{III}TPFPP\right] + H^+ + e^- \leftrightarrow \left[Fe^{II}TPFPP\right] + H_2O \tag{7-X}$$

$$\left[Fe^{II}TPFPP\right] + O_2 \rightarrow \left[O_2 - Fe^{II}TPFPP\right] \tag{7-XI}$$

$$\left[O_2Fe^{II}TPFPP\right] + 3H^+ + 3e^- \leftrightarrow \left[HO - Fe^{III}TPFPP\right] + H_2O \tag{7-XII}$$

Normally, Fe-based macrocycle catalyst such as $Fe^{III}TPFPP$ could catalyze the ORR with a 4-electron-transfer pathway from O_2 to H_2O, while those of Co-based could only catalyze the process with a 2-electron-transfer pathway to produce peroxide. For example, a monolayer of cobalt(II) 1,2,3,4, 8,9,10,11, 15,16,17,18, 22,23,24,25-hexadecafluoro-29H,31H-phthalocyanine (abbreviated as $Co^{II}HFPC$) was coated on a graphite electrode by spontaneously adsorption displayed a strong electrocatalytic activity

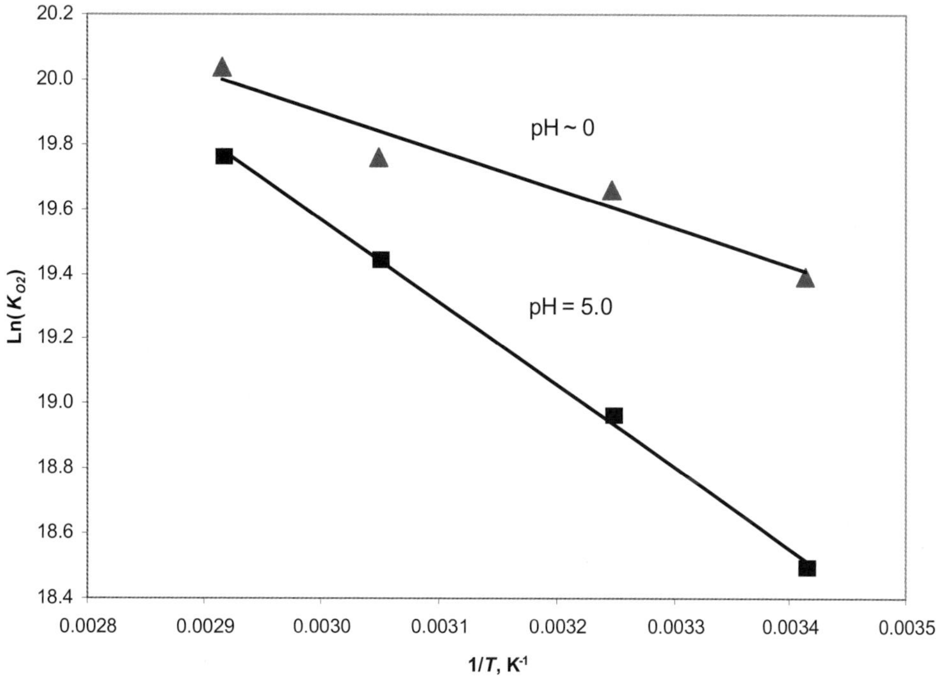

Figure 7.13 Arrhenius plots of O$_2$-reduction rate constant vs the reciprocal of temperature at two pH levels. Reprinted with permission from Ref. 46.

toward O_2 reduction. The kinetics of catalyzed ORR at different temperatures from 20 to 70 °C was measured by both cyclic voltammetry and RDE techniques.[47] Figure 7.14(A) shows the current–potential curves at different electrode rotating rates. The corresponding K–L plot shown in Figure 7.14(B) shows that this Co^{II}HFPC could only catalyze a 2-electron-transfer ORR process from O_2 to H_2O_2.

7.5. RDE/RRDE Study for ORR on the Surfaces of Supported Pt Particle- and Pt Alloy Particle-Based Catalyst Layer

One of the most practical applications of RDE/RRDE techniques is the evaluation of ORR electrocatalysts in terms of both their activity and stability for fuel cells or metal-air batteries. In order to increase the catalyst surface area and the catalyst utilization, all catalysts developed are normally supported on conductive support particles such as active carbon. In the

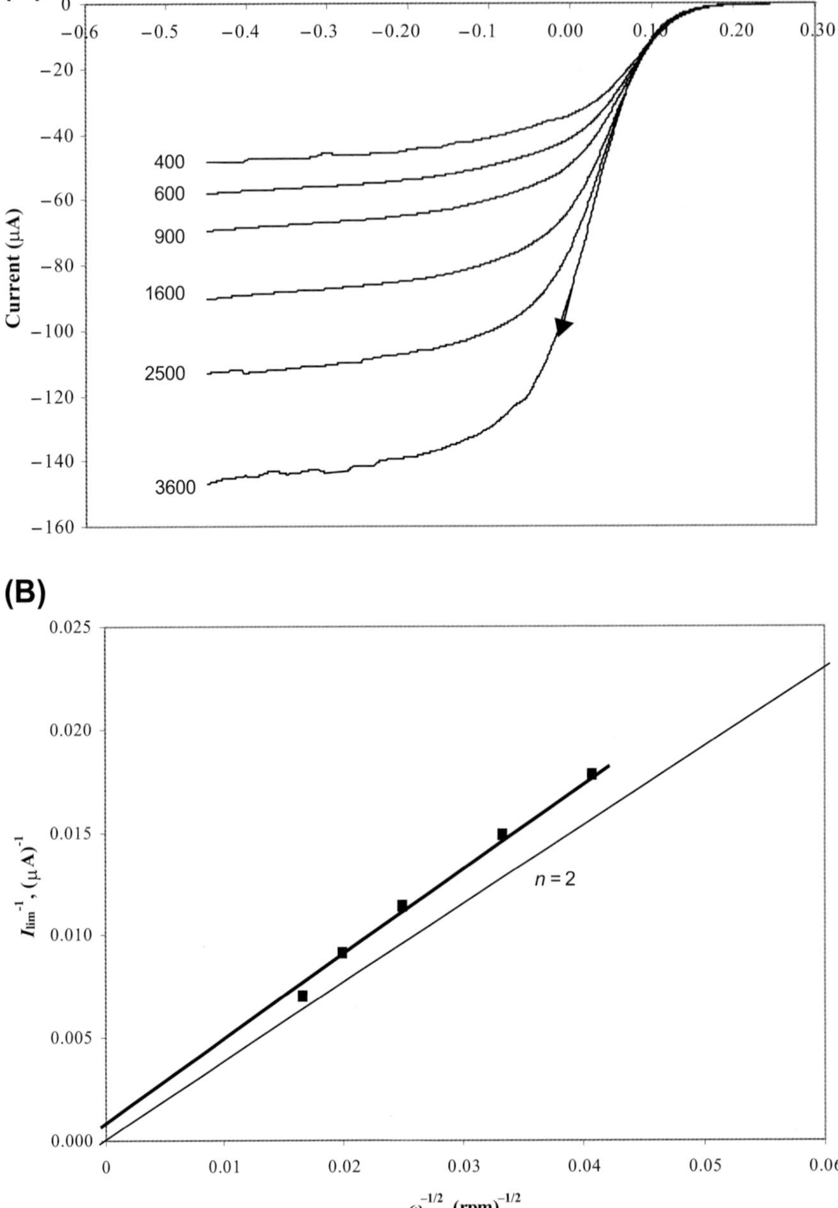

Figure 7.14 (A) Current–potential curves for $Co^{II}HFPC$ adsorbed on a rotating graphite disk electrode at different rates of rotation as (marked on each curve), recorded in an air-saturated 0.1 M Na_2SO_4 solution buffered to a pH 6. Temperature: 20 °C. Potential scan rate: 10 mV s^{-1}; (B) Koutecky–Levich plot, data from (B). The thinner solid line is calculated according to Levich theory for a 2-electron O_2 reduction process. The thicker solid line and data points are the experimental data. Reprinted with permission from Ref. 47.

evaluation, the catalyst-based catalyst layers will be coated onto the disk electrode substrates such as GC and gold, and then put into the N_2- or O_2-saturated electrolyte solution for measurement.

7.5.1. ORR on the Surfaces of Pt-Based Catalyst Layer

Pt particles supported on high surface-area carbon substrates to form supported catalyst (abbreviated as Pt/C) is the most widely used ORR catalyst at the current state of technology. The surface characteristics of Pt particles supported on carbon are similar to that of Pt polycrystalline electrode surface, which can be observed by their surface CVs. However, the ORR activity of Pt/C is different from what can be observed for polycrystalline Pt surface.[21] This difference is possibly due to some differences in the superficial structure, or a too strong adsorption of oxygenated species on very small particles. In addition, the ORR activity of carbon-supported Pt particles might also be affected by the electronic properties of Pt atoms from carbon.

Therefore, Pt-based catalysts are the most popular systems studied by RDE/RRDE techniques.[48] Figure 7.15 shows the RRDE curves on a GC disk electrode coated with a catalyst layer containing 20 wt% Pt/C catalyst and Nafion® ionomer, recorded at different electrode rotating rates in O_2-saturated 0.5 M H_2SO_4 solution. Their experiment results showed that the equivalent thickness of Nafion® ionomer ($\sim$0.1 μm) did not have a significant effect on the K−L plots, suggesting that its effect on the ORR kinetics could be negligible. From the ring currents, the percentages of H_2O_2 produced during the ORR catalyzed by Pt/C catalyst in the potential range of 0.2−1.0 V vs RHE were calculated, which were all within 1%, suggesting the catalyzed ORR was dominated by a 4-electron-transfer process to produce water.

7.5.2. ORR on the Surfaces of Pt Alloy-Based Catalyst Layer

It has been recognized that Pt-alloy catalyst-based catalyst layers can give better ORR activities than that of pure Pt-based catalyst layer. This may be due to the interaction between the Pt and alloy atom, altering the electron energy levels in the alloy, favoring the ORR process. However, the alloy's composition plays a major role in ORR activity of a Pt alloy.

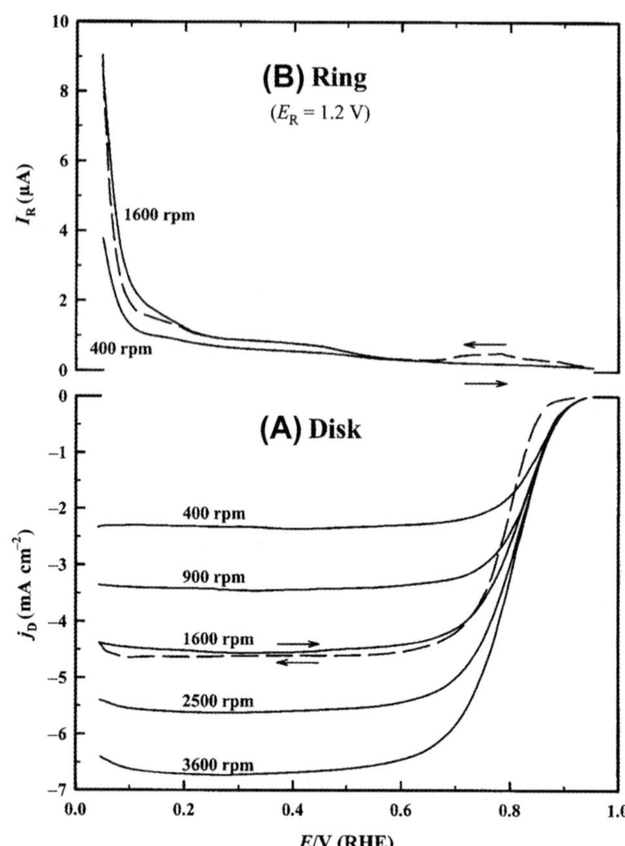

Figure 7.15 Current—potential curves on (A) glassy carbon disk for O_2 reduction and (B) Pt ring electrode for H_2O_2 oxidation obtained in an RRDE study. The disk electrode was coated by Pt/Vulcan with a Pt loading of 0.014 mg cm^{-2}. T = 60 °C; O_2-saturated 0.5 M H_2SO_4; scan rate = 5 mV s^{-1}; Ring potential = 1.2 V. Reprinted with permission from Ref. 48.

For example, Table 7.3 shows the comparison of half-wave potential, and the number of electrons transferred for the O_2 reduction in acidic solution at Au—Pt/C bimetallic nanoparticles of various compositions, measured by RDE technique. It can be clearly seen that a 4-electron-transfer pathway is mainly operative for ORR and the half-wave potential for ORR on bimetallic Au—Pt/C (20%:20%) is 100 mV less negative when compared with that of Pt/C (home-made and E-Tek). Au—Pt/C (30:10) shows a negative shift in a half-wave potential in comparison with both Pt/C samples, indicating that the catalytic activity decreases with increasing Au content. It is believed that an optimum content of Au will contribute to a reduction in the Pt-OH formation that is known to "poison" the catalyst for ORR.

It is worthwhile to point out that during the last decade, carbon-supported Au@Pt core—shell nanostructured catalysts were explored and synthesized to improve the ORR performance of the electrocatalyst.[50] The results showed that the limiting

Table 7.3. Comparison of ORR Half-Wave Potential in Acidic Solution, B Values, and the Overall Number of Electrons Transferred at Au–Pt/C Bimetallic Nanoparticles With Various Compositions

Composition (atomic wt%)	Half-Wave Potential of O_2 Reduction (mV) vs 0.5 M MSE (0.5 M H_2SO_4)	B (mA cm^{-2} rpm$^{-1/2}$)	n
0 Au:40Pt	96	0.130	3.96
10 Au:30Pt	98	0.124	3.76
20 Au:20Pt	200	0.127	3.92
30 Au:10Pt	50	0.114	3.47
E-tek-40 (Pt/C)	100	0.128	3.99

Reprinted with permission from Ref. 49.

current density exhibited a maximum value for the Au@Pt/C with Au:Pt at atomic ratios of 2:4 catalyst.

In addition to Pt–Au/C catalysts, several other Pt-based alloy catalysts, such as PtCo/C, PtNi/C, PtV/C, and PtPd/C were also reported, and the mechanism of enhancing ORR activity was investigated using both RDE and RRDE techniques. The mechanism of ORR improvement by alloying is ascribed to: (1) increase in the catalyst surface roughness,[15] (2) decrease in the coverage of surface oxides and an enrichment of the Pt-active sites of the catalyst surface,[51] (3) increase in the d-orbital vacancy, which strengthened the Pt–O_2^- interaction,[52] and (4) decrease in the Pt–Pt distance and the Pt–Pt coordination numbers.[53] Table 7.4 lists the properties of PtCo/C and PtNi/C catalysts and their ORR performance parameters measured by RDE techniques for comparison.

The RDE/RRDE curves of ORR catalyzed by PtNi/C catalysts are shown in Figure 7.16.[55] It can be seen that the catalytic ORR activities of all PtNi/C catalysts are higher than that of pure Pt/C. The amounts of hydrogen peroxide production on all PtNi/C catalysts are roughly similar to or lower than that on Pt/C. Besides, it is worth noting that the open-circuit potentials of PtNi/C catalysts in oxygen-saturated solution are about 30 mV higher than that of Pt/C catalyst, suggesting that the oxygen adsorption on the PtNi surface is more favored than that on a pure Pt surface.

Table 7.4. Summary of the Electrocatalytic Data for PtCo/C and PtNi/C Catalysts[a]. ECA: Electrochemical Surface Area, MA: ORR Mass Activity, and SA: ORR Specific Activity

Catalysts	Particle Size on Carbon (nm)	Metal Loading (% wt)	Lattice Parameter (nm)	ECA ($m^2\,g_{pt}^{-1}$)	MA ($A\,mg_{pt}^{-1}$)	SA ($mA\,cm^{-2}$)
Pt/C_Etek	2.5 ± 0.5	20	—	91	0.22	0.24
Pt/C	7.1 ± 1.6	23	0.3947	55	0.11	0.20
(A) PtCo/C						
$Pt_{73}Co_{27}/C$	2.8 ± 0.4	37	0.3910	73	0.33	0.45
$Pt_{75}Co_{25}/C$	4.6 ± 0.6	40	0.3904	58	0.15	0.26
$Pt_{70}Co_{30}/C$	7.0 ± 0.7	36	0.3874	38	0.12	0.32
$Pt_{52}Co_{48}/C$	3.9 ± 0.7	24	0.3844	44	0.25	0.57
(B) PtNi/C						
$Pt_{71}Ni_{29}/C$	6.6 ± 1.0	45	0.3920	27	0.17	0.62
$Pt_{78}Ni_{22}/C$	2.7 ± 1.0	38	0.3895	55	0.07	0.13
$Pt_{64}Ni_{36}/C$	8.7 ± 0.8	24	0.3890	24	0.14	0.57
$Pt_{56}Ni_{44}/C$	6.1 ± 0.8	44	0.3817	25	0.17	0.69

[a]Catalysts: 15 µg, electrode: 0.196 cm^2, electrolyte: 0.5 M H_2SO_4, scan rate: 5 mV s^{-1}, rotating speed: 2000 rpm.
Reprinted with permission from Ref. 54.

Pt-based trimetallic catalysts have also explored for ORR. For example, Ir, as the third metal, was added into PtCo alloy.[56] By the addition of Ir, a factor of 3–5 for the PtIrCo/C catalysts toward ORR in comparison with Pt/C catalysts was achieved. There is a clear trend that the mass activity increases with increasing Co% and decreasing Pt%. Since the fcc-type lattice constant for $Pt_{25}Ir_{20}Co_{55}/C$ (0.385 nm) is smaller than both those of $Pt_{65}Ir_{11}Co_{24}/C$ and $Pt_{40}Ir_{28}Co_{32}/C$ (0.389 nm), the specific activity of $Pt_{25}Ir_{20}Co_{55}/C$ is expected to be enhanced. Moreover, when Pt and Ir atoms are mixed on the surface of the catalyst, the OH adsorbed on Ir at low potentials has a repulsion effect on the OH and oxide formation on Pt, resulting in a higher ORR kinetics. It was also observed that the $Pt_{65}Ir_{11}Co_{24}/C$ catalyst had also a comparable durability to that of commercial catalysts upon the severe potential cycling (Figure 7.17).

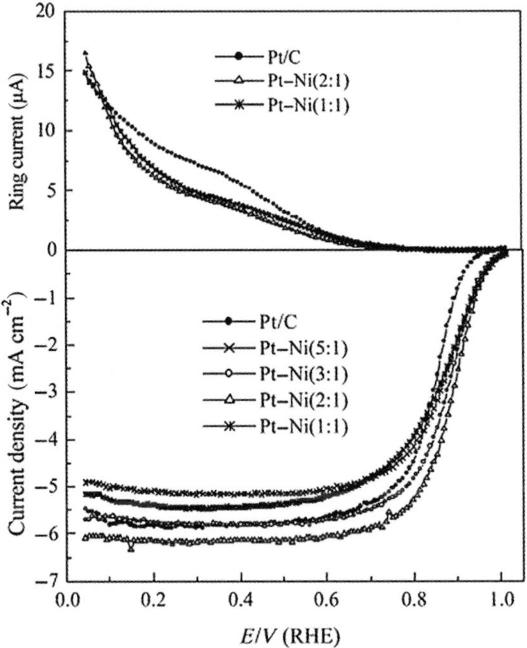

Figure 7.16 Current—potential curves (bottom) of the carbon-supported nanosized Pt and PtNi alloy catalysts coated on a glassy carbon disk electrode in O_2-saturated 0.5 M $HClO_4$ solution (scan rate of 5 mV s^{-1}, and rotating speed of 2000 rpm); and the corresponding ring currents (Pt at 1.2 V vs RHE) data for the hydrogen peroxide production (upper) on Pt/C, Pt Ni(2:1)/C and Pt Ni(1:1)/C systems. Reprinted with permission from Ref. 55.

Pt_3Co was exclusively under investigation due to its high specific activity, which was generally more than 100% higher than pure Pt.[57,58] But the Pt_3Co catalyst still needs more improvements to solve the problem of Co dissolution, which leads to reduced ORR activity. For Pt-alloy catalysts, the dissolution of the alloyed metals has been recognized as the major drawback of the Pt-alloy catalysts, particularly the catalysts of PtNi, PtCo, PtCu, and PtFe.

Furthermore, the corrosion of carbon support is also a challenge. The harsh condition at the cathode ORR includes oxidative oxygen, hot water, acidic environment, as well as high electrode potentials. These conditions are favorable to the oxidation of carbon support, especially in temperature range of 95—200 °C. To address the carbon support degradation issue, alternative supports are being developed with the objectives to increase both the support durability and the catalyst activity through improving the catalyst—support interaction. Titanium

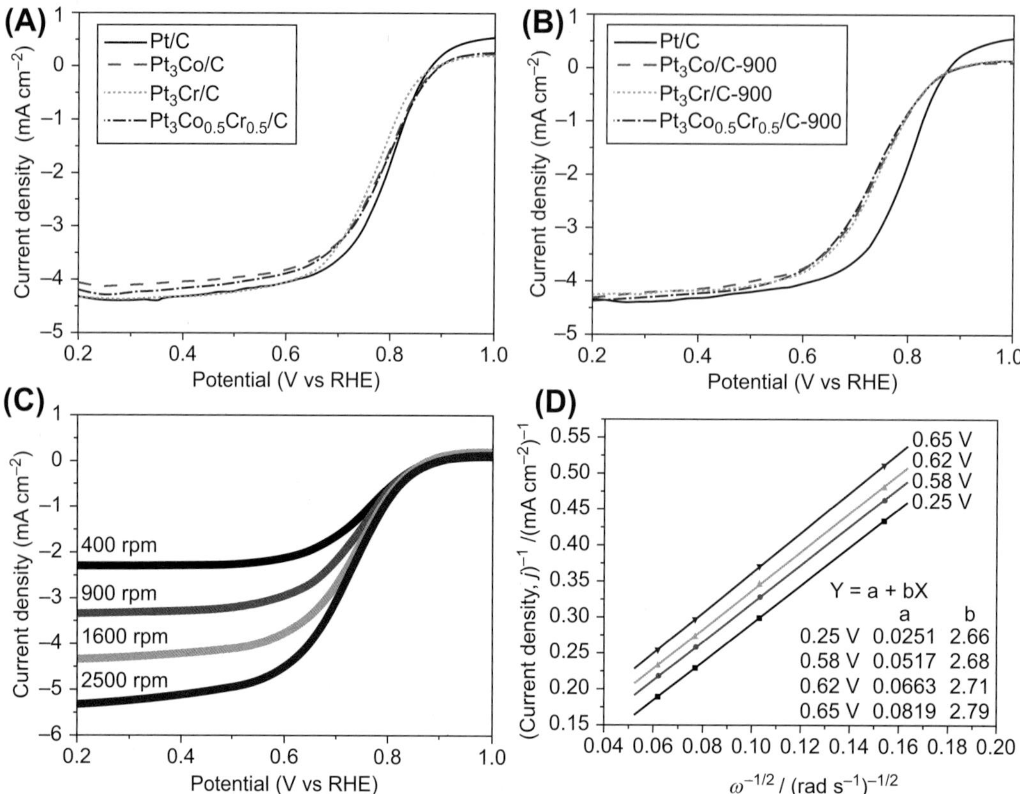

Figure 7.17 RDE testing results of Pt₃Co/C, Pt₃Cr/C, and Pt₃Co₀.₅Cr₀.₅/C catalysts. (A) before and (B) after 900 °C annealing. The results of Pt/C catalyst were included for comparison. (C) RDE testing results of the Pt₃Co₀.₅Cr₀.₅/C-900 catalyst at various rotation rates; and (D) Koutecky—Levich plots fitting results of the Pt₃Co₀.₅Cr₀.₅/C-900 catalyst. Electrolyte solution: Reprinted with permission from Ref. 56.

dioxide-carbon composite materials have been extensively investigated as a promising catalyst support candidate due to their high stability and low solubility.[58–61] For example, conductive NbO_2 and magneli phase titanium oxide (suboxides) materials, such as Ti_nO_{2n-1} ($4< n <10$) were explored as potential fuel cell catalyst supports, as shown in Figure 7.18 recorded using RDE technique at different electrode rotating rates. For stability test, after the 1000 cycles of potential cycling, the ORR current was reduced significantly, indicating the catalyst activity has decreased. The mass activity of Pt/NbO_2 catalyst dropped significantly, from 3.0 to 1.3 mA mgPt^{-1} after durability testing, and corresponded to a loss of 57% of ORR mass activity. Similarly, ORR activity of Pt/Ti_4O_7 catalysts also drastically drops 86% loss of ORR mass activity.[61] It is due to part of the NbO_2 nanofiber surface has been oxidized into more insulating

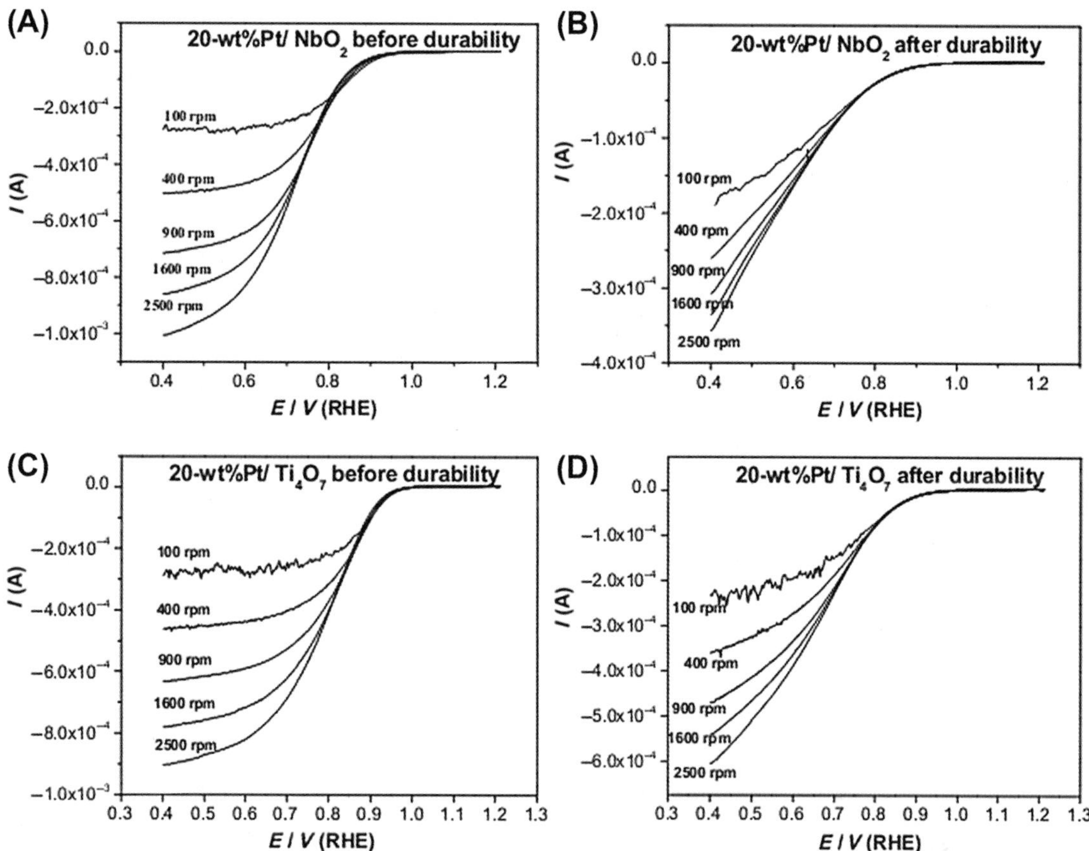

Figure 7.18 ORR curves on 20-wt% Pt/NbO$_2$ and 20-wt% Pt/Ti$_4$O$_7$ nanofiber catalysts coated on a glassy carbon electrode (0.196 cm^2), measured in O$_2$-saturated 0.1 M HClO$_4$ solution (30 °C). (A) Pt/NbO$_2$, before durability test of cyclic voltammetric cycling for 2000 cycles in N$_2$-saturated 0.1 M HClO$_4$ solution, (B) Pt/NbO$_2$, after durability test, (C) Pt/Ti$_4$O$_7$, before durability test, (D) Pt/Ti$_4$O$_7$, after durability test. Potential scan rate: 5 mV s^{-1}. Reprinted with permission from Ref. 61.

Nb$_2$O$_5$ even before subjecting to electrochemical potential cycling for Pt/NbO$_2$ catalyst. Additionally, thin TiO$_2$ layers formed due to durability tests for Pt/Ti$_4$O$_7$ catalysts. The effect of Nb$_2$O$_5$ and thin TiO$_2$ layers on electrical conductivity would be significant, which could limit the electron transport and create a poor electrical communication between the Pt catalyst particles with the nanofiber support, leading to a lower mass activity. Regarding the performance improvement of metal oxide or carbon-metal oxide-supported Pt catalysts, alloying Pt with a second metal or even third and fourth metals can further improve ORR activity. As support, carbon-Nb$_{0.07}$Ti$_{0.93}$O$_2$ (carbon-Nb$_{0.07}$Ti$_{0.93}$O$_2$) composite was prepared for Pt–Pd alloy.[60]

Table 7.5. Summary of ORR Mass Activities of Various Catalyst Prepared Using USP Synthesized Supports. Activity of Commercial Pt/C Catalyst is Also Given for Comparison

Catayst	ORR Mass Activity @ 0.9 V (mA mgPt^{-1})	
	Initial	% Loss after Durability
48 wt%Pt/C (TKK)	110	33.0
20 wt%Pt/25 wt% $(Nb_{0.07}Ti_{0.93}O_2)$-75 wt% C (600 °C)	129	39.0
20 wt%Pt$_{62}$Pd$_{38}$/TKK-E	132	40.0
20 wt%Pt$_{0.62}$Pd$_{0.38}$/25 wt% $(Nb_{0.07}Ti_{0.93}O_2)$-75 wt% C (600 °C)	157	51.0
20 wt%Pt$_{0.62}$Pd$_{0.38}$/25 wt% $(Nb_{0.07}Ti_{0.93}O_2)$-75 wt% C (700 °C)	155	52.0
20 wt%Pt$_{0.62}$Pd$_{0.38}$/25 wt% $(Nb_{0.07}Ti_{0.93}O_2)$-75 wt% C (800 °C)	157	49.0

Reprinted with permission from Ref. 62.

The formed $C_{67wt\%}$-$(Nb_{0.07}Ti_{0.93}O_2)_{33wt\%}$ composite support showed a high surface area of 176 m^2 g^{-1} and a similar porosity to that of carbon. The formed $Pt_{0.55}Pd_{0.45}/C_{67wt\%}$-$(Nb_{0.07}Ti_{0.93}O_2)_{33wt\%}$ catalyst showed a uniform distribution on the support, its catalytic activity toward ORR was much higher than other composition catalysts. It might be due to the electronic effect of the Pt–Pd alloying, and favorable adsorption and dissociation of O_2 on the Pt–Pd clusters. However, it is the most important that the interaction between Pt–Pd alloy and $C_{67wt\%}$-$(Nb_{0.07}Ti_{0.93}O_2)_{33wt\%}$ can be prevented Pd from electrochemical dissolution and enhanced the catalysts durability. Table 7.5 shows the measured ORR activities and stabilities for several Pt–Pd alloy catalyst supported on different carbon-$Nb_{0.07}Ti_{0.93}O_2$ hybrid materials. It can be seen that the best ORR mass activity of 157 mA mgPt^{-1}, which is much higher than the commercially available Pt/C catalyst.

Other non-carbon supports such as TiN, WC, and TiC have also been explored as the Pt catalyst supports to form Pt or Pt-alloy catalysts in the effort to improve the support's electric conductivity and durability.[63] For example, tungsten carbide (WC) appears to be attractive. The experiment results showed that WC showed a high stability and good electronic conductivity.

7.6. RDE/RRDE Study for ORR on the Surfaces of Non-Noble Metal Catalyst Layer

As a solution to provide a long-term solution to Pt cost and scarcity, a variety of non-noble metal-based catalysts has been explored as promising cathode catalysts for fuel cells.[38] These ORR catalysts include heat-treated metal-nitrogen-carbon complexes (M-N_x/C, M = Fe or Co),[64] carbon-supported chalcogenides, and carbon-supported metal oxides. These catalysts have been synthesized and showed considerable ORR activity and stability when compared to those of Pt/C catalyst. In the exploration, RDE/RRDE techniques are the most commonly employed tools in evaluating the catalysts' activity and stability toward ORR and its associated mechanism.

For example, a heat-treated cobalt tripyridyl triazine (Co-TPTZ) electrocatalysts was synthesized and tested in acidic medium for ORR using both RDE and RRDE techniques.[65] In this work, one of the Co-N/C catalysts developed with a Co loading of 5 wt% obtained via 700 °C heat-treatment was chosen for kinetic ORR investigation. Figure 7.19(A) shows the current–potential curves collected on a Co-N/C-coated GC electrode at different rotating rates from 100 to 1600 rpm in an O_2-saturated acidic solution. It can be seen that there is no well-defined limiting currents, indicating the ORR kinetics catalyzed by Co-N/C catalyst is not fast enough to fully exhaust the surface O_2 concentration.

The disk current density (i_d) for the ORR in the potential range of 0.7–0.8 V vs RHE in Figure 7.19(A) is almost independent of the electrode rotation rate, suggesting that the current densities in this narrow potential (or low overpotential) range are purely electrochemical kinetic current densities. The disk current density between 0.7 and 0.8 V, which is the pure kinetic current density, can be plotted as a function of electrode potential, as shown in Figure 7.19(B). From the Tafel slope and the intercept can be separately obtained, and two important ORR kinetic parameters (the electron-transfer coefficient, and the exchange current density, i^o). From the value of i^o, which is a kinetic rate constant (or electron-transfer rate constant) in the rate-determining step (RDS), k_e can be obtained using the relationship $i^o = nFk_eC_{O_2}$ (where the value of n is obtained by averaging the values for overall electron-transfer number at different electrode potentials), which is about $8.6 \times 10^{-10}\,A\,cm^{-2}$ as listed in Table 7.6.

Figure 7.19(A) also shows that when the potential is less than 0.7 V, the current density becomes dependent on the electrode rotation rate, suggesting that the current densities in the potential

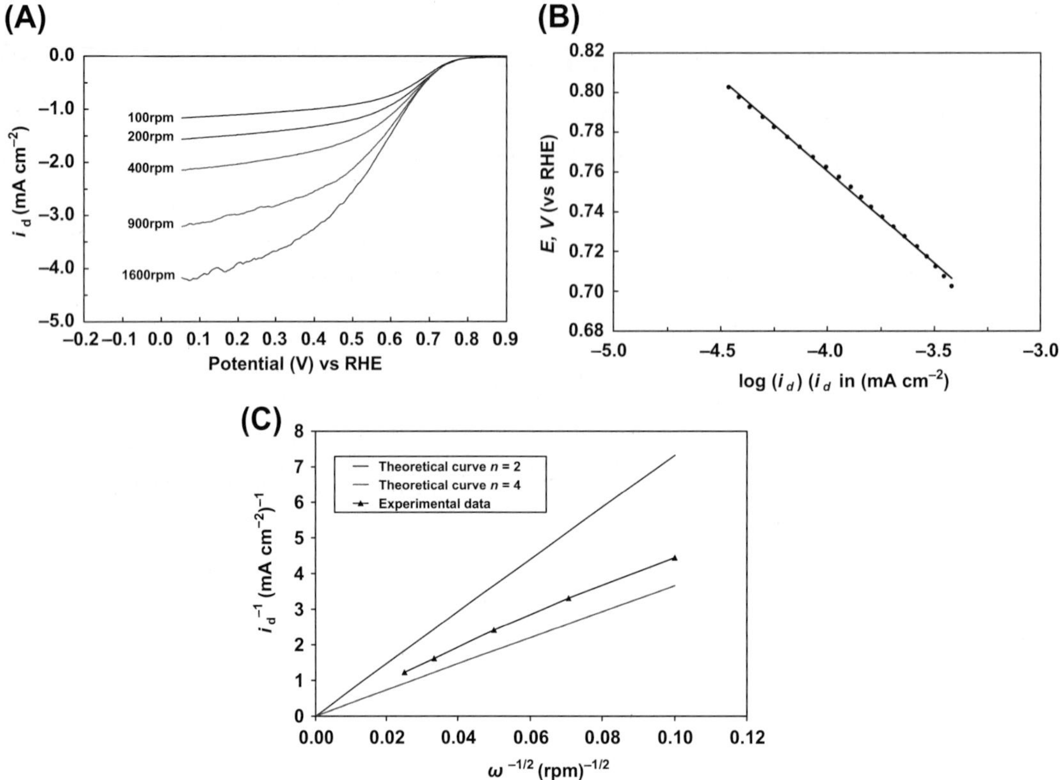

Figure 7.19 (A) Current–potential curves for the ORR on a glassy carbon electrode coated with Co-N/C. Recorded in a 0.5 M H_2SO_4 solution at various electrode rotation rates, as marked. Potential scan rate: 5 mV s^{-1}; Co-N/C loading in the catalyst coating: 0.10 mg cm^{-2}; (B) Koutecky–Levich plots for the ORR on a Co-N/C-coated glassy carbon electrode, data from (A); and (C) Tafel plot, data from (A). Reprinted with permission from Ref. 65.

range <0.7 V are affected by both the electrochemical kinetic current and O_2 diffusion. The K–L plots at different electrode potentials can be drawn, for example as shown in Figure 7.19(C), from the intercepts of which the kinetic current densities can be estimated as a function of electrode potential.

The kinetic current densities were believed to have two major contributors: (1) the chemical reaction between O_2 and the catalyst active site to form an adduct prior to the electron-transfer reaction, and (2) the electron-transfer reaction in the RDS. Therefore, the ORR kinetic current density (i_k) may be expressed as a function of electrode potential (E), as in Eqn (7.4):

$$\frac{1}{i_k} = \frac{1}{nFk_cC_{O_2}} + \frac{1}{nFk_eC_{O_2}\exp\left(-\frac{\alpha n_\alpha F}{RT}(E - E^o)\right)} \qquad (7.4)$$

Table 7.6. ORR Kinetic Current Densities and Overall Electron Transfer Numbers. Obtained Based on the Data in Figure 7.19, Collected on a Co-N/C Coated Glassy Carbon Electrode Rotating at Various Rotation Rates. Electrolyte: O_2-Saturated 0.5 M H_2SO_4 Solution; Co Loading in the Coating Layer: $5\ \mu g \cdot cm^{-2}$; Temperature: 25 °C

Electrode Potential, E(V vs. RHE)	Kinetic Current Density, ik (mA cm^{-2})	Overall Electron Number, n
0.10	5.1	3.4
0.20	4.0	3.3
0.30	3.4	3.2
0.40	2.2	3.1
0.50	1.2	3.1
0.55	0.74	3.1
0.60	0.42	3.2
0.65	0.21	3.5
0.70	0.086	3.7

Reprinted with permission from Ref. 65.

where k_c is the chemical reaction rate between O_2 and the active site on the electrode surface before electron transfer, k_e is the electron-transfer rate in the RDS of the overall ORR, and α, n_α, E, E^o, R, T, n, F, and C_{O_2} have their usual meanings. Based on the known values of n, F, k_e, α, n_a, and C_{O_2}, and the values of i_k in Table (7.6), according to Eqn (7.4), a plot of $\frac{1}{i_k}$ vs $\frac{1}{nFk_eC_{O_2}\exp\left(-\frac{\alpha n_\alpha F}{RT}(E-E^o)\right)}$ can give the intercept $\left(\frac{1}{nFk_cC_{O_2}}\right)$, from which the chemical reaction rate k_c can be roughly estimated, as listed in Table 7.7 with the values for exchange current density and the kinetic reaction constants.

To obtain the overall ORR electron-transfer numbers, the RRDE measurements were also carried out in this work. The obtained overall electron numbers and the percentages of H_2O_2 production are shown as a function of electrode potential in Figure 7.20(A) and (B), respectively. It can be seen that the ORR electron-transfer number varies in the range of 3.6–3.8,

Table 7.7. Electron-Transfer Coefficiency α (Assuming the Electron-Transfer Number n_α in ORR Rate-Determining Step is 1), the Exchange Current Density i^o, the Electron-Transfer Rate Constant k_e, and the Chemical Reaction Rate Constant k_c

Electron-Transfer Coefficiency α	Exchange Current Density i^o (mA cm^{-2})	Electron-Transfer Rate Constant in RDS k_e (cm s^{-1})	Chemical Reaction Rate Constant k_c (cm s^{-1})
0.277	8.18×10^{-4}	2.26×10^{-6}	1.30×10^{-2}

Reprinted with permission from Ref. 65.

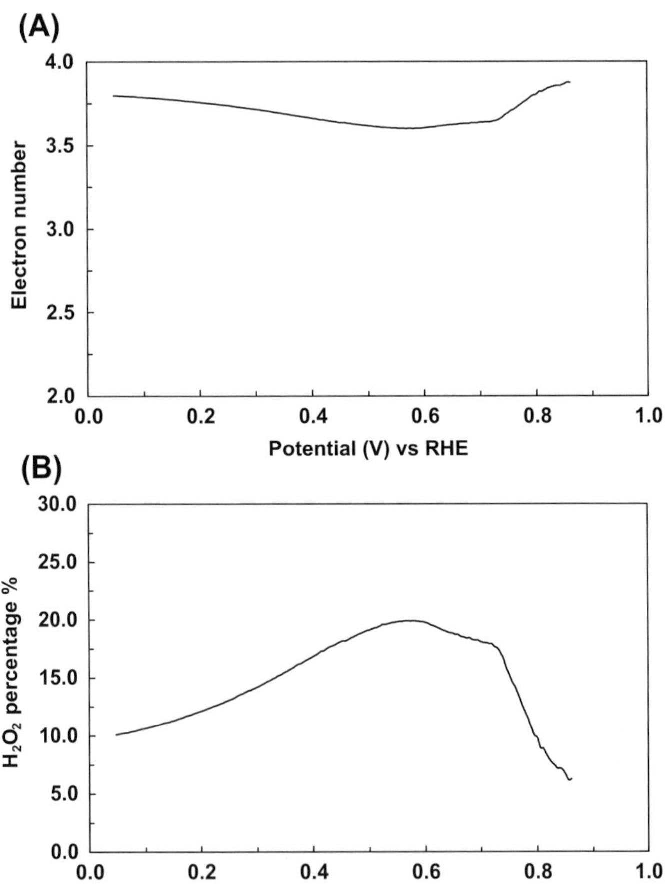

(A)

(B)

Figure 7.20 (A) Electron-transfer numbers for the ORR catalyzed by 5 wt% Co-N/C catalyst obtained after heat treating 5 wt % Co-TPTZ/C at 700 °C; (B) the corresponding %H$_2$O$_2$ produced. Data obtained using rotating ring (Pt)-disk electrode (glassy carbon coated by Co-N/C) in an O$_2$-saturated 0.5 M H$_2$SO$_4$ solution; Co-N/C catalyst loading in the disk coating: 0.1 mg cm^{-2}. Reprinted with permission from Ref. 65.

suggesting that the dominant reaction pathway is a 4-electron-transfer process from O_2 to H_2O, with 5–20% H_2O_2 production.

As discussed above, the ORR mechanism catalyzed by the Co-N/C catalyst was a mixed process of 2- and 4-electron-transfer pathways, dominated by the 4-electron pathway. An ORR mechanism was proposed in this work as follows:

$$Co(II) - N/C + O_2 \rightarrow O_2 - Co(II) - N/C \qquad (7\text{-}XIII)$$

$$O_2 - Co(II) - N/C + e^- \rightarrow O_2^- - Co(II) - N/C \quad (RDS) \quad (7\text{-}XIV)$$

$$O_2^- - Co(II) - N/C + e^- + 2H^+ \rightarrow H_2O_2 - Co(II) - N/C \quad (7\text{-}XV)$$

$$x(H_2O_2 - Co(II) - N/C) + 2xe^- + 2xH^+ \rightarrow xCo(II) - N/C$$
$$+ 2x(H_2O)$$
$$(7\text{-}XVI)$$

$$(1-x)(H_2O_2 - Co(II) - N/C) \rightarrow (1-x)Co(II) - N/C$$
$$+ (1-x)H_2O_2 \qquad (7\text{-}XVII)$$

In this mechanism, Reaction (7-XIII) is the chemical reaction discussed above. Reaction (7-XIV) is the ORR RDS on the Co-N/C catalyst surface. Reaction (7-XV) is peroxide formation. This H_2O_2 will go two ways: (1) further 2-electron reduction to H_2O through Reaction (7-XVI), or (2) chemical desorption through Reaction (7-XVII) to form free H_2O_2, and then entrance into the bulk solution, which can be detected by the ring electrode. The relative proportions of Reactions (7-XVI) and (7-XVII) determine the overall electron number and the amount of H_2O_2 produced. In Reaction (7-XVI), an x is used to express the proportion that is a 4-electron-transfer process, and in Reaction (7-XVII), $(1-x)$ to express the proportion of 2-electron process. It can be seen that when $x = 1$, the mechanism will be a completely 4-electron-transfer pathway, and when $x = 0$, the mechanism will be a completely 2-electron pathway. If x is in between, the ORR will be a mixture of 2- and 4-electron-transfer pathways.

In basic solution, some carbon-supported metal oxides, such as Ti_4O_7/C, $La_{0.6}Ca_{0.4}CoO_3$/C, Carbon-$La_{0.6}Ca_{0.4}CoO_3$/C, MnO_x/C, and Me-MnO_x/C (Me = Ni, Mg), have been explored for ORR catalysts, evaluated by RDE/RRDE techniques.[66,67]

For example, carbon-supported magneli phase Ti_4O_7 was synthesized and characterized for ORR catalysts, and evaluated cyclic voltammetry (CV), RDE, and RRDE techniques in electrolyte solutions containing various concentrations of KOH. The electrochemical stability of the catalyst was also evaluated by cyclic voltammetry scans and chronopotentiometric tests.

To obtain the kinetic information in 1, 4, and 6 M KOH solutions, Figure 7.21 a shows the typical RRDE disk and ring currents in 6 M KOH, collected at rotation rates from 400 to 2500 rpm. The ORR on the Ti_4O_7 electrode was under kinetic control in the potential range of c. -0.05 to -0.15 V, under combined kinetic-diffusion control from c. -0.15 to -0.20 V, and reached the limiting current at higher overpotentials (>-0.20 V). Using K–L plots shown in Figure 7.21(B), several ORR kinetic parameters can be obtained, from which the effect of KOH concentration on the ORR catalytic activity of Ti_4O_7/C can be understood. From the ring currents, the overall electron-transfer numbers and the percentages of H_2O_2 production at different potentials were also obtained, as plotted in Figure 7.22.

From Figure 7.22, it can be seen that the overall ORR electron-transfer numbers catalyzed by Ti_4O_7 in these KOH solutions are in

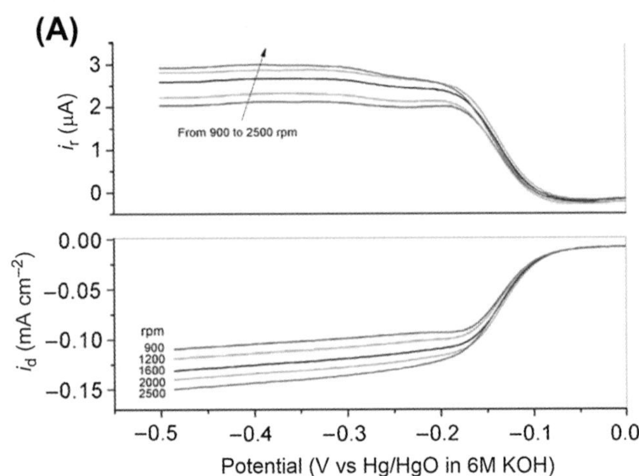

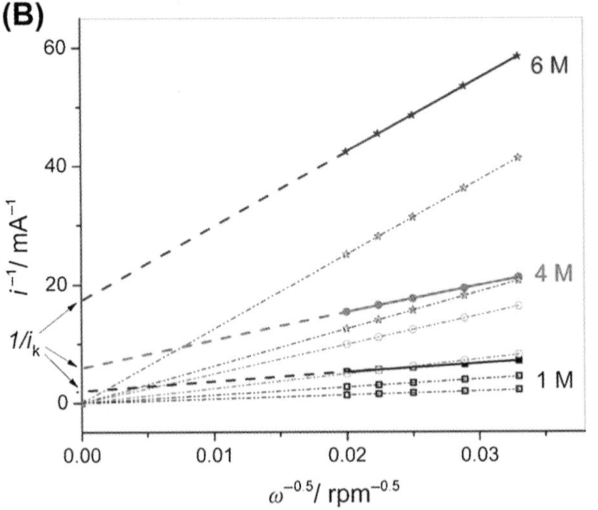

Figure 7.21 (A) Rotating ring-disk current–potential curves at different electrode rotation rates as marked in the figure. Disk currents were recorded on Ti_4O_7/C-coated glassy carbon disk electrode in potential range 0.0 to -0.50 V vs Hg/HgO, and ring currents were recorded on platinum ring electrode with fixed potential of $+0.5$ V in O_2-saturated 6 M KOH solution. Potential scan rate: 5 mV s^{-1}. Ti_4O_7 loading on disk electrode: 0.73 mg cm^{-2}; (B) Koutecky–Levich plots for ORR on Ti_4O_7 electrode at -0.40 V vs Hg/HgO in 1 M (solid black line), 4 M (solid red line), and 6 M KOH (solid blue line) solutions; theoretical plots for 2- and 4-electron-transfer process are also displayed as dashed lines for comparison. (For interpretation of the references to color in this figure legend, the reader is referred to the on-line version of this book.) Reprinted with permission from Ref. 66.

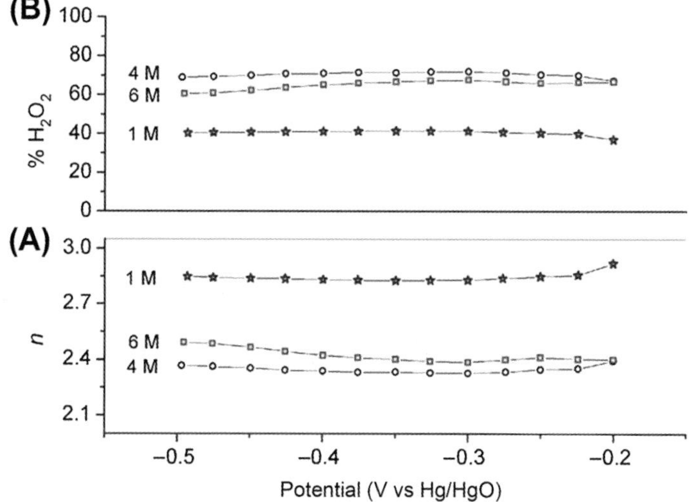

Figure 7.22 (A) Plots of overall electron-transfer number; and (B) percentage H_2O_2 produced as a function of electrode potential at three different concentrations of O_2-saturated KOH solution. Calculations based on current–potential data collected at rotation rate of 1600 rpm. Reprinted with permission from Ref. 66.

the range of 2.3–2.9, with the corresponding percentages of H_2O_2 being 35–72%, suggesting that the ORR on Ti_4O_7 is a mixed process of 2- and 4-electron-transfer pathways. It can also been seen that the higher the KOH concentration, the greater the amount of H_2O_2 detected (or the lower the overall electron-transfer number). It is understandable that in a more concentrated alkaline medium, H_2O_2 in the form of HO_2^- will be more stable than in a less concentrated OH^- solution.

Figure 7.23 shows the ORR Tafel plots catalyzed by Ti_4O_7/C catalyst at three KOH concentrations, from which the exchange current densities were obtained. The obtained results showed

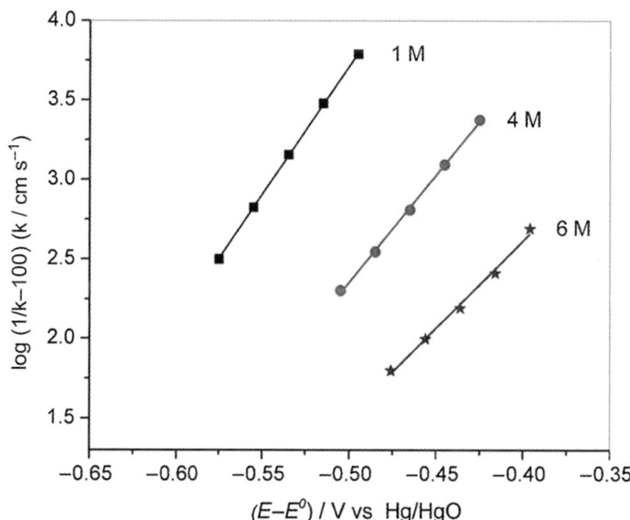

Figure 7.23 ORR kinetic constants catalyzed by Ti_4O_7/C catalyst as a function of applied electrode potential $(E - E^\circ)$ at three different KOH concentrations. Reprinted with permission from Ref. 66.

that the exchange current densities could be greatly increased with increasing alkaline concentration.

Another important kind of non-noble metal ORR catalysts are chalcogenides (M_xX_y, M = Co, Ni, Fe; X = S, Se), which have been paid great attention to because they show promising electrocatalytic activity and high selectivity toward ORR in both acid and alkaline media, evaluated by both RDE and RRDE techniques.[68] For example, carbon-supported cubic cobalt selenide ($CoSe_2$) nanoparticles showed a high activity and "complete" selectivity toward ORR in 0.5 M H_2SO_4. However, this catalyst is even more active in alkaline solution. As shown in Figure 7.24, the current density of 4.2 mA cm^{-2} at 0.4 V at 1600 rpm is much higher than that of 2.5 mA cm^{-2} at the same potential and rotating speed in

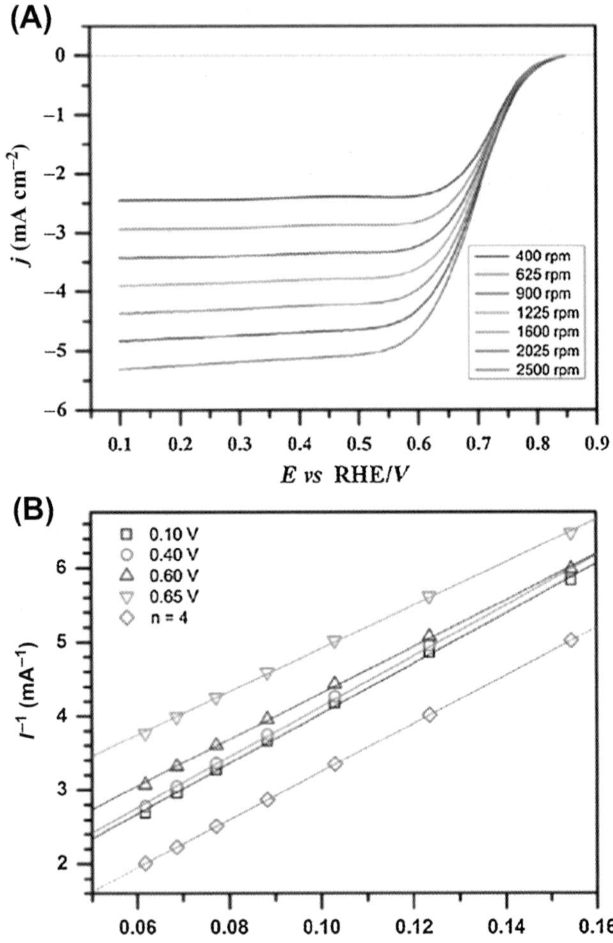

Figure 7.24 (A) Rotating-disk voltammograms of 20 wt% $CoSe_2$/C nanocatalyst in O_2- saturated 0.1 M KOH at 25 °C with a sweep rate of 5 mV s^{-1} at different rotating rates from 400 rpm to 2500 rpm from top to bottom. (B) The corresponding Koutecky—Levich plots as a function of $\omega^{-1/2}$ at four different potentials: 0.1, 0.4, 0.6 and 0.65 V vs RHE. Catalyst mass loading is c. 44 μg cm^{-2} for $CoSe_2$. The calculated slope for $n = 4$ electrons is shown in the same figure. Reprinted with permission from Ref. 68.

0.5 M H_2SO_4 reported in Ref. 69. The average ORR electron-transfer number obtained was very close to 4.0, suggesting that the 20 wt% $CoSe_2/C$ catalyst favors a 4-electron ORR process from O_2 to H_2O.

7.7. Chapter Summary

In exploring ORR catalysts, both RDE and RRDE techniques have been recognized as the most useful electrochemical techniques in evaluating the catalysts' activity and stability as well as performance optimization. Using RDE and RRDE techniques, ORR has been studied on Pt electrode, carbon electrode, monolayer metal catalyst, Pt-based catalyst, and non-noble metal-based catalysts, respectively. To give more detailed information, this chapter reviews the applications of RDE and RRDE techniques in ORR research and its associated catalyst evaluation. Some typical examples for RDE and RRDE analysis in obtaining the ORR kinetic information such as the overall electron-transfer number, electron-transfer coefficiency, and exchange current density are also given in this chapter.

References

1. Lipkowski J, Ross P. *Electrocatalysis*, vol. 3. Wiley-VCH; 1998.
2. Gasteiger HA, Ross PN. Oxygen reduction on platinum low-index single-crystal surfaces in alkaline solution: rotating ring DiskPt(hkl) studies. *J Phys Chem* 1996;**100**(16):6715−21.
3. Zhang J, Mo Y, Vukmirovic M, Klie R, Sasaki K, Adzic R. Platinum monolayer electrocatalysts for O_2 reduction: Pt monolayer on Pd(111) and on carbon-supported Pd nanoparticles. *J Phys Chem B* 2004;**108**(30):10955−64.
4. Kondo T, Song C, Hayashi N, Sakurai T, Shibata M, Notsu H, et al. Electrocatalytic activity for oxygen reduction reaction of Pseudomorphic Pt monolayer prepared electrochemically on a Au (111) surface. *Chem Lett* 2011;**40**(11):1235−7.
5. Kongkanand A, Kuwabata S. Oxygen reduction at platinum monolayer islands deposited on Au(111). *J Phys Chem B* 2005;**109**(49):23190−5.
6. Zhang J, Vukmirovic M, Xu Y, Mavrikakis M, Adzic R. Controlling the catalytic activity of platinum-monolayer electrocatalysts for oxygen reduction with different substrates. *Angew Chem Int Ed* 2005;**44**(14):2132−5.
7. Kitchin J, Norskov J, Barteau M, Chen J. Modification of the surface electronic and chemical properties of Pt(111) by subsurface 3d transition metals. *J Chem Phys* 2004;**120**:10240.
8. Nagaoka T, Sakai T, Ogura K, Yoshino T. Oxygen reduction at electrochemically treated glassy carbon electrodes. *Anal Chem* 1986;**58**(9):1953−5.
9. Sheng Z, Gao H, Bao W, Wang F, Xia X. Synthesis of boron doped graphene for oxygen reduction reaction in fuel cells. *J Mater Chem* 2012;**22**:390−5.

10. Li Y, Zhou W, Wang H, Xie L, Liang Y, Wei F, et al. An oxygen reduction electrocatalyst based on carbon nanotube-graphene complexes. *Nature Nanotechnology* 2012;**7**:394–400.

11. Hossain M, Tryk D, Yeager E. The electrochemistry of graphite and modified graphite surfaces: the reduction of O_2. *Electrochim Acta* 1989;**34**(12):1733–7.

12. Sarapuua A, Helsteina K, Vaika K, Schiffrinb D, Tammeveski K. Electrocatalysis of oxygen reduction by quinones adsorbed on highly oriented pyrolytic graphite electrodes. *Electrochim Acta* 2010;**55**(22):6376–82.

13. Vaik K, Sarapuu A, Tammeveski K, Mirkhalaf F, Schiffrin D. Oxygen reduction on phenanthrenequinone-modified glassy carbon electrodes in 0.1 M KOH. *J Electroanal Chem* 2004;**564**:159–66.

14. Lin R, Zhang H, Zhao T, Cao C, Yang D, Ma J. Investigation of Au@ Pt/C electro-catalysts for oxygen reduction reaction. *Electrochimica Acta* 2012;**62**:263–8.

15. Beard BC, Ross PN. The structure and activity of Pt-Co alloys as oxygen reduction electrocatalysts. *J Electrochem Soc* 1990;**137**(11):3368–74.

16. Aricòa A, Shuklab A, Kimc H, Parkc S, Minc M, Antonucci V. An XPS study on oxidation states of Pt and its alloys with Co and Cr and its relevance to electroreduction of oxygen. *Appl Surf Sci* 2001;**172**(1):33–40.

17. Toda T, Igarashi H, Watanabe M. Enhancement of the electrocatalytic O_2 reduction on Pt–Fe alloys. *J Electroanal Chem* 1999;**460**(1):258–62.

18. Shukla A, Neergat M, Bert P, Jayaram V, Hegde MS. An XPS study on binary and ternary alloys of transition metals with platinized carbon and its bearing upon oxygen electroreduction in direct methanol fuel cells. *J Electroanal Chem* 2001;**504**(1):111–9.

19. Marković N, Schmidt T, Stamenković V, Ross P. Oxygen reduction reaction on Pt and Pt bimetallic surfaces: a selective review. *Fuel Cells* 2001;**1**(2):105–16.

20. Markovic N, Gasteiger H, Ross PN. Kinetics of oxygen reduction on Pt(hkl) electrodes: implications for the crystallite size effect with supported Pt electrocatalysts. *J Electrochem Soc* 1997;**144**(5):1591–7.

21. Grgur BN, Markovic NM, Ross PN. Temperature dependent oxygen electrochemistry on platinum low index single crystal surfaces in acid solutions. *Can J Chem* 1997;**75**(11):1465–71.

22. Kitchin J, Norskov J, Barteau M, Chen J. Modification of the surface electronic and chemical properties of Pt(111) by subsurface 3d transition metals. *J Chem Phys* 2004;**120**(21):10240–6.

23. Kakinuma K, Wakasugi Y, Uchida M, Kamino T, Uchida H, Deki S, et al. Preparation of titanium nitride-supported platinum catalysts with well controlled morphology and their properties relevant to polymer electrolyte fuel cells. *Electrochim Acta* 2012;**77**:279–84.

24. Paliteiro C, Hamnett A, Goodenough JB. The electroreduction of oxygen on pyrolytic graphite. *J Electroanal Chem Interfacial Electrochem* 1987;**233**(1):147–59.

25. Xu J, Huang W, McCreery RL. Isotope and surface preparation effects on alkaline dioxygen reduction at carbon electrodes. *J Electroanal Chem* 1996;**410**(2):235–42.

26. Hu I-F, Karweik DH, Kuwana T. Activation and deactivation of glassy carbon electrodes. *J Electroanal Chem Interfacial Electrochem* 1985;**188**(1–2):59–72.

27. Hossain MS, Tryk D, Yeager E. The electrochemistry of graphite and modified graphite surfaces: the reduction of O_2. *Electrochim Acta* 1989;**34**(12):1733–7.

28. Gajendran P, Saraswathi R. Enhanced electrochemical growth and redox characteristics of poly(o-phenylenediamine) on a carbon nanotube modified

glassy carbon electrode and its application in the electrocatalytic reduction of oxygen. *J Phys Chem C* 2007;**111**(30):11320−8.

29. Yu D, Zhang Q, Dai L. Highly efficient metal-free growth of nitrogen-doped single-walled carbon nanotubes on plasma-etched substrates for oxygen reduction. *J Am Chem Soc* 2010;**132**(43):15127.

30. Geng D, Liu H, Chen Y, Li R, Sun X, Ye S, et al. Non-noble metal oxygen reduction electrocatalysts based on carbon nanotubes with controlled nitrogen contents. *J Power Sources* 2011;**196**(4):1795−801.

31. Wang SY, Yu DS, Dai LM, Chang DW, Baek JB. Polyelectrolyte-functionalized graphene as metal-free electrocatalysts for oxygen reduction. *ACS Nano* 2011;**5**(8):6202−9.

32. Lai L, Potts J, Zhan D, Wang L, Poh C, Tang C, et al. Exploration of the active center structure of nitrogen-doped graphene-based catalysts for oxygen reduction reaction. *Energy Environ Sci* 2012;**5**:7936−42.

33. Wang S, Zhang L, Xia Z, Roy A, Chang D, Baek J, et al. BCN graphene as efficient metal-free electrocatalyst for the oxygen reduction reaction. *Angew Chem Int Ed* 2012;**51**(17):4209−12.

34. Wang S, Iyyamperumal E, Roy A, Xue Y, Yu D, Dai L. Vertically aligned BCN nanotubes as efficient metal-free electrocatalysts for the oxygen reduction reaction: a synergetic effect by co-doping with boron and nitrogen. *Angew Chem Int Ed* 2011;**50**(49):11756−60.

35. Choi CH, Park SH, Woo SI. Phosphorus-nitrogen dual doped carbon as an effective catalyst for oxygen reduction reaction in acidic media: effects of the amount of P-doping on the physical and electrochemical properties of carbon. *J Mater Chem* 2012;**22**(24):12107−15.

36. Wilson T, Zhang J, Oloman C, Wayner D. Anthraquinone-2-carboxylic-allyl ester as a new electrocatalyst for dioxygen reduction to produce H_2O_2. *Int J Electrochem Sci* 2006;**1**:99−109.

37. Vaik K, Sarapuu A, Tammeveski K, Mirkhalaf F, Schiffrin DJ. Oxygen reduction on phenanthrenequinone-modified glassy carbon electrodes in 0.1 M KOH. *J Electroanal Chem* 2004;**564**(0):159−66.

38. Chen Z, Higgins D, Yu A, Zhang L, Zhang J. A review on non-precious metal electrocatalysts for PEM fuel cells. *Energy Environ Sci* 2011;**4**(9):3167−92.

39. Shigehara K, Anson FC. Electrocatalytic activity of three iron porphyrins in the reduction of dioxygen and hydrogen peroxide at graphite cathodes. *J Phys Chem* 1982;**86**(14):2776−83.

40. Shi C, Anson FC. Catalytic pathways for the electroreduction of oxygen by iron tetrakis(4-*N*-methylpyridyl)porphyrin or iron tetraphenylporphyrin adsorbed on edge plane pyrolytic graphite electrodes. *Inorg Chem* 1990;**29**(21):4298−305.

41. Mouahid O, Coutanceau C, Belgsir E, Crouigneau P, Léger J, Lamy C. Electrocatalytic reduction of dioxygen at macrocycle conducting polymer electrodes in acid media. *J Electroanal Chem* 1997;**426**(1−2):117−23.

42. Tse Y, Janan P, Lam H, Zhang J, Pietro W, Lever A. Monomeric and polymeric tetra-aminophthalocyanatocobalt(II) modified electrodes: electrocatalytic reduction of oxygen. *J Porphyrins Phthalocyanines* 1997;**01**(01):3−16.

43. Shi C, Anson FC. (5,10,15,20-Tetramethylporphyrinato)cobalt(II): a remarkably active catalyst for the electroreduction of O_2 to H_2O. *Inorg Chem* 1998;**37**(5):1037−43.

44. Anson FC, Shi C, Steiger B. Novel multinuclear catalysts for the electroreduction of dioxygen directly to water. *Acc Chem Res* 1997;**30**(11):437−44.

45. Steiger B, Anson FC. [5,10,15,20-Tetrakis(4-((pentaammineruthenio)- cyano)phenyl)porphyrinato]cobalt(II) immobilized on graphite electrodes catalyzes the electroreduction of O_2 to H_2O, but the corresponding 4-Cyano-2,6-dimethylphenyl derivative catalyzes the reduction only to H_2O_2. *Inorg Chem* 1997;**36**(18):4138−40.

46. Zhang L, Song C, Zhang J, Wang H, Wilkinson D. Temperature and pH dependence of oxygen reduction catalyzed by iron fluoroporphyrin adsorbed on a graphite electrode. *J Electrochem Soc* 2005;**152**(12):A2421−6.

47. Song C, Zhang L, Zhang J, Wilkinson D, Baker R. Temperature dependence of oxygen reduction catalyzed by cobalt fluoro-phthalocyanine adsorbed on a graphite electrode. *Fuel Cells* 2007;**7**(1):9−15.

48. Paulus U, Schmidt T, Gasteiger H, Behm R. Oxygen reduction on a high-surface area Pt/Vulcan carbon catalyst: a thin-film rotating ring-disk electrode study. *J Electroanal Chem* 2001;**495**(2):134−45.

49. Senthil Kumar S, Phani KLN. Exploration of unalloyed bimetallic Au−Pt/C nanoparticles for oxygen reduction reaction. *J Power Sources* 2009;**187**(1):19−24.

50. Lin R, Zhang H, Zhao T, Cao C, Yang D, Ma J. Investigation of Au@Pt/C electro-catalysts for oxygen reduction reaction. *Electrochim Acta* 2012;**62**(0):263−8.

51. AricoÁ A, Shukla A, Kim H, Park S, Min M, Antonucci V. An XPS study on oxidation states of Pt and its alloys with Co and Cr and its relevance to electroreduction of oxygen. *Appl Surf Sci* 2001;**172**(1−2):33−40.

52. Toda T, Igarashi H, Watanabe M. Enhancement of the electrocatalytic O_2 reduction on Pt−Fe alloys. *J Electroanal Chem* 1999;**460**(1−2):258−62.

53. Shukla A, Neergat M, Bera P, IJayaram V, Hegde M. An XPS study on binary and ternary alloys of transition metals with platinized carbon and its bearing upon oxygen electroreduction in direct methanol fuel cells. *J Electroanal Chem* 2001;**504**(1):111−9.

54. Loukrakpam R, Luo J, He T, Chen Y, Xu Z, Njoki P, et al. Nanoengineered PtCo and PtNi catalysts for oxygen reduction reaction: an assessment of the structural and electrocatalytic properties. *J Phys Chem C* 2011;**115**(5):1682−94.

55. Yang H, Vogel W, Lamy C, Nicolás AV. Structure and electrocatalytic activity of carbon-supported Pt−Ni alloy nanoparticles toward the oxygen reduction reaction. *J Phys Chem B* 2004;**108**(30):11024−34.

56. Loukrakpam R, Wanjala B, Yin J, Fang B, Luo J, Shao M, et al. Structural and electrocatalytic properties of PtIrCo/C catalysts for oxygen reduction reaction. *ACS Catal* 2011;**1**(5):562−72.

57. Chen S, Sheng W, Yabuuchi N, Ferreira P, Allard L, Yang S, et al. Origin of oxygen reduction reaction activity on "Pt3Co" nanoparticles: atomically resolved chemical compositions and structures. *J Phys Chem C* 2008;**113**(3):1109−25.

58. García BL, Fuentes R, Weidner JW. Low-temperature synthesis of a PtRu / Nb0.1Ti0.9O2 electrocatalyst for methanol oxidation. *Electrochem Solid-State Lett* 2007;**10**(7):B108−10.

59. Akalework N, Pan C, Su W, Rick J, Tsai M, Lee J, et al. Ultrathin TiO2-coated MWCNTs with excellent conductivity and SMSI nature as Pt catalyst support for oxygen reduction reaction in PEMFCs. *J Mater Chem* 2012;**22**(39):20977−85.

60. Ignaszak A, Song C, Zhu W, Wang Y, Zhang J, Bauer A, et al. Carbon−Nb0.07Ti0.93O2 composite supported Pt−Pd electrocatalysts for PEM fuel cell oxygen reduction reaction. *Electrochim Acta* 2012;**75**(0):220−8.

61. Senevirathne K, Hui R, Campbell S, Ye S, Zhang J. Electrocatalytic activity and durability of Pt/NbO2 and Pt/Ti4O7 nanofibers for PEM fuel cell oxygen reduction reaction. *Electrochim Acta* 2012;**59**(0):538−47.

62. Senevirathne K, Neburchilov V, Alzate V, Baker R, Neagu R, Zhang J, et al. Nb-doped TiO2/carbon composite supports synthesized by ultrasonic spray pyrolysis for proton exchange membrane (PEM) fuel cell catalysts. *J Power Sources* 2012;**220**(0):1—9.

63. Kakinuma K, Wakasugi Y, Uchida M, Kamino T, Uchida H, Deki S, et al. Preparation of titanium nitride-supported platinum catalysts with well controlled morphology and their properties relevant to polymer electrolyte fuel cells. *Electrochim Acta* 2012;**77**(0):279—84.

64. Michel L, Eric P, Frédéric J, Jean-Pol D. Iron-based catalysts with improved oxygen reduction activity in polymer electrolyte fuel cells. *Science* 2009;**324**(5923):71—4.

65. Li S, Zhang L, Liu H, Panb M, Zan L, Zhang J. Heat-treated cobalt—tripyridyl triazine (Co—TPTZ) electrocatalysts for oxygen reduction reaction in acidic medium. *Electrochim Acta* 2010;**55**(15):4403—11.

66. Li X, Zhu A, Qu W, Wang H, Hui R, Zhang L, et al. Magneli phase $Ti_4 O_7$ electrode for oxygen reduction reaction and its implication for zinc-air rechargeable batteries. *Electrochim Acta* 2010;**55**(20):5891—8.

67. Li X, Qu W, Zhang J, Wang H. Electrocatalytic activities of La0.6Ca0.4CoO3 and La0.6Ca0.4CoO3-carbon composites toward the oxygen reduction reaction in concentrated alkaline electrolytes. *J Electrochem Soc* 2011;**158**(5):A597—604.

68. Feng Y, Alonso-Vante N. Carbon-supported cubic CoSe2 catalysts for oxygen reduction reaction in alkaline medium. *Electrochim Acta* 2012;**72**(0):129—33.

69. Feng YJ, He T, Alonso-Vante N. Carbon-supported CoSe2 nanoparticles for oxygen reduction reaction in acid medium. *Fuel Cells* 2010;**10**(1):77—83.

INDEX

Note: Page numbers with "*f*" denote figures; "*t*" tables.

A

ACAE. *See* Anthraquinone-carboxylic-allyl ester
Acid medium
  bisulfate adsorption, 235
  ORR pathways in, 162t
Activation energies
  catalytic reactions and, 70
  as function of reaction coordination, 38f
Active center density, 191–192
Active site
  concentration, of electrocatalysts, 73–74
  formation
    active site structures and, 92
    high-temperature pyrolysis and, 158–159
  structures, 92
Adenosine triphosphate, 4–5
"ads" subscript, 138–139
Adsorption
  bisulfate, 235
  on disk electrode surface, 220
  of $Fe^{III}TPFPP$, 252f
  hydrogen, on Pt, 232–235
  of monolayer of catalyst, in smooth surface RDE, 187
  N_2, 86, 141f
  O_2
    energy, 144–145
    at Pt catalyst, 138–139
    with Pt surface, 141f
  sites, 235
Adsorption-desorption coulometry, of hydrogen, 189–190
AFM. *See* Atomic force microscope

Agglomeration, of Pt, 83–84
Ag-W_2C/C catalyst, 223
Alkaline fuel cells, 163–165
Allotropes, 3–4
Alloy metals. *See also* Pt alloy catalysts
  in catalysts, 85
ALS, 211f
American Society for Testing and Materials (ASTM), 18–19
Anodes
  electrode potential of, 136–137
  HOR at, 106
  reactions, 136
Anthraquinone-carboxylic-allyl ester (ACAE), 246
  coating on BPG, 247f
  current-potential curves of, 247f
  Koutecky-Levich plots and, 246, 247f
  as ORR catalyst, 246–247
Anthraquinone-modified carbon electrodes, 245–248
Anthraquinones, as ORR catalysts, 246–247
Aqueous solutions
  O_2 diffusion coefficients in, 11–18
  O_2 solubility in, 5–11
  ORR in, 134
  viscosity of, 18–24
Arrhenius activation energies, 254t
Arrhenius forms, 17
  of reaction rate constants, 37
Arrhenius model, 22

Arrhenius plots, of ORR constants, 255f
Arrhenius theory, 253–254
Assitanted precess, 95–96
ASTM. *See* American Society for Testing and Materials
Atomic force microscope (AFM), 101–102
Au. *See* Gold
AuPd, HAADF-STEM of, 102, 102f
AuPd/C catalysts, 104–105
AuPt, SEM of, 100f
Au@Pt core-shell nanostructured catalysts, 258–259
Au-Pt/C bimetallic nanoparticles, 258

B

Ball-milling method, 99
Basal plane graphite (BPG), 246
  ACAE-coated, 247f
Basic medium, ORR pathways in, 162t
Beginning of lifetime (BOL) test, 77
BET isotherm. *See* Brunauer-Emmett-Teller isotherm
Bimetallic alloy catalysts, 69
Bi-potentiostat, 213, 213f
Bisulfate adsorption, 235
BOL test. *See* Beginning of lifetime test
Boltzmann constant, 15
Bond order, of O-O, 160–161
Bonding orbitals, of O_2, 160–161
Boron, 245
Boundary conditions, for RRDE, 207–208